# The Picture Book of Quantum Mechanics

Siegmund Brandt • Hans Dieter Dahmen

# The Picture Book
# of Quantum Mechanics

Fourth Edition

 Springer

Siegmund Brandt
Physics Department
University of Siegen
D-57068 Siegen
Germany
brandt@physik.uni-siegen.de

Hans Dieter Dahmen
Physics Department
University of Siegen
D-57068 Siegen
Germany
dahmen@physik.uni-siegen.de

Please note that additional material for this book can be downloaded from
http://extras.springer.com

ISBN 978-1-4939-3695-3          ISBN 978-1-4614-3951-6 (eBook)
DOI 10.1007/978-1-4614-3951-6
Springer New York Heidelberg Dordrecht London

Library of Congress Control Number: 2012937850

Printed on acid-free paper

Springer is part of Springer Science+Business Media (www.springer.com)

# Preface to the Fourth Edition

In the present edition the illustrations of the *Picture Book* appear in full color. The scope of the book was extended again. There is now a chapter on hybridization and sections on bound states and on scattering in piecewise linear potentials in one dimension.

From the web page www.extras.springer.com all illustrations can be downloaded for easy use in lectures and seminars. The inclusion of a CD-ROM with with that material (as in the third edition) is no longer necessary.

To generate the computer graphics of the first edition of the *Picture Book*, we developed an interactive program on quantum mechanics. A modernized version, which we call INTERQUANTA (abbreviated **IQ**), together with an accompanying text has been published by Springer in various editions. The most recent one[1] can be regarded as companion to this *Picture Book*. It allows interactive manipulation of of a host of physics and graphics parameters and produces output in the form of static and moving pictures. The program runs under Windows, Linux and Mac OS X. It is a pleasure to acknowledge the generous help provided by IBM Germany in the development of **IQ**. In particular, we want to thank Dr. U. Groh for his competent help in the early phase of the work.

All computer-drawn figures in the present edition were produced using the published version of **IQ** or extensions realized with the help of Mr. Anli Shundi, Dr. Sergei Boris, and Dr. Tilo Stroh.

Siegen, January 2012

Siegmund Brandt
Hans Dieter Dahmen

---

[1] S. Brandt, H. D. Dahmen, and T. Stroh *Interactive Quantum Mechanics – Quantum Experiments on the Computer*, 2nd ed., Springer, New York, 2011

# Foreword

Students of classical mechanics can rely on a wealth of experience from everyday life to help them understand and apply mechanical concepts. Even though a stone is not a mass point, the experience of throwing stones certainly helps them to understand and analyze the trajectory of a mass point in a gravitational field. Moreover, students can solve many mechanical problems on the basis of Newton's laws and, in doing so, gain additional experience. When studying wave optics, they find that their knowledge of water waves, as well as experiments in a ripple tank, is very helpful in forming an intuition about the typical wave phenomena of interference and diffraction.

In quantum mechanics, however, beginners are without any intuition. Because quantum-mechanical phenomena happen on an atomic or a subatomic scale, we have no experience of them in daily life. The experiments in atomic physics involve more or less complicated apparatus and are by no means simple to interpret. Even if students are able to take Schrödinger's equation for granted, as many students do Newton's laws, it is not easy for them to acquire experience in quantum mechanics through the solution of problems. Only very few problems can be treated without a computer. Moreover, when solutions in closed form are known, their complicated structure and the special mathematical functions, which students are usually encountering for the first time, constitute severe obstacles to developing a heuristic comprehension. The most difficult hurdle, however, is the formulation of a problem in quantum-mechanical language, for the concepts are completely different from those of classical mechanics. In fact, the concepts and equations of quantum mechanics in Schrödinger's formulation are much closer to those of optics than to those of mechanics. Moreover, the quantities that we are interested in – such as transition probabilities, cross sections, and so on – usually have nothing to do with mechanical concepts such as the position, momentum, or trajectory of a particle. Nevertheless, actual insight into a process is a prerequisite for understanding its quantum-mechanical description and for interpreting basic properties in quantum mechanics like position, linear and angular momentum, as well as cross sections, lifetimes, and so on.

Actually, students must develop an intuition of how the concepts of classical mechanics are altered and supplemented by the arguments of optics in order to acquire a roughly correct picture of quantum mechanics. In particular, the time evolution of microscopic physical systems has to be studied to establish how it corresponds to classical mechanics. Here, computers and computer graphics offer incredible help, for they produce a large number of examples that are very detailed and that can be looked at in any phase of their time development. For instance, the study of wave packets in motion, which is practically impossible without the help of a computer, reveals the limited validity of intuition drawn from classical mechanics and gives us insight into phenomena like the tunnel effect and resonances, which, because of the importance of interference, can be understood only through optical analogies. A variety of systems in different situations can be simulated on the computer and made accessible by different types of computer graphics.

Some of the topics covered are

- scattering of wave packets and stationary waves in one dimension,
- the tunnel effect,
- decay of metastable states,
- bound states in various potentials,
- energy bands,
- distinguishable and indistinguishable particles,
- angular momentum,
- three-dimensional scattering,
- cross sections and scattering amplitudes,
- eigenstates in three-dimensional potentials, for example, in the hydrogen atom, partial waves, and resonances,
- motion of wave packets in three dimensions,
- spin and magnetic resonance.

Conceptual tools that bridge the gap between classical and quantum concepts include

- the phase-space probability density of statistical mechanics,
- the Wigner phase-space distribution,
- the absolute square of the analyzing amplitude as probability or probability density.

The graphical aids comprise

- time evolutions of wave functions for one-dimensional problems,
- parameter dependences for studying, for example, the scattering over a range of energies,
- three-dimensional surface plots for presenting two-particle wave functions or functions of two variables,
- polar (antenna) diagrams in two and three dimensions,
- plots of contour lines or contour surfaces, that is, constant function values, in two and three dimensions,
- ripple-tank pictures to illustrate three-dimensional scattering.

Whenever possible, how particles of a system would behave according to classical mechanics has been indicated by their positions or trajectories. In passing, the special functions typical for quantum mechanics, such as Legendre, Hermite, and Laguerre polynomials, spherical harmonics, and spherical Bessel functions, are also shown in sets of pictures.

The text presents the principal ideas of wave mechanics. The introductory Chapter 1 lays the groundwork by discussing the particle aspect of light, using the fundamental experimental findings of the photoelectric and Compton effects and the wave aspect of particles as it is demonstrated by the diffraction of electrons. The theoretical ideas abstracted from these experiments are introduced in Chapter 2 by studying the behavior of wave packets of light as they propagate through space and as they are reflected or refracted by glass plates.

Chapter 3 introduces material particles as wave packets of de Broglie waves. The ability of de Broglie waves to describe the mechanics of a particle is explained through a detailed discussion of group velocity, Heisenberg's uncertainty principle, and Born's probability interpretation. The Schrödinger equation is found to be the equation of motion.

Chapters 4 through 9 are devoted to the quantum-mechanical systems in one dimension. Study of the scattering of a particle by a potential helps us understand how it moves under the influence of a force and how the probability interpretation operates to explain the simultaneous effects of transmission and reflection. We study the tunnel effect of a particle and the excitation and decay of a metastable state. A careful transition to a stationary bound state is carried out. Quasi-classical motion of wave packets confined to the potential range is also examined.

The velocity of a particle experiencing the tunnel effect has been a subject of controversial discussion in the literature. In Chapter 7, we introduce the concepts of quantile position and quantile velocity with which this problem can be treated.

Chapters 8 and 9 cover two-particle systems. Coupled harmonic oscillators are used to illustrate the concept of indistinguishable particles. The striking differences between systems composed of different particles, systems of identical bosons, and systems of identical fermions obeying the Pauli principle are demonstrated.

Three-dimensional quantum mechanics is the subject of Chapters 10 through 16. We begin with a detailed study of angular momentum and discuss methods of solving the Schrödinger equation. The scattering of plane waves is investigated by introducing partial-wave decomposition and the concepts of differential cross sections, scattering amplitudes, and phase shifts. Resonance scattering, which is the subject of many fields of physics research, is studied in detail in Chapter 15. Bound states in three dimensions are dealt with in Chapter 13. The hydrogen atom and the motion of wave packets under the action of a harmonic force as well as the Kepler motion on elliptical orbits are among the topics covered. In Chapter 14 we discuss and illustrate hybridization, a model used in some cases to describe chemical binding. Chapter 16 is devoted to Coulomb scattering in terms of stationary wave functions as well as wave-packet motion on hyperbolic orbits.

Spin is treated in Chapter 17. After the introduction of spin states and operators, the Pauli equation is used for the description of the precession of a magnetic moment in a homogeneous magnetic field. The discussion of Rabi's magnetic resonance concludes this chapter.

The last chapter is devoted to results obtained through experiments in atomic, molecular, solid-state, nuclear, and particle physics. They can be qualitatively understood with the help of the pictures and the discussion in the body of the book. Thus, examples for

- typical scattering phenomena,
- spectra of bound states and their classifications with the help of models,
- resonance phenomena in total cross sections,
- phase-shift analyses of scattering and Regge classification of resonances,
- radioactivity as decay of metastable states,
- magnetic resonance phenomena,

taken from the fields of atomic and subatomic physics, are presented. Comparing these experimental results with the computer-drawn pictures of the book and their interpretation gives the reader a glimpse of the vast fields of science that can be understood only on the basis of quantum mechanics.

In Appendix A, the simplest aspects of the structure of quantum mechanics are discussed, and the matrix formulation in an infinite-dimensional vector space is juxtaposed to the more conventional formulation in terms of wave functions and differential operators. Appendix B gives a short account of two-level systems that is helpful for the discussion of spin. In Appendix C, we introduce the analyzing amplitude using as examples the free particle and the harmonic oscillator. Appendix D discusses Wigner's phase-space distribution. Appendixes E through G give short accounts of the gamma, Bessel, and Airy functions, as well as the Poisson distribution.

There are more than a hundred problems at the ends of the chapters. Many are designed to help students extract the physics from the pictures. Others will give them practice in handling the theoretical concepts. On the endpapers of the book are a list of frequently used symbols, a list of basic equations, a short list of physical constants, and a brief table converting SI units to particle-physics units. The constants and units will make numerical calculations easier.

We are particularly grateful to Professor Eugen Merzbacher for his kind interest in our project and for many valuable suggestions he gave before the publication of the first edition that helped to improve the book. We are indebted to Ms. Ute Smolik and Mr. Anli Shundi who helped with the computer typesetting of several editions and to Drs. Sergei Boris, Erion Gjonaj and Tilo Stroh who contributed greatly to the computer programs used to generate the pictures of this book. Last but not least, we would like to thank Drs. Jeanine Burke, Hans-Ulrich Daniel, Thomas von Foerster, and Hans Kölsch of Springer New York for their constant interest and support.

Siegmund Brandt
Hans Dieter Dahmen

# Contents

# 1. Introduction

The basic fields of classical physics are mechanics and heat on the one hand and electromagnetism and optics on the other. Mechanical and heat phenomena involve the motion of particles as governed by Newton's equations. Electromagnetism and optics deal with fields and waves, which are described by Maxwell's equations. In the classical description of particle motion, the position of the particle is exactly determined at any given moment. Wave phenomena, in contrast, are characterized by interference patterns which extend over a certain region in space. The strict separation of particle and wave physics loses its meaning in atomic and subatomic processes.

Quantum mechanics goes back to Max Planck's discovery in 1900 that the energy of an oscillator of *frequency* $\nu$ is quantized. That is, the energy emitted or absorbed by an oscillator can take only the values $0$, $h\nu$, $2h\nu$,.... . Only multiples of *Planck's quantum of energy*

$$E = h\nu$$

are possible. *Planck's constant*

$$h = 6.262 \times 10^{-34} \,\text{J s}$$

is a fundamental constant of nature, the central one of quantum physics. Often it is preferable to use the *angular frequency* $\omega = 2\pi\nu$ of the oscillator and to write Planck's quantum of energy in the form

$$E = \hbar\omega \quad .$$

Here

$$\hbar = \frac{h}{2\pi}$$

is simply Planck's constant divided by $2\pi$. Planck's constant is a very small quantity. Therefore the quantization is not apparent in macroscopic systems. But in atomic and subatomic physics Planck's constant is of fundamental importance. In order to make this statement more precise, we shall look at experiments showing the following fundamental phenomena:

S. Brandt and H.D. Dahmen, *The Picture Book of Quantum Mechanics*, DOI 10.1007/978-1-4614-3951-6_1, © Springer Science+Business Media New York 2012

- the photoelectric effect,

- the Compton effect,

- the diffraction of electrons,

- the orientation of the magnetic moment of electrons in a magnetic field.

## 1.1 The Photoelectric Effect

The photoelectric effect was discovered by Heinrich Hertz in 1887. It was studied in more detail by Wilhelm Hallwachs in 1888 and Philipp Lenard in 1902. We discuss here the quantitative experiment, which was first carried out in 1916 by R. A. Millikan. His apparatus is shown schematically in Figure 1.1a. Monochromatic light of variable frequency falls onto a photocathode in a vacuum tube. Opposite the photocathode there is an anode – we assume cathode and anode to consist of the same metal – which is at a negative voltage $U$ with respect to the cathode. Thus the electric field exerts a repelling force on the electrons of charge $-e$ that leave the cathode. Here $e = 1.609 \times 10^{-19}$ Coulomb is the elementary charge. If the electrons reach the anode, they flow back to the cathode through the external circuit, yielding a measurable current $I$. The kinetic energy of the electrons can therefore be determined by varying the voltage between anode and cathode. The experiment yields the following findings.

1. The electron current sets in, independent of the voltage $U$, at a frequency $\nu_0$ that is characteristic for the material of the cathode. There is a current only for $\nu > \nu_0$.

2. The voltage $U_s$ at which the current stops flowing depends linearly on the frequency of the light (Figure 1.1b). The kinetic energy $E_{kin}$ of the electrons leaving the cathode then is equal to the potential energy of the electric field between cathode and anode,

$$E_{kin} = eU_s \quad .$$

If we call $h/e$ the proportionality factor between the frequency of the light and the voltage,

$$U_s = \frac{h}{e}(\nu - \nu_0) \quad ,$$

we find that light of frequency $\nu$ transfers the kinetic energy $eU_s$ to the electrons kicked out of the material of the cathode. When light has a frequency less than $\nu_0$, no electrons leave the material. If we call

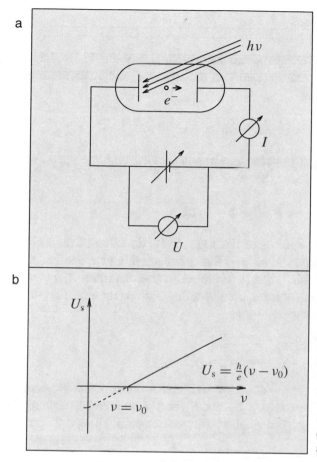

a

$h\nu$

$e^-$

$I$

$U$

b

$U_s$

$U_s = \frac{h}{e}(\nu - \nu_0)$

$\nu$

$\nu = \nu_0$

**Fig. 1.1. Photoelectric effect.
(a) The apparatus to measure
the effect consists of a vacuum tube containing two elec-
trodes. Monochromatic light
of frequency $\nu$ shines on the
cathode and liberates electrons which may reach the
anode and create a current
$I$ in the external circuit. The
flow of electrons in the vacuum
tube is hindered by the external voltage $U$. It stops once the
voltage exceeds the value $U_s$.
(b) There is a linear dependence between the frequency $\nu$
and the voltage $U_s$.**

$$h\nu_0 = eU_k$$

the ionization energy of the material that is needed to free the electrons,
we must conclude that light of frequency $\nu$ has energy

$$E = h\nu = \hbar\omega$$

with

$$\omega = 2\pi\nu \quad , \qquad \hbar = \frac{h}{2\pi} \quad .$$

3. The number of electrons set free is proportional to the intensity of the
   light incident on the photocathode.

In 1905 Albert Einstein explained the photoelectric effect by assuming
that light consists of quanta of energy $h\nu$ which act in single elementary processes. The *light quanta* are also called *photons* or $\gamma$ quanta. The number of
quanta in the light wave is proportional to its intensity.

## 1.2 The Compton Effect

If the light quanta of energy $E = h\nu = \hbar\omega$ are particles, they should also have momentum. The relativistic relation between the energy $E$ and momentum $p$ of a particle of rest mass $m$ is

$$p = \frac{1}{c}\sqrt{E^2 - m^2 c^4} \quad ,$$

where $c$ is the speed of light in vacuum. Quanta moving with the speed of light must have rest mass zero, so that we have

$$p = \frac{1}{c}\sqrt{\hbar^2\omega^2} = \hbar\frac{\omega}{c} = \hbar k \quad ,$$

where $k = \omega/c$ is the wave number of the light. If the direction of the light is $\mathbf{k}/k$, we find the vectorial relation $\mathbf{p} = \hbar\mathbf{k}$. To check this idea one has to perform an experiment in which light is scattered on free electrons. The conservation of energy and momentum in the scattering process requires that the following relations be fulfilled:

$$E_\gamma + E_e = E'_\gamma + E'_e \quad ,$$
$$\mathbf{p}_\gamma + \mathbf{p}_e = \mathbf{p}'_\gamma + \mathbf{p}'_e \quad ,$$

where $E_\gamma$, $\mathbf{p}_\gamma$ and $E'_\gamma$, $\mathbf{p}'_\gamma$ are the energies and the momenta of the incident and the scattered photon, respectively. $E_e$, $\mathbf{p}_e$, $E'_e$, and $\mathbf{p}'_e$ are the corresponding quantities of the electron. The relation between electron energy $E_e$ and momentum $\mathbf{p}_e$ is

$$E_e = c\sqrt{\mathbf{p}_e^2 + m_e^2 c^2} \quad ,$$

where $m_e$ is the rest mass of the electron. If the electron is initially at rest, we have $\mathbf{p}_e = \mathbf{0}$, $E_e = m_e c^2$. Altogether, making use of these relations, we obtain

$$c\hbar k + m_e c^2 = c\hbar k' + c\sqrt{\mathbf{p}_e'^2 + m_e^2 c^2} \quad ,$$
$$\hbar\mathbf{k} = \hbar\mathbf{k}' + \mathbf{p}'_e$$

as the set of equations determining the wavelength $\lambda' = 2\pi/k'$ of the scattered photon as a function of the wavelength $\lambda = 2\pi/k$ of the initial photon and the scattering angle $\vartheta$ (Figure 1.2a). Solving for the difference $\lambda' - \lambda$ of the two wavelengths, we find

$$\lambda' - \lambda = \frac{h}{m_e c}(1 - \cos\vartheta) \quad .$$

This means that the angular frequency $\omega' = ck' = 2\pi c/\lambda'$ of the light scattered at an angle $\vartheta > 0$ is smaller than the angular frequency $\omega = ck = 2\pi c/\lambda$ of the incident light.

a

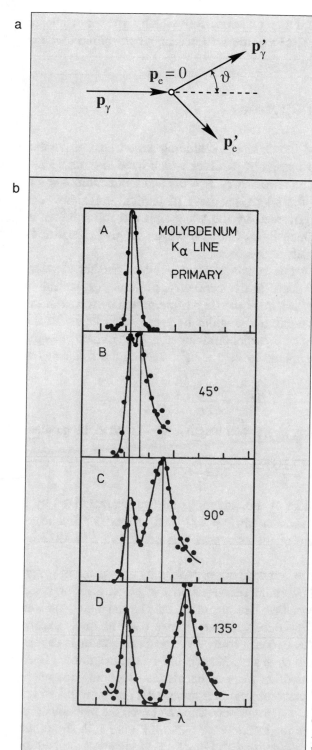

b

**Fig. 1.2. The Compton effect.**
**(a) Kinematics of the process.**
**A photon of momentum $p_\gamma$**
**is scattered by a free elec-**
**tron at rest, one with momen-**
**tum $p_e = 0$. After the scat-**
**tering process the two par-**
**ticles have the momenta $p'_\gamma$**
**and $p'_e$, respectively. The di-**
**rection of the scattered pho-**
**ton forms an angle $\vartheta$ with its**
**original direction. From en-**
**ergy and momentum conser-**
**vation in the collision, the ab-**
**solute value $p'_\gamma$ of the mo-**
**mentum of the scattered pho-**
**ton and the corresponding**
**wavelength $\lambda' = h/p'_\gamma$ can be**
**computed. (b) Compton's re-**
**sults. Compton used mono-**
**chromatic X-rays from the $K_\alpha$**
**line of molybdenum to bom-**
**bard a graphite target. The**
**wavelength spectrum of the**
**incident photons shows the**
**rather sharp $K_\alpha$ line at the top.**
**Observations of the photons**
**scattered at three different an-**
**gles $\vartheta$ (45°, 90°, 135°) yielded**
**spectra showing that most of**
**them had drifted to the longer**
**wavelength $\lambda'$. There are also**
**many photons at the original**
**wavelength $\lambda$, photons which**
**were not scattered by single**
**electrons in the graphite.** From
A. H. Compton, *The Physical Review* **22**
(1923) 409, copyright © 1923 by the
American Physical Society, reprinted by
permission.

Arthur Compton carried out an experiment in which light was scattered on electrons; he reported in 1923 that the scattered light had shifted to lower frequencies $\omega'$ (Figure 1.2b).

## 1.3 The Diffraction of Electrons

The photoelectric effect and the Compton scattering experiment prove that light must be considered to consist of particles which have rest mass zero, move at the speed of light, and have energy $E = \hbar\omega$ and momentum $\mathbf{p} = \hbar\mathbf{k}$. They behave according to the relativistic laws of particle collisions. The propagation of photons is governed by the wave equation following from Maxwell's equations. The intensity of the light wave at a given location is a measure of the photon density at this point.

Once we have arrived at this conclusion, we wonder whether classical particles such as electrons behave in the same way. In particular, we might conjecture that the motion of electrons should be determined by waves. If the relation $E = \hbar\omega$ between energy and angular frequency also holds for the kinetic energy $E_{kin} = \mathbf{p}^2/2m$ of a particle moving at nonrelativistic velocity, that is, at a speed small compared to that of light, its angular frequency is given by

$$\omega = \frac{1}{\hbar}\frac{p^2}{2m} = \frac{\hbar k^2}{2m}$$

provided that its wave number $k$ and wavelength $\lambda$ are related to the momentum $p$ by

$$k = \frac{p}{\hbar} \quad , \quad \lambda = \frac{h}{p} \quad .$$

Thus the motion of a particle of momentum $p$ is then characterized by a wave with the *de Broglie wavelength* $\lambda = h/p$ and an angular frequency $\omega = p^2/(2m\hbar)$. The concept of matter waves was put forward in 1923 by Louis de Broglie.

If the motion of a particle is indeed characterized by waves, the propagation of electrons should show interference patterns when an electron beam suffers diffraction. This was first demonstrated by Clinton Davisson and Lester Germer in 1927. They observed interference patterns in an experiment in which a crystal was exposed to an electron beam. In their experiment the regular lattice of atoms in a crystal acts like an optical grating. Even simpler conceptually is diffraction from a sharp edge. Such an experiment was performed by Hans Boersch in 1943. He mounted a platinum foil with a sharp edge in the beam of an electron microscope and used the magnification of the microscope to enlarge the interference pattern. Figure 1.3b shows his result. For comparison it is juxtaposed to Figure 1.3a indicating the pattern

a

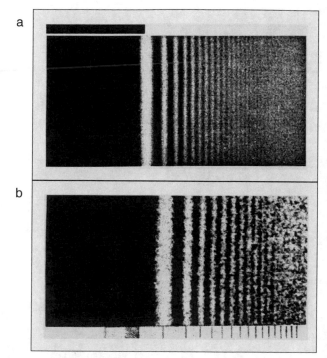

b

**Fig. 1.3. (a) Interference pattern caused by the scattering of red light on a sharp edge. The edge is the border line of an absorbing half-plane, the position of which is indicated at the top of the figure. (b) Interference pattern caused by the scattering of electrons on a sharp edge.** *Sources*: (a) From R. W. Pohl, *Optik und Atomphysik*, ninth edition, copyright © 1954 by Springer-Verlag, Berlin, Göttingen, Heidelberg, reprinted by permission. (b) From H. Boersch, *Physikalische Zeitschrift*, **44** (1943) 202, copyright © 1943 by S.-Hirzel-Verlag, Leipzig, reprinted by permission.

produced by visible light diffracted from a sharp edge. The wavelength determined in electron diffraction experiments is in agreement with the formula of de Broglie.

## 1.4 The Stern–Gerlach Experiment

In 1922 Otto Stern and Walther Gerlach published the result of an experiment in which they measured the magnetic moment of silver atoms. By evaporating silver in an oven with a small aperture they produced a beam of silver atoms which was subjected to a magnetic induction field **B**. In the coordinate system shown in Figure 1.4 together with the principal components of the experiment the beam travels along the $x$ axis. In the $x, z$ plane the field $\mathbf{B} = (B_x, B_y, B_z)$ has only a $z$ component $B_z$. Caused by the form of the pole shoes the field is inhomogeneous. The magnitude of $B_z$ is larger near the upper pole shoe which has the shape of a wedge. In the $x, z$ plane the derivative of the field is

$$\frac{\partial \mathbf{B}}{\partial z} = \frac{\partial B_z}{\partial z} \mathbf{e}_z \quad , \qquad \frac{\partial B_z}{\partial z} > 0 \quad .$$

Here $\mathbf{e}_z$ is the unit vector in $z$ direction. In the field a silver atom with the *magnetic moment $\boldsymbol{\mu}$* experiences the force

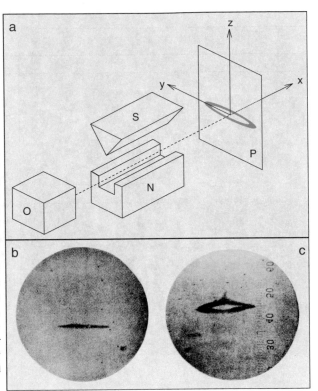

**Fig. 1.4. Stern–Gerlach experiment. Experimental setup with oven O, magnet pole shoes N and S, and glass screen P (a). Silver deposit on screen without field (b) and with field (c)** as shown in Stern's and Gerlach's original publication. The splitting is largest in the middle and gets smaller to the left and the right of the picture because the field inhomogeneity is largest in the $x, z$ plane.
*Source:* (b) and (c) from W. Gerlach and O. Stern, *Zeitschrift für Physik* **9** (1922) 349 © 1922 by Springer-Verlag, Berlin, reprinted by permission.

$$\mathbf{F} = \left( \boldsymbol{\mu} \cdot \frac{\partial \mathbf{B}}{\partial z} \right) \mathbf{e}_z = (\boldsymbol{\mu} \cdot \mathbf{e}_z) \frac{\partial B_z}{\partial z} \mathbf{e}_z \quad .$$

Since the scalar product of $\boldsymbol{\mu}$ and $\mathbf{e}_z$ is

$$\boldsymbol{\mu} \cdot \mathbf{e}_z = \mu \cos \alpha \quad ,$$

where $\alpha$ is the angle between the direction of the magnetic moment and the $z$ direction and $\mu$ is the magnitude of the magnetic moment, the force has its maximum strength in the $z$ direction if $\boldsymbol{\mu}$ is parallel to $\mathbf{e}_z$ and its maximum strength in the opposite direction if $\boldsymbol{\mu}$ is antiparallel to $\mathbf{e}_z$. For intermediate orientations the force has intermediate values. In particular, the force vanishes if $\boldsymbol{\mu}$ is perpendicular to $\mathbf{e}_z$, i.e., if $\boldsymbol{\mu}$ is parallel to the $x, y$ plane.

Stern and Gerlach measured the deflection of the silver atoms by this force by placing a glass plate behind the magnet perpendicular to the $x$ axis. In those areas where atoms hit the glass a thin but visible layer of silver formed after some time. Along the $z$ axis they observed two distinct areas of silver indicating that the magnetic moments $\boldsymbol{\mu}$ were oriented preferentially parallel ($\alpha = 0$) or antiparallel ($\alpha = \pi$) to the field $\mathbf{B}$. This finding is contrary to the classical expectation that all orientations of $\boldsymbol{\mu}$ are equally probable.

It remains to be said that the magnetic moment of a silver atom is practically identical to the magnetic moment of a single free electron. A silver atom has 47 electrons but the contributions of 46 electrons to the total magnetic moment cancel. The contribution of the nucleus to the magnetic moment of the atom is very small. The quantitative result of the Stern–Gerlach experiment is

1. The magnetic moment of the electron is

$$\mu = -\frac{e}{m}\frac{\hbar}{2} \quad .$$

2. In the presence of a magnetic field the magnetic moment is found to be oriented parallel or antiparallel to the field direction.

## Problems

1.1. Thirty percent of the 100 W power consumption of a sodium lamp goes into the emission of photons with the wavelength $\lambda = 589\,\text{nm}$. How many photons are emitted per second? How many hit the eye of an observer – the diameter of the pupil is 5 mm – stationed 10 km from the lamp?

1.2. The minimum energy $E_0 = h\nu_0$ needed to set electrons free is called the work function of the material. For cesium it is $3.2 \times 10^{-19}\,\text{J}$. What is the minimum frequency and the corresponding maximum wavelength of light that make the photoelectric effect possible? What is the kinetic energy of an electron liberated from a cesium surface by a photon with a wavelength of 400 nm?

1.3. The energy $E = h\nu$ of a light quantum of frequency $\nu$ can also be interpreted in terms of Einstein's formula $E = Mc^2$, where $c$ is the velocity of light in a vacuum. (See also the introduction to Chapter 18.) What energy does a blue quantum ($\lambda = 400\,\text{nm}$) lose by moving 10 m upward in the earth's gravitational field? How large is the shift in frequency and wavelength?

1.4. Many radioactive nuclei emit high-energy photons called $\gamma$ rays. Compute the recoil momentum and velocity of a nucleus possessing 100 times the proton mass and emitting a photon of 1 MeV energy.

1.5. Calculate the maximum change in wavelength experienced by a photon in a Compton collision with an electron initially at rest. The initial wavelength of the photon is $\lambda = 2 \times 10^{-12}\,\text{m}$. What is the kinetic energy of the recoil electron?

1.6. Write the equations for energy and momentum conservation in the Compton scattering process when the electron is not at rest before the collision.

1.7. Use the answer to problem 1.6 to calculate the maximum change of energy and wavelength of a photon of red light ($\lambda = 8 \times 10^{-7}$ m) colliding head on with an electron of energy $E_e = 20\,\mathrm{GeV}$. (Collisions of photons from a laser with electrons from the Stanford linear accelerator are in fact used to prepare monochromatic high-energy photon beams.)

1.8. Electron microscopes are chosen for very fine resolution because the de Broglie wavelength $\lambda = h/p$ can be made much shorter than the wavelength of visible light. The resolution is roughly $\lambda$. Use the relativistic relation $E^2 = p^2c^2 + m^2c^4$ to determine the energy of electrons needed to resolve objects of the size $10^{-6}$ m (a virus), $10^{-8}$ m (a DNA molecule), and $10^{-15}$ m (a proton). Determine the voltage $U$ needed to accelerate the electrons to the necessary kinetic energy $E - mc^2$.

1.9. What are the de Broglie frequency and wavelength of an electron moving with a kinetic energy of $20\,\mathrm{keV}$, which is typical for electrons in the cathode-ray tube of a color television set?

# 2. Light Wavess

## 2.1 Harmonic Plane Waves, Phase Velocity

Many important aspects and phenomena of quantum mechanics can be visualized by means of *wave mechanics*, which was set up in close analogy to
*wave optics*. Here the simplest building block is the harmonic plane wave of
light in a vacuum describing a particularly simple configuration in space and
time of the *electric field* **E** and the *magnetic induction field* **B**. If the $x$ axis
of a rectangular coordinate system has been oriented parallel to the direction
of the wave propagation, the $y$ axis can always be chosen to be parallel to
the electric field strength so that the $z$ axis is parallel to the magnetic field
strength. With this choice the field strengths can be written as

$$
\begin{aligned}
E_y &= E_0 \cos(\omega t - kx) \quad , & B_z &= B_0 \cos(\omega t - kx) \quad , \\
E_x &= E_z = 0 \quad , & B_x &= B_y = 0 \quad .
\end{aligned}
$$

They are shown in Figures 2.1 and 2.2. The quantities $E_0$ and $B_0$ are the
maximum values reached by the electric and magnetic fields, respectively.
They are called *amplitudes*. The *angular frequency* $\omega$ is connected to the *wave
number* $k$ by the simple relation

$$ \omega = ck \quad . $$

The points where the field strength is maximum, that is, has the value $E_0$,
are given by the *phase* of the cosine function

$$ \delta = \omega t - kx = 2\ell\pi \quad , $$

where $\ell$ takes the integer values $\ell = 0, \pm 1, \pm 2, \ldots$ . Therefore such a point
moves with the velocity

$$ c = \frac{x}{t} = \frac{\omega}{k} \quad . $$

Since this velocity describes the speed of a point with a given phase, $c$ is
called the *phase velocity* of the wave. For light waves in a vacuum, it is independent of the wavelength. For positive, or negative, $k$ the propagation is in
the direction of the positive, or negative, $x$ axis, respectively.

S. Brandt and H.D. Dahmen, *The Picture Book of Quantum Mechanics*,
DOI 10.1007/978-1-4614-3951-6_2, © Springer Science+Business Media New York 2012

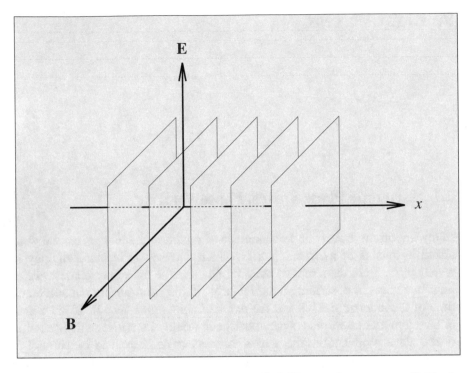

**Fig. 2.1. In a plane wave the electric and magnetic field strengths are perpendicular to the direction of propagation. At any moment in time, the fields are constant within planes perpendicular to the direction of motion. As time advances, these planes move with constant velocity.**

At a fixed point in space, the field strengths **E** and **B** oscillate in time with the angular frequency $\omega$ (Figures 2.3a and c). The *period* of the oscillation is

$$T = \frac{2\pi}{\omega} \quad .$$

For fixed time the field strengths exhibit a periodic pattern in space with a spatial period, the *wavelength*

$$\lambda = \frac{2\pi}{|k|} \quad .$$

The whole pattern moves with velocity $c$ along the $x$ direction. Figures 2.3b and 2.3d present the propagation of waves by a set of curves showing the field strength at a number of consecutive equidistant moments in time. Earlier moments in time are drawn in the background of the picture, later ones toward the foreground. We call such a representation a *time development*.

For our purpose it is sufficient to study only the electric field of a light wave,

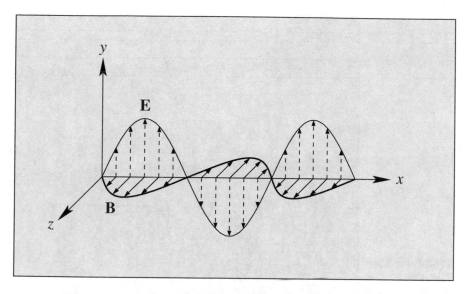

**Fig. 2.2.** For a given moment in time, the electric field strength **E** and the magnetic field strength **B** are shown along a line parallel to the direction of motion of the harmonic plane wave.

$$E_y = E = E_0 \cos(\omega t - kx - \alpha) \quad .$$

We have included an additional *phase* $\alpha$ to allow for the fact that the maximum of $E$ need not be at $x = 0$ for $t = 0$. To simplify many calculations, we now make use of the fact that cosine and sine are equal to the real and imaginary parts of an exponential,

$$\cos \beta + i \sin \beta = e^{i\beta} \quad ,$$

that is,

$$\cos \beta = \mathrm{Re}\, e^{i\beta} \quad , \qquad \sin \beta = \mathrm{Im}\, e^{i\beta} \quad .$$

The wave is then written as

$$E = \mathrm{Re}\, E_c \quad ,$$

where $E_c$ is the complex field strength:

$$E_c = E_0 e^{-i(\omega t - kx - \alpha)} = E_0 e^{i\alpha} e^{-i\omega t} e^{ikx} \quad .$$

It factors into a complex amplitude

$$A = E_0 e^{i\alpha}$$

and two exponentials containing the time and space dependences, respectively. As mentioned earlier, the wave travels in the positive or negative $x$

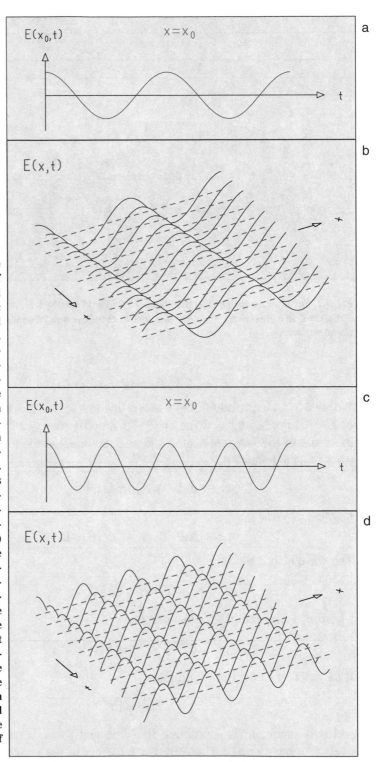

**Fig. 2.3. (a) Time dependence of the electric field of a harmonic wave at a fixed point in space. (b) Time development of the electric field of a harmonic wave. The field distribution along the** $x$ **direction is shown for several moments in time. Early moments are in the background, later moments in the foreground. (c, d) Here the wave has twice the frequency. We observe that the period** $T$ **and the wavelength** $\lambda$ **are halved, but that the phase velocity** $c$ **stays the same. The time developments in parts b and d are drawn for the same interval of time.**

direction, depending on the sign of $k$. Such waves with different amplitudes are

$$E_{c+} = A e^{-i\omega t} e^{ikx} \quad , \qquad E_{c-} = B e^{-i\omega t} e^{-ikx} \quad .$$

The factorization into a time- and a space-dependent factor is particularly convenient in solving Maxwell's equations. It allows the separation of time and space coordinates in these equations. If we divide by $\exp(-i\omega t)$, we arrive at the time-independent expressions

$$E_{s+} = A e^{ikx} \quad , \qquad E_{s-} = B e^{-ikx} \quad ,$$

which we call *stationary waves*.

The *energy density* in an electromagnetic wave is equal to a constant, $\varepsilon_0$, times the square of the field strength,

$$w(x,t) = \varepsilon_0 E^2 \quad .$$

Because the plane wave has a cosine structure, the energy density varies twice as fast as the field strength. It remains always a positive quantity; therefore the variation occurs around a nonzero average value. This average taken over a period $T$ of the wave can be written in terms of the complex field strength as

$$w = \frac{\varepsilon_0}{2} E_c E_c^* = \frac{\varepsilon_0}{2} |E_c|^2 \quad .$$

Here $E_c^*$ stands for the complex conjugate,

$$E_c^* = \mathrm{Re}\, E_c - i \,\mathrm{Im}\, E_c \quad ,$$

of the complex field strength,

$$E_c = \mathrm{Re}\, E_c + i \,\mathrm{Im}\, E_c \quad .$$

For the average energy density in the plane wave, we obtain

$$w = \frac{\varepsilon_0}{2} |A|^2 = \frac{\varepsilon_0}{2} E_0^2 \quad .$$

## 2.2 Light Wave Incident on a Glass Surface

The effect of glass on light is to reduce the phase velocity by a factor $n$ called the *refractive index*,

$$c' = \frac{c}{n} \quad .$$

Although the frequency $\omega$ stays constant, wave number and wavelength are changed according to

$$k' = nk \quad , \qquad \lambda' = \frac{\lambda}{n} \quad .$$

The Maxwell equations, which govern all electromagnetic phenomena, demand the continuity of the electric field strength and its first derivative at the boundaries of the regions with different refractive indices. We consider a wave traveling in the $x$ direction and encountering at position $x = x_1$ the surface of a glass block filling half of space (Figure 2.4a). The surface is oriented perpendicular to the direction of the light. The complex expression

$$E_{1+} = A_1 e^{ik_1 x}$$

describes the incident stationary wave to the left of the glass surface, that is, for $x < x_1$, where $A_1$ is the known amplitude of the incident light wave. At the surface only a part of the light wave enters the glass block; the other part will be reflected. Thus, in the region to the left of the glass block, $x < x_1$, we find in addition to the incident wave the reflected stationary wave

$$E_{1-} = B_1 e^{-ik_1 x}$$

propagating in the opposite direction. Within the glass the transmitted wave

$$E_2 = A_2 e^{ik_2 x}$$

propagates with the wave number

$$k_2 = n_2 k_1$$

altered by the refractive index $n = n_2$ of the glass. The waves $E_{1+}$, $E_{1-}$, and $E_2$ are called *incoming*, *reflected*, and *transmitted constituent waves*, respectively. The continuity for the field strength $E$ and its derivative $E'$ at $x = x_1$ means that

$$E_1(x_1) = E_{1+}(x_1) + E_{1-}(x_1) = E_2(x_1)$$

and

$$E'_1(x_1) = ik_1 \left[ E_{1+}(x_1) - E_{1-}(x_1) \right] = ik_2 E_2(x_1) = E'_2(x_1) \quad .$$

The two unknown amplitudes, $B_1$ of the reflected wave, and $A_2$ of the transmitted, can now be calculated from these two continuity equations. The electric field in the whole space is determined by two expressions incorporating these amplitudes,

$$E_s = \begin{cases} A_1 e^{ik_1 x} + B_1 e^{-ik_1 x} & \text{for } x < x_1 \\ A_2 e^{ik_2 x} & \text{for } x > x_1 \end{cases} \quad .$$

The electric field in the whole space is obtained as a superposition of constituent waves physically existing in regions 1 and 2. By multiplication with the time-dependent phase $\exp(-i\omega t)$, we obtain the complex field strength $E_c$, the real part of which is the physical electric field strength.

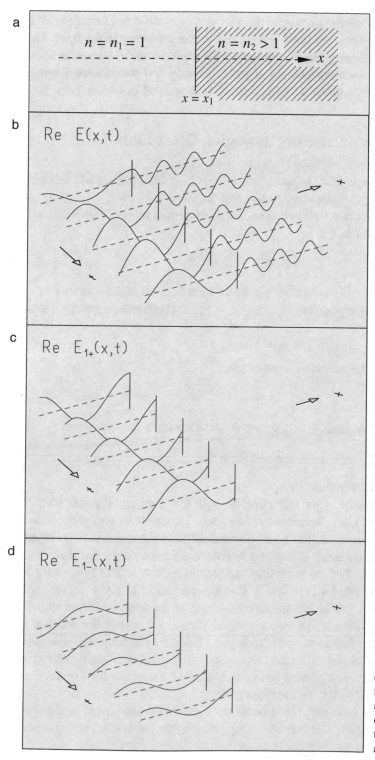

Fig. 2.4. (a) To the right of the plane $x = x_1$, a glass block extends with refractive index $n = n_2$; to the left there is empty space, $n = 1$. (b) Time development of the electric field strength of a harmonic wave which falls from the left onto a glass surface, represented by the vertical line, and is partly reflected by and partly transmitted into the glass. (c) Time development of the incoming wave alone. (d) Time development of the reflected wave alone.

Figure 2.4b gives the time development of this electric field strength. It is easy to see that in the glass there is a harmonic wave moving to the right. The picture in front of the glass is less clear. Figures 2.4c and d therefore show separately the time developments of the incoming and the reflected waves which add up to the total wave to the left of $x_1$, observed in Figure 2.4b.

## 2.3 Light Wave Traveling through a Glass Plate

It is now easy to see what happens when light falls on a glass plate of finite thickness. When the light wave penetrates the front surface at $x = x_1$, again reflection occurs so that we have as before the superposition of two stationary waves in the region $x < x_1$:

$$E_1 = A_1 e^{ik_1 x} + B_1 e^{-ik_1 x} \quad .$$

The wave moving within the glass plate suffers reflection at the rear surface at $x = x_2$, so that the second region, $x_1 < x < x_2$, also contains a superposition of two waves,

$$E_2 = A_2 e^{ik_2 x} + B_2 e^{-ik_2 x} \quad ,$$

which now have the refracted wave number

$$k_2 = n_2 k_1 \quad .$$

Only in the third region, $x_2 < x$, do we observe a single stationary wave

$$E_3 = A_3 e^{ik_1 x}$$

with the original wave number $k_1$.

As a consequence of the reflection on both the front and the rear surface of the glass plate, the reflected wave in region 1 consists of two parts which interfere with each other. The most prominent phenomenon observed under appropriate circumstances is the destructive interference between these two reflected waves, so that no reflection remains in region 1. The light wave is completely transmitted into region 3. This phenomenon is called a *resonance of transmission*. It can be illustrated by looking at the *frequency dependence* of the stationary waves. The upper plot of Figure 2.5 shows the stationary waves for different fixed values of the angular frequency $\omega$, with its magnitude rising from the background to the foreground. A resonance of transmission is recognized through a maximum in the amplitude of the transmitted wave, that is, in the wave to the right of the glass plate.

The signature of a resonance becomes even more prominent in the frequency dependence of the average energy density in the wave. As discussed in Section 2.1, in a vacuum the average energy density has the form

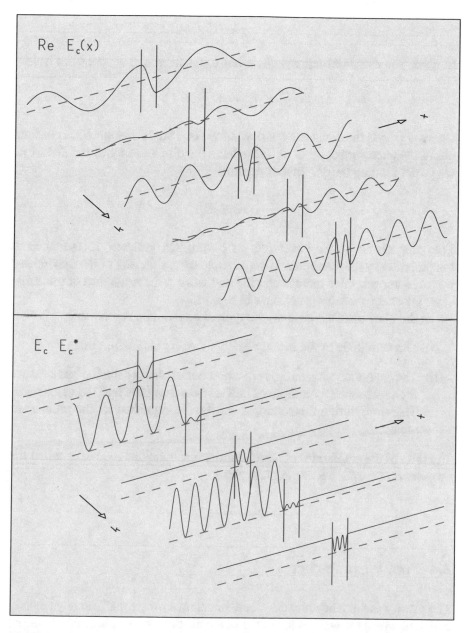

**Fig. 2.5.** Top: Frequency dependence of stationary waves when a harmonic wave is incident from the left on a glass plate. The two vertical lines indicate the thickness of the plate. Small values of the angular frequency $\omega$ are given in the background, large values in the foreground of the picture. Bottom: Frequency dependence of the quantity $E_c E_c^*$ (which except for a factor $n_2$ is proportional to the average energy density) of a harmonic wave incident from the left on a glass plate. The parameters are the same as in part a. At a resonance of transmission, the average energy density is constant in the left region, indicating through the absence of interference wiggles that there is no reflection.

$$w = \frac{\varepsilon_0}{2} E_c E_c^* \quad .$$

In glass, where the refractive index $n$ has to be taken into account, we have

$$w = \frac{\varepsilon \varepsilon_0}{2} E_c E_c^* = n^2 \frac{\varepsilon_0}{2} E_c E_c^* \quad ,$$

where $\varepsilon = n^2$ is the dielectric constant of glass. Thus, although $E_c$ is continuous at the glass surface, $w$ is not. It reflects the discontinuity of $n^2$. Therefore we prefer plotting the continuous quantity

$$\frac{2}{n^2 \varepsilon_0} w = E_c E_c^* \quad .$$

This plot, shown in the lower plot of Figure 2.5, indicates a resonance of transmission by the maximum in the average energy density of the transmitted wave. Moreover, since there is no reflected wave at the resonance of transmission, the energy density is constant in region 1.

In the glass plate we observe the typical pattern of a resonance.

(i)  The amplitude of the average energy density is maximum.

(ii)  The energy density vanishes in a number of places called *nodes* because for a resonance a multiple of half a wavelength fits into the glass plate. Therefore different resonances can be distinguished by the number of nodes.

The ratio of the amplitudes of the transmitted and incident waves is called the *transmission coefficient* of the glass plate,

$$T = \frac{A_3}{A_1} \quad .$$

## 2.4 Free Wave Packet

The plane wave extends into all space, in contrast to any realistic physical situation in which the wave is localized in a finite domain of space. We therefore introduce the concept of a wave packet. It can be understood as a *superposition*, that is, a sum of plane waves of different frequencies and amplitudes. As a first step we concentrate the wave only in the $x$ direction. It still extends through all space in the $y$ and the $z$ direction. For simplicity we start with the sum of two plane waves with equal amplitudes, $E_0$:

$$E = E_1 + E_2 = E_0 \cos(\omega_1 t - k_1 x) + E_0 \cos(\omega_2 t - k_2 x) \quad .$$

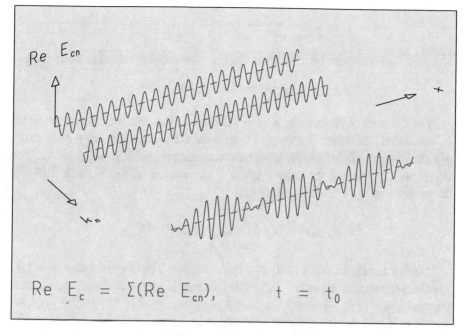

**Fig. 2.6. Superposition of two harmonic waves of slightly different angular frequencies** $\omega_1$ **and** $\omega_2$ **at a fixed moment in time.**

For a fixed time this sum represents a plane wave with two periodic structures. The slowly varying structure is governed by a spatial period,

$$\lambda_- = \frac{4\pi}{|k_2 - k_1|} \quad ,$$

the rapidly varying structure by a wavelength,

$$\lambda_+ = \frac{4\pi}{|k_2 + k_1|} \quad .$$

The resulting wave can be described as the product of a "carrier wave" with the short wavelength $\lambda_+$ and a factor modulating its amplitude with the wavelength $\lambda_-$:

$$E = 2E_0 \cos(\omega_- t - k_- x) \cos(\omega_+ t - k_+ x) \quad ,$$

$$k_\pm = |k_2 \pm k_1|/2 \quad , \qquad \omega_\pm = ck_\pm \quad .$$

Figure 2.6 plots for a fixed moment in time the two waves $E_1$ and $E_2$, and the resulting wave $E$. Obviously, the field strength is now concentrated for the most part in certain regions of space. These regions of great field strength propagate through space with the velocity

$$\frac{\Delta x}{\Delta t} = \frac{\omega_-}{k_-} = c \quad .$$

Now we again use complex field strengths. The superposition is written as

$$E_c = E_0 e^{-i(\omega_1 t - k_1 x)} + E_0 e^{-i(\omega_2 t - k_2 x)} \quad .$$

For the sake of simplicity, we have chosen in this example a superposition of two harmonic waves with equal amplitudes. By constructing a more complicated "sum" of plane waves, we can concentrate the field in a single region of space. To this end we superimpose a continuum of waves with different frequencies $\omega = ck$ and amplitudes:

$$E_c(x,t) = E_0 \int_{-\infty}^{+\infty} dk \, f(k) e^{-i(\omega t - kx)} \quad .$$

Such a configuration is called a *wave packet*. The *spectral function* $f(k)$ specifies the amplitude of the harmonic wave with wave number $k$ and circular frequency $\omega = ck$. We now consider a particularly simple spectral function which is significantly different from zero in the neighborhood of the wave number $k_0$. We choose the *Gaussian* function

$$f(k) = \frac{1}{\sqrt{2\pi}\sigma_k} \exp\left[-\frac{(k-k_0)^2}{2\sigma_k^2}\right] \quad .$$

It describes a bell-shaped spectral function which has its maximum value at $k = k_0$; we assume the value of $k_0$ to be positive, $k_0 > 0$. The width of the region in which the function $f(k)$ is different from zero is characterized by the parameter $\sigma_k$. In short, one speaks of a Gaussian with *width* $\sigma_k$. The Gaussian function $f(k)$ is shown in Figure 2.7a. The factors in front of the exponential are chosen so that the area under the curve equals one. We illustrate the construction of a wave packet by replacing the integration over $k$ by a sum over a finite number of terms,

$$E_c(x,t) \approx \sum_{n=-N}^{N} E_n(x,t) \quad ,$$

$$E_n(x,t) = E_0 \Delta k \, f(k_n) e^{-i(\omega_n t - k_n x)} \quad ,$$

where

$$k_n = k_0 + n\,\Delta k \quad , \qquad \omega_n = ck_n \quad .$$

In Figure 2.7b the different terms of this sum are shown for time $t = 0$, together with their sum, which is depicted in the foreground. The term with the lowest wave number, that is, the longest wavelength, is in the background of the picture. The variation in the amplitudes of the different terms reflects the Gaussian form of the spectral function $f(k)$, which has its maximum, for $k = k_0$, at the center of the picture. On the different terms, the partial waves, the point $x = 0$ is marked by a circle. We observe that the sum over all terms is concentrated around a rather small region near $x = 0$.

Figure 2.7c shows the same wave packet, similarly made up of its partial waves, for later time $t_1 > 0$. The wave packet as well as all partial waves have moved to the right by the distance $ct_1$. The partial waves still carry marks at the phases that were at $x = 0$ at time $t = 0$. The picture makes it clear that all partial waves have the same velocity as the wave packet, which maintains the same shape for all moments in time.

If we perform the integral explicitly, the wave packet takes the simple form

$$
\begin{aligned}
E_c(x,t) &= E_c(ct - x) \\
&= E_0 \exp\left[-\frac{\sigma_k^2}{2}(ct - x)^2\right] \exp\left[-i(\omega_0 t - k_0 x)\right] \quad ,
\end{aligned}
$$

that is,

$$
E(x,t) = \mathrm{Re}\, E_c = E_0 \exp\left[-\frac{\sigma_k^2}{2}(ct - x)^2\right] \cos(\omega_0 t - k_0 x) \quad .
$$

It represents a plane wave propagating in the positive $x$ direction, with a field strength concentrated in a region of the spatial extension $1/\sigma_k$ around point $x = ct$. The time development of the field strength is shown in Figure 2.8b. Obviously, the maximum of the field strength is located at $x = ct$; thus the wave packet moves with the velocity $c$ of light. We call this configuration a *Gaussian wave packet* of spatial width

$$
\Delta x = \frac{1}{\sigma_k} \quad ,
$$

and of wave-number width

$$
\Delta k = \sigma_k \quad .
$$

We observe that a spatial concentration of the wave in the region $\Delta x$ necessarily requires a spectrum of different wave numbers in the interval $\Delta k$ so that

$$
\Delta x\, \Delta k = 1 \quad .
$$

This is tantamount to saying that the sharper the localization of the wave packet in $x$ space, the wider is its spectrum in $k$ space. The original harmonic wave $E = E_0 \cos(\omega t - kx)$ was perfectly sharp in $k$ space ($\Delta k = 0$) and therefore not localized in $x$ space. The time development of the average energy density $w$ shown in Figure 2.8c appears even simpler than that of the field

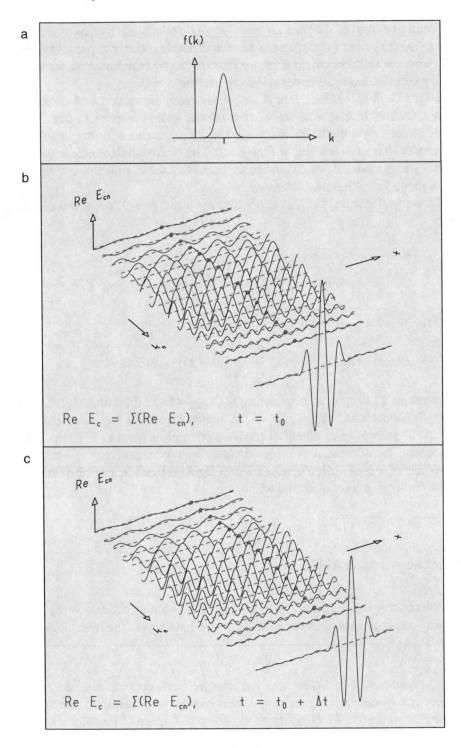

a

$f(k)$

$k$

b

$Re\ E_{cn}$

$k_n$

$Re\ E_c = \Sigma(Re\ E_{cn}),\qquad t = t_0$

c

$Re\ E_{cn}$

$k_n$

$Re\ E_c = \Sigma(Re\ E_{cn}),\qquad t = t_0 + \Delta t$

**Fig. 2.7.**

strength. It is merely a Gaussian traveling with the velocity of light along the $x$ direction. The Gaussian form is easily explained if we remember that

$$w = \frac{\varepsilon_0}{2} E_c E_c^* = \frac{\varepsilon_0}{2} E_0^2 e^{-\sigma_k^2 (ct-x)^2} \quad .$$

We demonstrate the influence of the spectral function on the wave packet by showing in Figure 2.8 spectral functions with two different widths $\sigma_k$. For both we show the time development of the field strength and of the average energy density.

## 2.5 Wave Packet Incident on a Glass Surface

The wave packet, like the plane waves of which it is composed, undergoes reflection and transmission at the glass surface. The upper plot of Figure 2.9 shows the time development of the average energy density in a wave packet moving in from the left. As soon as it hits the glass surface, the already reflected part interferes with the incident wave packet, causing the wiggly structure at the top of the packet. Part of the packet enters the glass, moving with a velocity reduced by the refractive index. For this reason it is compressed in space. The remainder is reflected and moves to the left as a regularly shaped wave packet as soon as it has left the region in front of the glass where interference with the incident packet occurs.

We now demonstrate that the wiggly structure in the interference region is caused by the fast spatial variation of the carrier wave characterized by its wavelength. To this end let us examine the time development of the field strength in the packet, shown in the lower plot of Figure 2.9. Indeed, the spatial variation of the field strength has twice the wavelength of the average energy density in the interference region.

Another way of studying the reflection and transmission of the packet is to look separately at the average energy densities of the constituent waves, namely the incoming, transmitted, and reflected waves. We show these constituent waves in both regions 1, a vacuum, and 2, the glass, although they contribute physically only in either the one or the other. Figure 2.10 gives

**Fig. 2.7.** (a) Gaussian spectral function describing the amplitudes of harmonic waves of different wave numbers $k$. (b) Construction of a light wave packet as a sum of harmonic waves of different wavelengths and amplitudes. For time $t = 0$ the different terms of the sum are plotted, starting with the contribution of the longest wavelength in the background. Points $x = 0$ are indicated as circles on the partial waves. The resulting wave packet is shown in the foreground. (c) The same as part b, but for time $t_1 > 0$. The phases that were at $x = 0$ for $t = 0$ have moved to $x_1 = ct_1$ for all partial waves. The wave packet has consequently moved by the same distance and retained its shape.

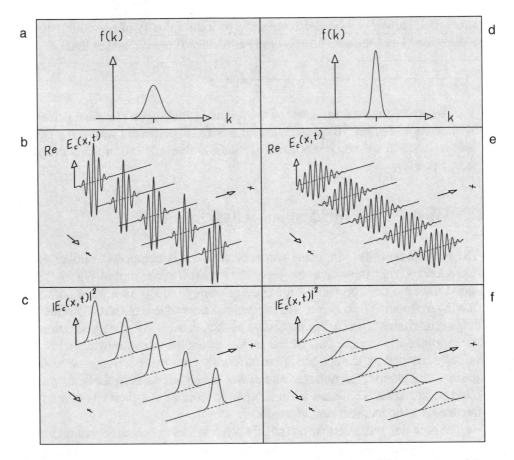

**Fig. 2.8. (a, d) Spectral functions, (b, e) time developments of the field strength, and (c, f) time developments of the average energy density for two different Gaussian wave packets.**

their time developments. All three have a smooth bell-shaped form and no wiggles, even in the interference region. The time developments of the field strengths of the constituent waves are shown in Figure 2.11. The observed average energy density of Figure 2.9 corresponds to the absolute square of the sum of the incoming and reflected field strengths in the region in front of the glass and, of course, not to the sum of the average energy densities of these two constituent fields. Their interference pattern shows half the wavelength of the carrier waves.

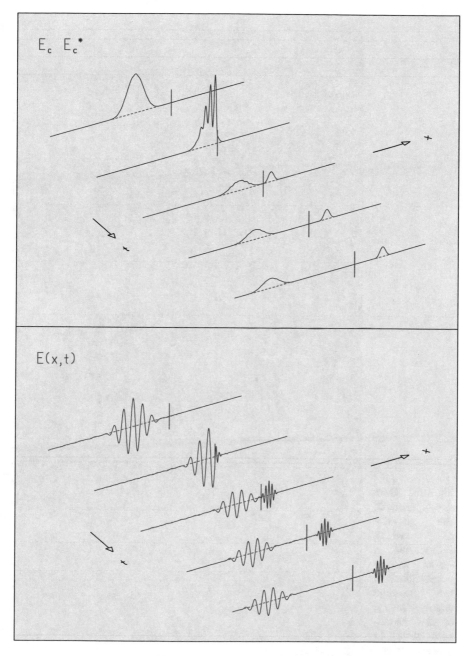

**Fig. 2.9.** Time developments of the quantity $E_c E_c^*$ (which except for a factor $n^2$ is proportional to the average energy density) and of the field strength in a wave packet of light falling onto a glass surface where it is partly reflected and partly transmitted through the surface. The glass surface is indicated by the vertical line.

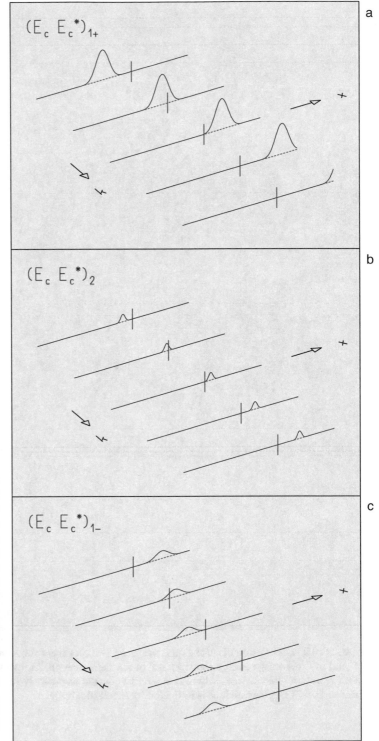

**Fig. 2.10. Time developments of the quantity $E_c E_c^*$ (which except for a factor $n^2$ is proportional to the average energy density) of the constituent waves in a wave packet of light incident on a glass surface: (a) incoming wave, (b) transmitted wave, and (c) reflected wave.**

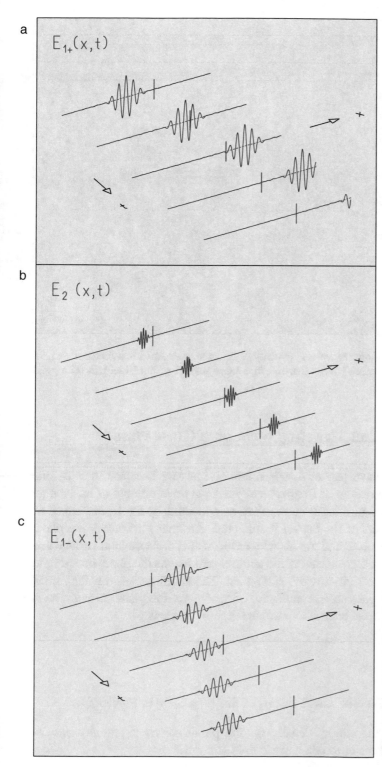

**Fig. 2.11.** Time developments of the electric field strengths of the constituent waves in a wave packet of light incident on a glass surface: (a) incoming wave, (b) transmitted wave, and (c) reflected wave.

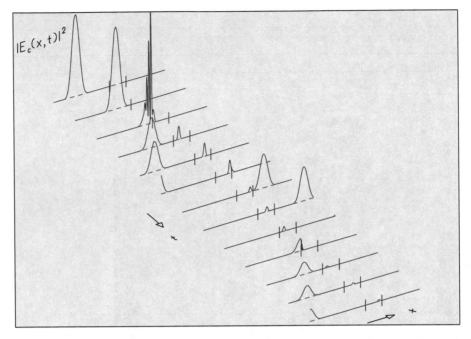

$|E_c(x,t)|^2$

**Fig. 2.12. Time development of the quantity $E_c E_c^*$ (which except for a factor $n^2$ is proportional to the average energy density) in a wave packet of light incident on a glass plate.**

## 2.6 Wave Packet Traveling through a Glass Plate

Let us study a wave packet that is relatively narrow in space, that is, one containing a wide range of frequencies. The time development of its average energy density (Figure 2.12) shows that, as expected, at the front surface of the glass plate part of the packet is reflected. Another part enters the plate, where it is compressed and travels with reduced speed. At the rear surface this packet is again partly reflected while another part leaves the plate, traveling to the right with the original width and speed. The small packet traveling back and forth in the glass suffers multiple reflections on the glass surfaces, each time losing part of its energy to packets leaving the glass.

## Problems

2.1. Estimate the refractive index $n_2$ of the glass plate in Figure 2.4b.

2.2. Calculate the energy density for the plane electromagnetic wave described by the complex electric field strength

$$E_{\mathrm{c}} = E_0 \mathrm{e}^{-\mathrm{i}(\omega t - kx)}$$

and show that its average over a temporal period $T$ is $\omega = (\varepsilon_0/2)E_{\mathrm{c}}E_{\mathrm{c}}^*$.

2.3. Give the qualitative reason why the resonance phenomena in Figure 2.5 (top) occurs for the wavelengths

$$\lambda = \ell\frac{nd}{2} \quad , \qquad \ell = 1, 2, 3 \ldots \quad .$$

Use the continuity condition of the electric field strength and its derivative. Here $n$ is the refractive index of the glass plate of thickness $d$.

2.4. Calculate the ratio of the frequencies of the two electric field strengths, as they are plotted in Figure 2.6, from the beat in their superposition.

2.5. The one-dimensional wave packet of light does not show any dispersion, that is, spreading with time. What causes the dispersion of a wave packet of light confined in all three spatial dimensions?

2.6. Estimate the refractive index of the glass, using the change in width or velocity of the light pulse in Figure 2.9 (top).

2.7. Verify in Figure 2.12 that the stepwise reduction of the amplitude of the pulse within the glass plate proceeds with approximately the same reduction factor, thus following on the average an exponential decay law.

2.8. Calculate energy $E$ and momentum $p$ of a photon of blue ($\lambda = 450 \times 10^{-9}\,\mathrm{m}$), green ($\lambda = 530 \times 10^{-9}\,\mathrm{m}$), yellow ($\lambda = 580 \times 10^{-9}\,\mathrm{m}$), and red ($\lambda = 700 \times 10^{-9}\,\mathrm{m}$) light. Use Einstein's formula $E = Mc^2$ to calculate the relativistic mass of the photon. Give the results in SI units.

# 3. Probability Waves of Matter

## 3.1 de Broglie Waves

For a particle with a finite rest mass $m$, which moves with a velocity $v$ slow compared to the velocity of light, the relation between energy and momentum is

$$E = \frac{p^2}{2m} \quad , \qquad p = mv \quad .$$

In Section 1.3 we saw that such a particle possesses wave properties, in particular an angular frequency $\omega$ and a wave number $k$, which are related to its energy and its momentum, respectively,

$$E = \hbar\omega \quad , \qquad p = \hbar k \quad .$$

In analogy to the electromagnetic wave $\mathrm{Re}\, E_c$ with $E_c = A\,e^{-i(\omega t - kx)}$ of Section 2.1 we can write down a wave function for the particle of momentum $p$,

$$
\begin{aligned}
\psi_p(x,t) &= \frac{1}{(2\pi\hbar)^{1/2}} \exp[-i(\omega t - kx)] \\
&= \frac{1}{(2\pi\hbar)^{1/2}} \exp\left[-\frac{i}{\hbar}(Et - px)\right] \quad ,
\end{aligned}
$$

which we call a *de Broglie wave* of matter. The factor in front of the exponential is chosen for convenience. The *phase velocity* of a de Broglie wave is

$$v_\mathrm{p} = \frac{E}{p} = \frac{p}{2m}$$

and is thus different from the particle velocity $v = p/m$.

## 3.2 Wave Packet, Dispersion

The harmonic de Broglie waves, like the harmonic electric waves, are not localized in space and therefore are not suited to describing a particle. To

S. Brandt and H.D. Dahmen, *The Picture Book of Quantum Mechanics*,
DOI 10.1007/978-1-4614-3951-6_3, © Springer Science+Business Media New York 2012

localize a particle in space, we again have to superimpose harmonic waves to form a wave packet. To keep things simple, we first restrict ourselves to discussing a one-dimensional wave packet.

For the spectral function we again choose a Gaussian function,[1]

$$f(p) = \frac{1}{(2\pi)^{1/4}\sqrt{\sigma_p}} \exp\left[-\frac{(p-p_0)^2}{4\sigma_p^2}\right] \quad .$$

The corresponding de Broglie wave packet is then

$$\psi(x,t) = \int_{-\infty}^{+\infty} f(p)\psi_p(x-x_0,t)\,dp \quad .$$

For the de Broglie wave packet, as for the light wave packet, we first approximate the integral by a sum,

$$\psi(x,t) \approx \sum_{n=-N}^{N} \psi_n(x,t) \quad ,$$

where the $\psi_n(x,t)$ are harmonic waves for different values $p_n = p_0 + n\,\Delta p$ multiplied by the spectral weight $f(p_n)\Delta p$,

$$\psi_n(x,t) = f(p_n)\psi(x-x_0,t)\,\Delta p \quad .$$

Figure 3.1a shows the real parts $\mathrm{Re}\,\psi_n(x,t)$ of the harmonic waves $\psi_n(x,t)$ as well as their sum being equal to the real part $\mathrm{Re}\,\psi(x,t)$ of the wave function $\psi(x,t)$ for the wave packet at time $t = t_0 = 0$. The point $x = x_0$ is marked on each harmonic wave. In Figure 3.1b the real parts $\mathrm{Re}\,\psi_n(x,t)$ and their sum $\mathrm{Re}\,\psi(x,t)$ are shown at later time $t = t_1$. Because of their different phase velocities, the partial waves have moved by different distances $\Delta x_n = v_n(t_1 - t_0)$ where $v_n = p_n/(2m)$ is the phase velocity of the harmonic wave of momentum $p_n$. This effect broadens the extension of the wave packet.

The integration over $p$ can be carried out so that the explicit expression for the wave packet has the form

$$\psi(x,t) = M(x,t)e^{i\phi(x,t)} \quad .$$

Here the exponential function represents the carrier wave with a *phase* $\phi$ varying rapidly in space and time. The bell-shaped amplitude function

---

[1]We have chosen this spectral function to correspond to the square root of the spectral function that was used in Section 2.4 to construct a wave packet of light. Since the area under the spectral function $f(k)$ of Section 2.4 was equal to one, the area under $[f(p)]^2$ is now equal to one. This guarantees that the normalization condition of the wave function $\psi$ in the next section will be fulfilled.

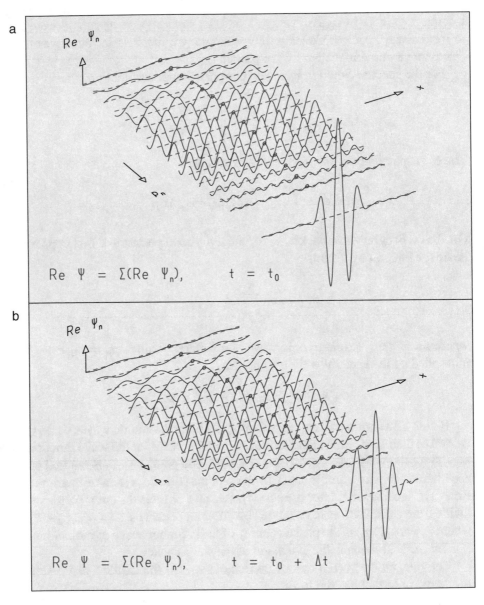

**Fig. 3.1. Construction of a wave packet as a sum of harmonic waves** $\psi_n$ **of different momenta and consequently of different wavelengths. Plotted are the real parts of the wave functions. The terms of different momenta and different amplitudes begin with the one of longest wavelength in the background. In the foreground is the wave packet resulting from the summation. (a) The situation for time** $t = t_0$. **All partial waves are marked by a circle at point** $x = x_0$. **(b) The same wave packet and its partial waves at time** $t_1 > t_0$. **The partial waves have moved different distances** $\Delta x_n = v_n(t_1 - t_0)$ **because of their different phase velocities** $v_n$, **as indicated by the circular marks which have kept their phase with respect to those in part a. Because of the different phase velocities, the wave packet has changed its form and width.**

$$M(x,t) = \frac{1}{(2\pi)^{1/4}\sqrt{\sigma_x}} \exp\left[-\frac{(x - x_0 - v_0 t)^2}{4\sigma_x^2}\right]$$

travels in $x$ direction with the *group velocity*

$$v_0 = \frac{p_0}{m} \quad .$$

The group velocity is indeed the particle velocity and different from the phase velocity. The localization in space is given by

$$\sigma_x^2 = \frac{\hbar^2}{4\sigma_p^2}\left(1 + \frac{4\sigma_p^4\, t^2}{\hbar^2\, m^2}\right) \quad .$$

This formula shows that the spatial extension $\sigma_x$ of the wave packet increases with time. This phenomenon is called *dispersion*. Figure 3.2 shows the time developments of the real and imaginary parts of two wave packets with different group velocities and widths. We easily observe the dispersion of the wave packets in time. The fact that a wave packet comprises a whole range of momenta is the physical reason why it disperses. Its components move with different velocities, thus spreading the packet in space.

The function $\phi(x,t)$ determines the phase of the carrier wave. It has the form

$$\phi(x,t) = \frac{1}{\hbar}\left[p_0 + \frac{\sigma_p^2}{\sigma_x^2}\frac{v_0 t}{2p_0}(x - x_0 - v_0 t)\right](x - x_0 - v_0 t) + \frac{p_0}{2\hbar}v_0 t - \frac{\alpha}{2}$$

with

$$\tan\alpha = \frac{2\,\sigma_p^2}{\hbar\, m}t \quad .$$

For fixed time $t$ it represents the phase of a harmonic wave modulated in wave number. The effective wave number $k_{\text{eff}}$ is the factor in front of $x - x_0 - v_0 t$ and is given by

$$k_{\text{eff}}(x) = \frac{1}{\hbar}\left[p_0 + \frac{\sigma_p^2}{\sigma_x^2}\frac{v_0 t}{2p_0}(x - x_0 - v_0 t)\right] \quad .$$

At the value $x = \langle x \rangle$ corresponding to the maximum value of the bell-shaped amplitude modulation $M(x,t)$, that is, its position average

$$\langle x \rangle = x_0 + v_0 t \quad ,$$

the effective wave number is simply equal to the wave number that corresponds to the average momentum $p_0$ of the spectral function,

$$k_0 = \frac{1}{\hbar}p_0 = \frac{1}{\hbar}mv_0 \quad .$$

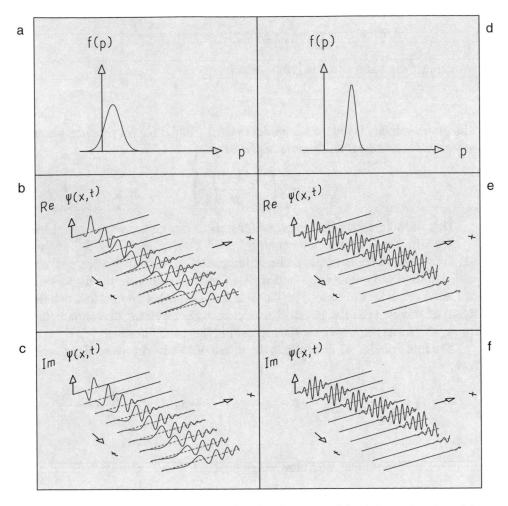

**Fig. 3.2. (a, d) Spectral functions and time developments of (b, e) the real parts and (c, f) the imaginary parts of the wave functions for two different wave packets. The two packets have different group velocities and different widths and spread differently with time.**

For values $x > x_0 + v_0 t$, that is, in front of the average position $\langle x \rangle$ of the moving wave packet, the effective wave number increases,

$$k_{\text{eff}}(x > x_0 + v_0 t) > k_0 \quad ,$$

so that the local wavelength

$$\lambda_{\text{eff}}(x) = \frac{2\pi}{|k_{\text{eff}}(x)|}$$

decreases.

For values $x < x_0 + v_0 t$, that is, behind the average position $\langle x \rangle$, the effective wave number decreases,

$$k_{\text{eff}}(x < x_0 + v_0 t) < k_0 \quad .$$

This decrease leads to negative values of $k_{\text{eff}}$ of large absolute value, which, far behind the average position, makes the wavelengths $\lambda_{\text{eff}}(x)$ short again. This wave number modulation can easily be verified in Figures 3.1 and 3.2. For a wave packet at rest, that is, $p_0 = 0$, $v_0 = p_0/m = 0$, the effective wave number

$$k_{\text{eff}}(x) = \frac{1}{\hbar} \frac{\sigma_p^2}{\sigma_x^2} \frac{t}{2m}(x - x_0)$$

has the same absolute value to the left and to the right of the average position $x_0$. This implies a decrease of the effective wavelength that is symmetric on both sides of $x_0$. Figure 3.4 corroborates this statement.

## 3.3 Probability Interpretation, Uncertainty Principle

Following Max Born (1926), we interpret the wave function $\psi(x,t)$ as follows. Its absolute square

$$\rho(x,t) = |\psi(x,t)|^2 = M^2(x,t)$$

is identified with the *probability density* for observing the particle at position $x$ and time $t$, that is, the probability of observing the particle at a given time $t$ in the space region between $x$ and $x + \Delta x$ is $\Delta P = \rho(x,t)\Delta x$. This is plausible since $\rho(x,t)$ is positive everywhere. Furthermore, its integral over all space is equal to one for every moment in time so that the *normalization condition*

$$\int_{-\infty}^{+\infty} |\psi(x,t)|^2 \, dx = \int_{-\infty}^{+\infty} \psi^*(x,t)\psi(x,t)\, dx = 1$$

holds.

Notice, that there is a strong formal similarity between the average energy density $w(x,t) = \varepsilon_0 |E_c(x,t)|^2/2$ of a light wave and the probability density $\rho(x,t)$. Because of the probability character, the wave function $\psi(x,t)$ is not a field strength, since the effect of a field strength must be measurable wherever the field is not zero. A probability density, however, determines the probability that a particle, which can be point-like, will be observed at a given position. This probability interpretation is, however, restricted to normalized wave functions. Since the integral over the absolute square of a harmonic plane wave is

$$\frac{1}{2\pi\hbar} \int_{-\infty}^{+\infty} \exp\left[\frac{i}{\hbar}(Et - px)\right] \exp\left[-\frac{i}{\hbar}(Et - px)\right] dx = \frac{1}{2\pi\hbar} \int_{-\infty}^{+\infty} dx$$

and diverges, the absolute square $|\psi(x,t)|^2$ of a harmonic plane wave cannot be considered a probability density. We shall call the absolute square of a wave function that cannot be normalized its *intensity*. Even though wave functions that cannot be normalized have no immediate physical significance, they are of great importance for the solution of problems. We have already seen that normalizable wave packets can be composed of these wave functions. This situation is similar to the one in classical electrodynamics in which the plane electromagnetic wave is indispensable for the solution of many problems. Nevertheless, a harmonic plane wave cannot exist physically, for it would fill all of space and consequently have infinite energy.

Figure 3.3 shows the time developments of the probability densities of the two Gaussian wave packets given in Figure 3.2. Underneath the two time developments the motion of a classical particle with the same velocity is presented. We see that the center of the Gaussian wave packet moves in the exact same way as the classical particle. But whereas the classical particle at every instant in time occupies a well-defined position in space, the quantum-mechanical wave packet has a finite width $\sigma_x$. It is a measure for the size of the region in space surrounding the classical position in which the particle will be found. The fact that the wave packet disperses in time means that the location of the particle becomes more and more uncertain with time.

The dispersion of a wave packet with zero group velocity is particularly striking. Without changing position it becomes wider and wider as time goes by (Figure 3.4a).

It is interesting to study the behavior of the real and imaginary parts of the wave packet at rest. Their time developments are shown in Figures 3.4b and 3.4c. Starting from a wave packet that at initial time $t = 0$ was chosen to be a real Gaussian packet, waves travel in both positive and negative $x$ directions. Obviously, the harmonic waves with the highest phase velocities, those whose wiggles escape the most quickly from the original position $x = 0$, possess the shortest wavelengths. The spreading of the wave packet can be explained in another way. Because the original wave packet at $t = 0$ contains spectral components with positive and negative momenta, it spreads in space as time elapses.

The probability interpretation of the wave function now suggests that we use standard concepts of probability calculus, in particular the expectation value and variance. The *expectation value* or *average value of the position* of a particle described by a wave function $\psi(x,t)$ is

$$\langle x \rangle = \int_{-\infty}^{+\infty} x\rho(x,t)\,dx = \int_{-\infty}^{+\infty} \psi^*(x,t)x\psi(x,t)\,dx \quad,$$

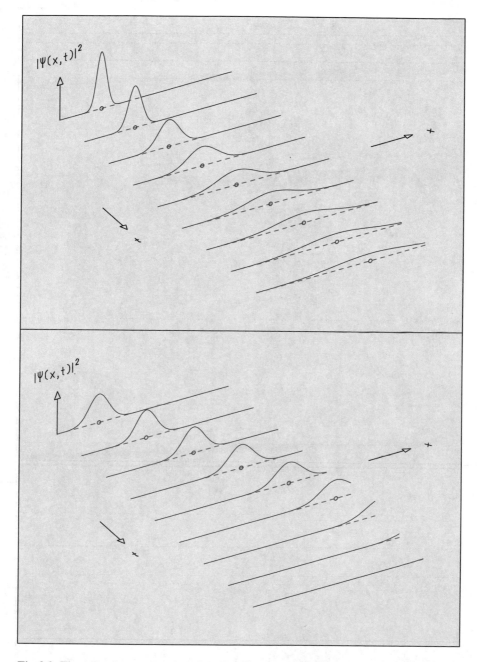

**Fig. 3.3. Time developments of the probability densities for the two wave packets of Figure 3.2. The two packets have different group velocities and different widths. Also shown, by the small circles, is the position of a classical particle moving with a velocity equal to the group velocity of the packet.**

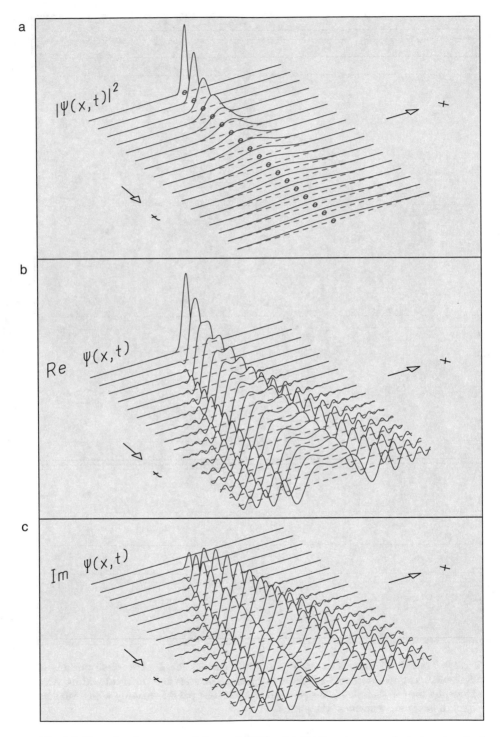

**Fig. 3.4. Time developments of the probability density for a wave packet at rest and of the real part and the imaginary part of its wave function.**

which, in general, remains a function of time. For a Gaussian wave packet the integration indeed yields

$$\langle x \rangle = x_0 + v_0 t \quad , \qquad v_0 = \frac{p_0}{m} \quad ,$$

corresponding to the trajectory of classical unaccelerated motion. We shall therefore interpret the Gaussian wave packet of de Broglie waves as a quantum-mechanical description of the unaccelerated *motion of a particle*, that is, a particle moving with constant velocity. Actually, the Gaussian form of the spectral function $f(k)$ allows the explicit calculation of the wave packet. With this particular spectral function, the wave function $\psi(x,t)$ can be given in closed form.

The *variance of the position* is the expectation value of the square of the difference between the position and its expectation:

$$
\begin{aligned}
\mathrm{var}(x) &= \left\langle (x - \langle x \rangle)^2 \right\rangle \\
&= \int_{-\infty}^{+\infty} \psi^*(x,t)(x - \langle x \rangle)^2 \psi(x,t)\,dx \quad .
\end{aligned}
$$

Again, for the Gaussian wave packet the integral can be carried out to give

$$\mathrm{var}(x) = \sigma_x^2 = \frac{\hbar^2}{4\sigma_p^2}\left(1 + \frac{4\sigma_p^4}{\hbar^2}\frac{t^2}{m^2}\right) \quad ,$$

which agrees with the formula quoted in Section 3.2.

Calculation of the expectation value of the momentum of a wave packet

$$\psi(x,t) = \int_{-\infty}^{+\infty} f(p)\psi_p(x - x_0, t)\,dp$$

is carried out with the direct help of the spectral function $f(p)$, that is,

$$\langle p \rangle = \int_{-\infty}^{+\infty} p|f(p)|^2\,dp \quad .$$

For the spectral function $f(p)$ of the Gaussian wave packet given at the beginning of Section 3.2, we find

$$\langle p \rangle = \int_{-\infty}^{+\infty} p\frac{1}{\sqrt{2\pi}\sigma_p}\exp\left[-\frac{(p - p_0)^2}{2\sigma_p^2}\right]dp \quad .$$

We replace the factor $p$ by the identity

$$p = p_0 + (p - p_0) \quad .$$

Since the exponential in the integral above is an even function in the variable $p - p_0$, the integral

$$\int_{-\infty}^{+\infty} (p - p_0) \frac{1}{\sqrt{2\pi}\sigma_p} \exp\left[-\frac{(p - p_0)^2}{2\sigma_p^2}\right] dp = 0$$

vanishes, for the contributions in the intervals $-\infty < p < p_0$ and $p_0 < p < \infty$ cancel. The remaining term is the product of the constant $p_0$ and the normalization integral,

$$\int_{-\infty}^{+\infty} |f(p)|^2 dp = 1 \quad ,$$

so that we find

$$\langle p \rangle = p_0 \quad .$$

This result is not surprising, for the Gaussian spectral function gives the largest weight to momentum $p_0$ and decreases symmetrically to the left and right of this value. At the end of Section 3.2, we found $v_0 = p_0/m$ as the group velocity of the wave packet. Putting the two findings together, we have discovered that the momentum expectation value of a free, unaccelerated Gaussian wave packet is the same as the momentum of a free, unaccelerated particle of mass $m$ and velocity $v_0$ in classical mechanics:

$$\langle p \rangle = p_0 = m v_0 \quad .$$

The expectation value of momentum can also be calculated directly from the wave function $\psi(x,t)$. We have the simple relation

$$\frac{\hbar}{i} \frac{\partial}{\partial x} \psi_p(x - x_0, t) = \frac{\hbar}{i} \frac{\partial}{\partial x} \left\{ \frac{1}{(2\pi\hbar)^{1/2}} \exp\left[-\frac{i}{\hbar}(Et - px)\right] \right\}$$
$$= p \psi_p(x - x_0, t) \quad .$$

This relation translates the momentum variable $p$ into the *momentum operator*

$$p \rightarrow \frac{\hbar}{i} \frac{\partial}{\partial x} \quad .$$

The momentum operator allows us to calculate the expectation value of momentum from the following formula:

$$\langle p \rangle = \int_{-\infty}^{+\infty} \psi^*(x,t) \frac{\hbar}{i} \frac{\partial}{\partial x} \psi(x,t) \, dx \quad .$$

It is completely analogous to the formula for the expectation value of position given earlier. We point out that the operator appears between the functions $\psi^*(x,t)$ and $\psi(x,t)$, thus acting on the second factor only. To verify this formula, we replace the wave function $\psi(x,t)$ by its representation in terms of the spectral function:

$$\langle p \rangle = \int_{-\infty}^{+\infty} \psi^*(x,t) \frac{\hbar}{i} \frac{\partial}{\partial x} \int_{-\infty}^{+\infty} f(p)\psi_p(x - x_0, t)\, dp\, dx$$

$$= \int_{-\infty}^{+\infty} \int_{-\infty}^{+\infty} \psi^*(x,t)\psi_p(x - x_0, t)\, dx\, pf(p)\, dp \quad .$$

The inner integral

$$\int_{-\infty}^{+\infty} \psi^*(x,t)\psi_p(x - x_0, t)\, dx$$

$$= \int_{-\infty}^{+\infty} \psi^*(x,t) \frac{1}{(2\pi\hbar)^{1/2}} \exp\left\{ -\frac{i}{\hbar} \left[ Et - p(x - x_0) \right] \right\} dx$$

is by Fourier's theorem the inverse of the representation

$$\psi^*(x,t) = \int_{-\infty}^{+\infty} f^*(p)\psi_p^*(x - x_0, t)\, dp$$

$$= \frac{1}{(2\pi\hbar)^{1/2}} \int_{-\infty}^{+\infty} f^*(p)\exp\left\{ \frac{i}{\hbar} \left[ Et - p(x - x_0) \right] \right\} dp$$

of the complex conjugate of the wave packet $\psi(x,t)$. Thus we have

$$\int_{-\infty}^{+\infty} \psi^*(x,t)\psi_p(x - x_0, t)\, dx = f^*(p) \quad .$$

Substituting this result for the inner integral of the expression for $\langle p \rangle$, we rediscover the expectation value of momentum in the form

$$\langle p \rangle = \int_{-\infty}^{+\infty} f^*(p)pf(p)\, dp = \int_{-\infty}^{+\infty} p|f(p)|^2\, dp \quad .$$

This equation justifies the identification of momentum $p$ with the operator $(\hbar/i)(\partial/\partial x)$ acting on the wave function. The *variance of the momentum* for a wave packet is

$$\mathrm{var}(p) = \langle (p - \langle p \rangle)^2 \rangle = \int_{-\infty}^{+\infty} \psi^*(x,t) \left( \frac{\hbar}{i} \frac{\partial}{\partial x} - p_0 \right)^2 \psi(x,t)\, dx \quad .$$

For our Gaussian packet we have

$$\mathrm{var}(p) = \sigma_p^2$$

again independent of time because momentum is conserved.

The square root of the variance of the position,

$$\Delta x = \sqrt{\mathrm{var}(x)} = \sigma_x \quad ,$$

determines the width of the wave packet in the position variable $x$ and therefore is a measure of the *uncertainty* about where the particle is located. By the same token, the corresponding uncertainty about the momentum of the particle is

$$\Delta p = \sqrt{\mathrm{var}(p)} = \sigma_p \quad .$$

For our Gaussian wave packet we found the relation

$$\sigma_x = \frac{\hbar}{2\sigma_p}\left(1 + \frac{4\sigma_p^4\, t^2}{\hbar^2\, m^2}\right)^{1/2} \quad .$$

For time $t = 0$ this relation reads

$$\sigma_x \sigma_p = \frac{\hbar}{2} \quad .$$

For later moments in time, the product becomes even larger so that, in general,

$$\Delta x \cdot \Delta p \geq \frac{\hbar}{2} \quad .$$

This relation expresses the fact that the product of uncertainties in position and momentum cannot be smaller than the fundamental Planck's constant $h$ divided by $4\pi$.

We have just stated the *uncertainty principle*, which is valid for wave packets of all forms. It was formulated by Werner Heisenberg in 1927. This relation says, in effect, that a small uncertainty in localization can only be achieved at the expense of a large uncertainty in momentum and vice versa. Figure 3.5 illustrates this principle by comparing the time development of the probability density $\rho(x,t)$ and the square of the spectral function $f^2(p)$. The latter, in fact, is the probability density in momentum. Looking at the spreading of the wave packets with time, we see that the initially narrow wave packet (Figure 3.5, top right) becomes quickly wide in space, whereas the initially wide wave packet (Figure 3.5, bottom right) spreads much more slowly. Actually, this behavior is to be expected. The spatially narrow wave packet requires a wide spectral function in momentum space. Thus it comprises components with a wide range of velocities. They, in turn, cause a quick dispersion of the packet in space compared to the initially wider packet with a narrower spectral function (Figures 3.5, bottom left and bottom right).

At its initial time $t = 0$ the Gaussian wave packet discussed at the beginning of Section 3.2 has the smallest spread in space and momentum because Heisenberg's uncertainty principle is fulfilled in the equality form $\sigma_x \cdot \sigma_p = \hbar/2$. The wave function at $t = 0$ takes the simple form

$$\begin{aligned}
\psi(x,0) &= \frac{1}{(2\pi)^{1/4}\sqrt{\sigma_x}}\exp\left[-\frac{(x-x_0)^2}{4\sigma_x^2}\right]\exp\left[\frac{i}{\hbar}p_0(x-x_0)\right] \\
&= M(x,0)\exp\left[i\phi(x,0)\right] \quad .
\end{aligned}$$

The bell-shaped amplitude function $M(x,0)$ is centered around the position $x_0$ with the width $\sigma_x$; $\phi$ is the phase of the wave function at $t = 0$ and has the simple linear dependence

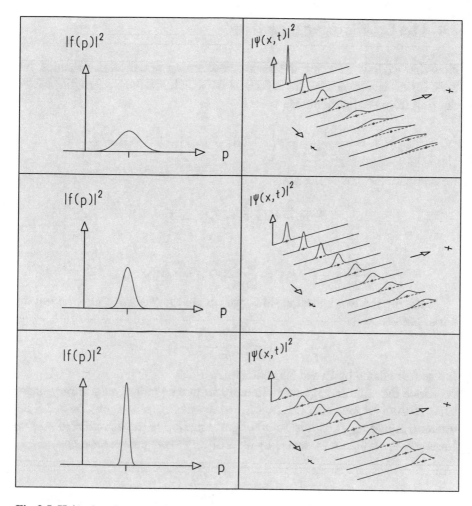

**Fig. 3.5. Heisenberg's uncertainty principle. For three different Gaussian wave packets the square $f^2(p)$ of the spectral function is shown on the left, the time development of the probability density in space on the right. All three packets have the same group velocity but different widths $\sigma_p$ in momentum. At $t = 0$ the widths $\sigma_x$ in space and $\sigma_p$ in momentum fulfill the equality $\sigma_x \sigma_p = \hbar/2$. For later moments in time the wave packets spread in space so that $\sigma_x \sigma_p > \hbar/2$.**

$$\phi(x,0) = \frac{1}{\hbar} p_0(x - x_0) \quad .$$

This phase ensures that the wave packet at $t = 0$ stands for a particle with an average momentum $p_0$. We shall use this observation when we have to prepare wave functions for the initial state of a particle with the initial conditions $\langle x \rangle = x_0$, $\langle p \rangle = p_0$ at the initial moment of time $t = t_0$.

## 3.4 The Schrödinger Equation

Now that we have introduced the wave description of particle mechanics, we look for a *wave equation*, the solutions of which are the de Broglie waves. Starting from the harmonic wave

$$\psi_p(x,t) = \frac{1}{(2\pi\hbar)^{1/2}} \exp\left[-\frac{i}{\hbar}(Et - px)\right] \quad , \qquad E = \frac{p^2}{2m} \quad ,$$

we compare the two expressions

$$i\hbar\frac{\partial}{\partial t}\psi_p(x,t) = E\psi_p(x,t)$$

and

$$-\frac{\hbar^2}{2m}\frac{\partial^2}{\partial x^2}\psi_p(x,t) = \frac{p^2}{2m}\psi_p(x,t) = E\psi_p(x,t) \quad .$$

Equating the two left-hand sides, we obtain the *Schrödinger equation* for a free particle,

$$i\hbar\frac{\partial}{\partial t}\psi_p(x,t) = -\frac{\hbar^2}{2m}\frac{\partial^2}{\partial x^2}\psi_p(x,t) \quad .$$

It was formulated by Erwin Schrödinger in 1926.

Since the solution $\psi_p$ occurs linearly in this equation, an arbitrary linear superposition of solutions, that is, any wave packet, is also a solution of Schrödinger's equation. Thus this Schrödinger equation is the *equation of motion* for any free particle represented by an arbitrary wave packet $\psi(x,t)$:

$$i\hbar\frac{\partial}{\partial t}\psi(x,t) = -\frac{\hbar^2}{2m}\frac{\partial^2}{\partial x^2}\psi(x,t) \quad .$$

In the spirit of representing physical quantities by differential operators, as we did for momentum, we can now represent kinetic energy $T$, which is equal to the total energy of the free particle $T = p^2/(2m)$, by

$$T \rightarrow \frac{1}{2m}\left(\frac{\hbar}{i}\frac{\partial}{\partial x}\right)\left(\frac{\hbar}{i}\frac{\partial}{\partial x}\right) = -\frac{\hbar^2}{2m}\frac{\partial^2}{\partial x^2} \quad .$$

The equation can be generalized to describe the motion of a particle in a force field represented by a potential energy $V(x)$. This is done by replacing the kinetic energy $T$ with the total energy,

$$E = T + V \rightarrow -\frac{\hbar^2}{2m}\frac{\partial^2}{\partial x^2} + V(x) \quad .$$

With this substitution we obtain the *Schrödinger equation for the motion of a particle in a potential* $V(x)$:

$$i\hbar\frac{\partial}{\partial t}\psi(x,t) = -\frac{\hbar^2}{2m}\frac{\partial^2}{\partial x^2}\psi(x,t) + V(x)\psi(x,t) \quad .$$

We now denote the operator of total energy by the symbol

$$H = -\frac{\hbar^2}{2m}\frac{\partial}{\partial x} + V(x) \quad .$$

In analogy to the Hamilton function of classical mechanics, operator $H$ is called the *Hamilton operator* or *Hamiltonian*. With its help the Schrödinger equation for the motion of a particle under the influence of a potential takes the form

$$i\hbar\frac{\partial}{\partial t}\psi(x,t) = H\psi(x,t) \quad .$$

At this stage we should point out that the Schrödinger equation, generalized to three spatial dimensions and many particles, is the fundamental law of nature for all of nonrelativistic particle physics and chemistry. The rest of this book will be dedicated to the pictorial study of the simple phenomena described by the Schrödinger equation.

## 3.5 Bivariate Gaussian Probability Density

To facilitate the physics discussion in the next section we now introduce a *Gaussian probability density of two variables* $x_1$ and $x_2$ and demonstrate its properties. The bivariate Gaussian probability density is defined by

$$\rho(x_1,x_2) = A\exp\left\{-\frac{1}{2(1-c^2)}\left[\frac{(x_1-\langle x_1\rangle)^2}{\sigma_1^2}\right.\right.$$
$$\left.\left. -2c\frac{(x_1-\langle x_1\rangle)}{\sigma_1}\frac{(x_2-\langle x_2\rangle)}{\sigma_2} + \frac{(x_2-\langle x_2\rangle)^2}{\sigma_2^2}\right]\right\} \quad .$$

The normalization constant

$$A = \frac{1}{2\pi\sigma_1\sigma_2\sqrt{1-c^2}}$$

ensures that the probability density is properly normalized:

$$\int_{-\infty}^{+\infty}\int_{-\infty}^{+\infty}\rho(x_1,x_2)\,\mathrm{d}x_1\,\mathrm{d}x_2 = 1 \quad .$$

The bivariate Gaussian is completely described by five parameters. They are the *expectation values* $\langle x_1\rangle$ and $\langle x_2\rangle$, the *widths* $\sigma_1$ and $\sigma_2$, and the *correlation coefficient* $c$. The *marginal distributions* defined by

$$\rho_1(x_1) = \int_{-\infty}^{+\infty} \rho(x_1, x_2) \, dx_2 \quad ,$$

$$\rho_2(x_2) = \int_{-\infty}^{+\infty} \rho(x_1, x_2) \, dx_1$$

are for the bivariate Gaussian distribution simply Gaussians of a single variable,

$$\rho_1(x_1) = \frac{1}{\sqrt{2\pi}\,\sigma_1} \exp\left[-\frac{(x_1 - \langle x_1 \rangle)^2}{2\sigma_1^2}\right] \quad ,$$

$$\rho_2(x_2) = \frac{1}{\sqrt{2\pi}\,\sigma_2} \exp\left[-\frac{(x_2 - \langle x_2 \rangle)^2}{2\sigma_2^2}\right] \quad .$$

Each marginal distribution depends on two parameters only, the expectation value and the width of its variable.

Lines of constant probability density in $x_1$, $x_2$ are the lines of intersection between the surface $\rho(x_1, x_2)$ and a plane $\rho = a = \text{const.}$

One particular ellipse, for which

$$\rho(x_1, x_2) = A \exp\left\{-\frac{1}{2}\right\} \quad ,$$

i.e., the one for which the exponent in the bivariate Gaussian is simply equal to $-1/2$, is called the *covariance ellipse*. Points $x_1$, $x_2$ on the covariance ellipse fulfill the equation

$$\frac{1}{1-c^2}\left\{\frac{(x_1 - \langle x_1 \rangle)^2}{\sigma_1^2} - 2c\frac{(x_1 - \langle x_1 \rangle)\,(x_2 - \langle x_2 \rangle)}{\sigma_1\,\sigma_2} + \frac{(x_2 - \langle x_2 \rangle)^2}{\sigma_2^2}\right\} = 1 \quad .$$

Projected on the $x_1$ axis and the $x_2$ axis, it yields lines of lengths $2\sigma_1$ and $2\sigma_2$, respectively.

The plots in Figure 3.6 differ only by the value $c$ of the covariance. The covariance ellipses are shown as lines of constant probability on the surfaces $\rho(x_1, x_2)$. For $c = 0$ the principal axes of the covariance ellipse are parallel to the coordinate axes. In this situation variables $x_1$ and $x_2$ are *uncorrelated*, that is, knowledge that $x_1 > \langle x_1 \rangle$ holds true does not tell us whether it is more probable to observe $x_2 > \langle x_2 \rangle$ or $x_2 < \langle x_2 \rangle$. For uncorrelated variables the relation between the joint probability density and the marginal distribution is simple, $\rho(x_1, x_2) = \rho_1(x_1)\rho_2(x_2)$. The situation is different for correlated variables, that is, for $c \neq 0$. For a positive correlation, $c > 0$, the major axis of the ellipse lies along a direction between those of the $x_1$ axis and the $x_2$ axis. If we know that $x_1 > \langle x_1 \rangle$ is valid it is more probable to have $x_2 > \langle x_2 \rangle$ than to have $x_2 < \langle x_2 \rangle$. If, on the other hand, the correlation is negative, $x < 0$, the major axis has a direction between those of the $x_1$ axis and the negative

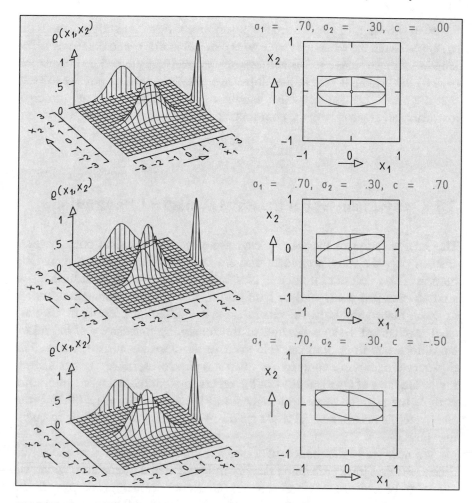

**Fig. 3.6. Bivariate Gaussian probability density** $\rho(x_1, x_2)$ **drawn as a surface over the** $x_1, x_2$ **plane and marginal distributions** $\rho_1(x_1)$ **and** $\rho_2(x_2)$**. The latter are drawn as curves over the margins parallel to the** $x_1$ **axis and the** $x_2$ **axis, respectively. Also shown is the covariance ellipse corresponding to the distribution. The rectangle circumscribing the ellipse has the sides** $2\sigma_1$ **and** $2\sigma_2$**, respectively. The pairs of plots in the three rows of the figure differ only by the correlation coefficient** $c$**.**

$x_2$ axis. In this situation, once it is known that $x_1 > \langle x_1 \rangle$ is valid, $x_2 < \langle x_2 \rangle$ is more probable than $x_2 > \langle x_2 \rangle$.

The amount of correlation is measured by the numerical value of $c$, which can vary in the range $-1 < c < 1$. In the limiting case of total correlation, $c = \pm 1$, the covariance ellipse degenerates to a line, the principal axis. The joint probability density is completely concentrated along this line. That is, knowing the value $x_1$ of one variable, we also know the value $x_2$ of the other.

Also shown in Figure 3.6 are the covariance ellipses directly drawn in the $x_1, x_2$ plane and the rectangles with sides parallel to the $x_1$ and the $x_2$ axes. The lengths of these sides are $2\sigma_1$ and $2\sigma_2$, respectively. If there is no correlation ($c = 0$) the principal axes of the ellipse are parallel to the coordinate axes so that the principal semi-axes have lengths $\sigma_1$ and $\sigma_2$. For $c \neq 0$ the principal axes form an angle $\alpha$ with the coordinate axes. The angle $\alpha$ is given by

$$\tan 2\alpha = \frac{2c\sigma_1\sigma_2}{\sigma_1^2 - \sigma_2^2} \quad .$$

## 3.6 Comparison with a Classical Statistical Description

The interpretation of the wave-packet solution as a classical point particle catches only the most prominent and simplest classical features of particle motion. To exploit our intuition of classical mechanics somewhat further, we study a classical point particle with initial position and momentum known to some inaccuracy only. In principle, such a situation prevails in all classical mechanical systems because of the remaining inaccuracy of the initial conditions due to errors inevitable even in all classical measurements. The difference in principle compared to quantum physics is, however, that according to the laws of classical physics the errors in location and momentum of a particle both can be made arbitrarily small independent of each other. From Heisenberg's uncertainty principle we know that this is not possible in quantum physics.

We now study the motion of a classical particle described at the initial time $t = 0$ by a joint probability density in location and momentum which we choose to be a bivariate Gaussian about the average values $x_0$ and $p_0$ with the widths $\sigma_{x0}$ and $\sigma_p$. We assume that at the initial time $t = 0$ there is no correlation between position and momentum. The initial joint probability density is then

$$\rho_i^{cl}(x, p) = \frac{1}{\sqrt{2\pi}\,\sigma_{x0}} \exp\left\{ -\frac{(x - x_0)^2}{2\sigma_{x0}^2} \right\} \frac{1}{\sqrt{2\pi}\,\sigma_p} \exp\left\{ -\frac{(p - p_0)^2}{2\sigma_p^2} \right\} \quad .$$

For force-free motion the particle does not suffer a change in momentum as time elapses, e.g., also at a later time $t > 0$ the particle still moves with its initial momentum, i.e., $p = p_i$. Thus, the momentum distribution does not change with time. The position of a particle of momentum $p_i$ at time $t$ initially having the position $x_i$ is given by

$$x = x_i + v_i t \quad , \qquad v_i = p_i/m \quad .$$

The probability density initially described by $\rho_i^{cl}(x_i, p_i)$ can be expressed at time $t$ by the positions $x$ at time $t$ by inserting

$$x_i = x - (p/m)t \quad ,$$

yielding the *classical phase-space probability density*

$$\rho^{cl}(x,p,t) = \rho_i^{cl}(x - pt/m, p)$$

$$= \frac{1}{2\pi\sigma_{x0}\sigma_p} \exp\left\{-\frac{1}{2}\left[\frac{(x - x_0 - pt/m)^2}{\sigma_{x0}^2} + \frac{(p - p_0)^2}{\sigma_p^2}\right]\right\}$$

$$= \frac{1}{2\pi\sigma_{x0}\sigma_p} \exp L \quad .$$

The exponent is a quadratic polynomial in $x$ and $p$ which can be written as

$$L = -\frac{1}{2}\left\{\frac{(x - [x_0 + p_0 t/m] - (p - p_0)t/m)^2}{\sigma_{x0}^2} + \frac{(p - p_0)^2}{\sigma_p^2}\right\}$$

$$= -\frac{1}{2}\frac{\sigma_{x0}^2 + \sigma_p^2 t^2/m^2}{\sigma_{x0}^2}\left\{\frac{(x - [x_0 + p_0 t/m])^2}{\sigma_{x0}^2 + \sigma_p^2 t^2/m^2}\right.$$

$$\left. - \frac{2(x - [x_0 + p_0 t/m])(p - p_0)}{(\sigma_{x0}^2 + \sigma_p^2 t^2/m^2)m/t} + \frac{(p - p_0)^2}{\sigma_p^2}\right\} \quad .$$

Comparing this expression with the exponent of the general expression for a bivariate probability density in Section 3.5 we find that $\rho^{cl}(x,p,t)$ is a bivariate Gaussian with the expectation values

$$\langle x(t)\rangle = x_0 + p_0 t/m \quad , \qquad \langle p(t)\rangle = p_0 \quad ,$$

the widths

$$\sigma_x(t) = \sqrt{\sigma_{x0}^2 + \sigma_p^2 t^2/m^2} \quad , \qquad \sigma_p(t) = \sigma_p \quad ,$$

and the correlation coefficient

$$c = \frac{\sigma_p t}{\sigma_x(t)m} = \frac{\sigma_p t/m}{\sqrt{\sigma_{x0}^2 + \sigma_p^2 t^2/m^2}} \quad .$$

In particular this means that the marginal distribution $\rho_x^{cl}(x,t)$, i.e., the spatial probability density for the classical particle with initial uncertainties $\sigma_{x0}$ in position and $\sigma_p$ momentum is

$$\rho_x^{cl}(x,t) = \frac{1}{\sqrt{2\pi}\sigma_x(t)} \exp\left\{-\frac{(x - [x_0 + p_0 t/m])^2}{2\sigma_x^2(t)}\right\} \quad .$$

Let us now study the classical probability density $\rho^{cl}(x,p,t)$ of a particle with initial uncertainties $\sigma_{x0}$ in position and $\sigma_p$ in momentum which satisfy the minimal uncertainty requirement of quantum mechanics:

$$\sigma_{x0}\sigma_p = \hbar/2 \quad .$$

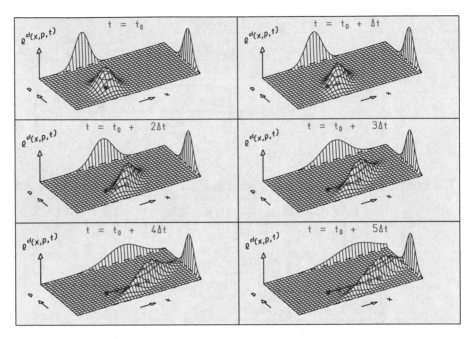

**Fig. 3.7. Time development of the classical phase-space probability density** $\rho^{\mathrm{cl}}(x,p,t)$ **for a free particle with uncertainty in position and momentum. Also shown are the marginal distributions** $\rho_x^{\mathrm{cl}}(x,t)$ **in the back and** $\rho_p^{\mathrm{cl}}(p,t)$ **on the right-hand side of the plots.**

In that case the spatial width of the classical probability distribution is

$$\sigma_x(t) = \frac{\hbar}{2\sigma_p}\sqrt{1 + \frac{4\sigma_p^2}{\hbar^2}\frac{t^2}{m^2}}$$

and thus identical to the width of the corresponding quantum-mechanical wave packet. Also the expectation values for $x$ and $p$ and the width in $p$ are identical for the classical and the quantum-mechanical case.

In Figure 3.7 we show the time development of the classical phase-space probability density. At the initial time $t = t_0 = 0$ there is no correlation between position and momentum. With increasing time the structure moves in $x$ direction and develops an increasing positive correlation between $x$ and $p$. The marginal distribution $\rho_p^{\mathrm{cl}}(p,t)$ remains unchanged whereas the marginal distribution $\rho_x^{\mathrm{cl}}(x,t)$ shows the motion in $x$ direction and the dispersion already well known from Section 3.5. The same information is presented in different form in Figure 3.8 which shows the covariance ellipse of $\rho^{\mathrm{cl}}(x,p,t)$ for several times. Its center moves with constant velocity $v_0 = p_0/m$ on a straight line parallel to the $x$ axis. The width in $p$ stays constant, the width in $x$ increases. It also becomes clear from the figure that the correlation coefficient, vanishing at $t = t_0 = 0$, tends towards $c = 1$ for $t \to \infty$, since in that

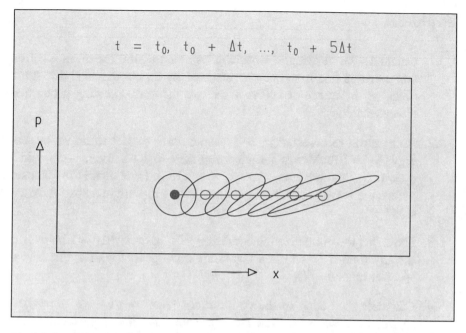

**Fig. 3.8. Motion of the covariance ellipse in phase space which characterizes the classical space probability density $\rho^{cl}(x, p, t)$ of a free particle. The ellipse is shown for six moments of time corresponding to the six plots in Figure 3.7. The center of the ellipse (indicated by a small circle) moves with constant velocity on a straight trajectory. The rectangle circumscribing the covariance ellipse has sides of lengths $2\sigma_x(t)$ and $2\sigma_p(t)$, respectively. While $\sigma_p$ stays constant, $\sigma_x$ increases with time. For $t = t_0 = 0$ there is no correlation between position and momentum (ellipse on the far left) but with increasing time a strong positive correlation develops.**

limit the covariance ellipse degenerates towards a line along the diagonal of the circumscribing rectangle.

The lesson we have learned so far is that the force-free motion of a classical particle described by a Gaussian probability distribution in phase space of position and momentum yields the same time evolution of the local probability density as in quantum mechanics if the initial widths $\sigma_{x0}, \sigma_p$ in position and momentum fulfill the relation

$$\sigma_{x0}\sigma_p = \frac{\hbar}{2} \quad .$$

In the further development of the quantum-mechanical description of particles we shall see that this finding does not remain true for particles under the action of forces other than constant in space or linear in the coordinates.

## Problems

3.1. Calculate the de Broglie wavelengths and frequencies of an electron and a proton that have been accelerated by an electric field through a potential difference of $100\,\mathrm{V}$. What are the corresponding group and phase velocities?

3.2. An electron represented by a Gaussian wave packet with average energy $E_0 = 100\,\mathrm{eV}$ was initially prepared to have momentum width $\sigma_p = 0.1\,p_0$ and position width $\sigma_x = \hbar/(2\sigma_p)$. How much time elapses before the wave packet has spread to twice the original spatial extension?

3.3. Show that the normalization condition $\int_{-\infty}^{+\infty} |\psi(x,t)|^2\,dx = 1$ holds true for any time if $\psi(x,t)$ is a Gaussian wave packet with a normalized spectral function $f(p)$.

3.4. Calculate the action of the commutator $[p,x] = px - xp$, $p = (\hbar/i)$ $(\partial/\partial x)$ on a wave function $\psi(x,t)$. Show that it is equivalent to the multiplication of $\psi(x,t)$ by $\hbar/i$ so that we may write $[p,x] = \hbar/i$.

3.5. Express the expectation value of the kinetic energy of a Gaussian wave packet in terms of the expectation value of the momentum and the width $\sigma_p$ of the spectral function.

3.6. Given a Gaussian wave packet of energy expectation value $\langle E \rangle$ and momentum expectation value $\langle p \rangle$, write its normalized spectral function $f(p)$.

3.7. A large virus may for purposes of this problem be approximated by a cube whose sides measure one micron and which has the density of water. Assuming as an upper estimate an uncertainty of one micron in position, calculate the minimum uncertainty in velocity of the virus.

3.8. The radius of both the proton and the neutron is measured to be of the order of $10^{-15}\,\mathrm{m}$. A free neutron decays spontaneously into a proton, an electron, and a neutrino. The momentum of the emitted electron is typically $1\,\mathrm{MeV}/c$. If the neutron were, as once thought, a bound system consisting of a proton and an electron, how large would be the position uncertainty of the electron and hence the size of the neutron? Take as the momentum uncertainty of the electron the value $1\,\mathrm{MeV}/c$.

3.9. Show that the solutions of the Schrödinger equation satisfy the continuity equation

$$\frac{\partial \rho(x,t)}{\partial t} + \frac{\partial j(x,t)}{\partial x} = 0$$

for the probability density

$$\rho(x,t) = \psi^*(x,t)\psi(x,t)$$

and the probability current density

$$j(x,t) = \frac{\hbar}{2im}\left[\psi^*(x,t)\frac{\partial}{\partial x}\psi(x,t) - \psi(x,t)\frac{\partial}{\partial x}\psi^*(x,t)\right] \ .$$

To this end, multiply the Schrödinger equation by $\psi^*(x,t)$ and its complex conjugate

$$i\hbar\frac{\partial \psi^*(x,t)}{\partial t} = \frac{\hbar^2}{2m}\frac{\partial^2}{\partial x^2}\psi^*(x,t) - V(x)\psi^*(x,t)$$

by $\psi(x,t)$, and add the two resulting equations.

3.10. Convince yourself with the help of the continuity equation that the normalization integral

$$\int_{-\infty}^{+\infty} \psi^*(x,t)\psi(x,t)\,dx$$

is independent of time if $\psi(x,t)$ is a normalized solution of the Schrödinger equation. To this end, integrate the continuity equation over all $x$ and use the vanishing of the wave function for large $|x|$ to show the vanishing of the integral over the probability current density.

3.11. Calculate the probability current density for the free Gaussian wave packet as given at the end of Section 3.2. Interpret the result for $t = 0$ in terms of the probability density and the group velocity of the packet.

3.12. Show that the one-dimensional Schrödinger equation possesses spatial reflection symmetry, that is, is invariant under the substitution $x \to -x$ if the potential is an even function, that is, $V(x) = V(-x)$.

3.13. Show that the *ansatz* for the Gaussian wave packet of Section 3.2 fulfills the Schrödinger equation for a free particle.

# 4. Solution of the Schrödinger Equation in One Dimension

## 4.1 Separation of Time and Space Coordinates, Stationary Solutions

The simple structure of the Schrödinger equation allows a particular *ansatz* in which the time and space dependences occur in separate factors,

$$\psi_E(x,t) = \exp\left(-\frac{i}{\hbar}Et\right)\varphi_E(x) \quad .$$

As in the case of electromagnetic waves, we call the factor $\varphi_E(x)$ that is independent of time a *stationary solution*. Inserting our *ansatz* into the Schrödinger equation yields an equation for the stationary wave,

$$-\frac{\hbar^2}{2m}\frac{d^2}{dx^2}\varphi_E(x) + V(x)\varphi_E(x) = E\varphi_E(x) \quad ,$$

which is often called the *time-independent Schrödinger equation*. It is characterized by the parameter $E$, which is called an *eigenvalue*. The left-hand side represents the sum of the kinetic and the potential energy, so that $E$ is the total energy of the stationary solution. The solution $\varphi_E(x)$ is called an *eigenfunction* of the Hamilton operator

$$H = -\frac{\hbar^2}{2m}\frac{d^2}{dx^2} + V(x) \quad ,$$

since the time-independent Schrödinger equation can be put into the form

$$H\varphi_E(x) = E\varphi_E(x) \quad .$$

We also say that the solution $\varphi_E(x)$ describes an *eigenstate* of the system specified by the Hamilton operator. This eigenstate is characterized by the eigenvalue $E$ of the total energy. Often the stationary solution $\varphi_E(x)$ is also called a *stationary state* of the system.

S. Brandt and H.D. Dahmen, *The Picture Book of Quantum Mechanics*,
DOI 10.1007/978-1-4614-3951-6_4, © Springer Science+Business Media New York 2012

The time-independent Schrödinger equation has a large manifold of solutions. It is supplemented by *boundary conditions* that have to be imposed on a particular solution. These boundary conditions must be abstracted from the physical process that the solution should describe. The boundary conditions on the solution for the elastic scattering in one dimension of a particle under the action of a force will be discussed in the next section. Because of the boundary conditions, solutions $\varphi_E(x)$ exist for particular values of the energy eigenvalues or for particular energy intervals only.

As a first example, we look at the de Broglie waves,

$$\psi_p(x - x_0, t) = \frac{1}{(2\pi\hbar)^{1/2}} \exp\left[-\frac{i}{\hbar}(Et - px + px_0)\right] \quad .$$

The function $\psi_p(x - x_0, t)$ factors into $\exp[-(i/\hbar)Et]$ and the stationary wave

$$\frac{1}{(2\pi\hbar)^{1/2}} \exp\left[\frac{i}{\hbar}p(x - x_0)\right] \quad .$$

It is a solution of the time-independent Schrödinger equation with a vanishing potential for the energy eigenvalue $E = p^2/2m$. A superposition of de Broglie waves fulfilling the normalization condition of Section 3.3 forms a wave packet describing an unaccelerated particle. Here $x_0$ is the position expectation value of the wave packet at time $t = 0$.

Since the momentum $p$ is a real parameter, the energy eigenvalue of a de Broglie wave is always positive. Thus, for the case of de Broglie waves, we have found the restriction $E \geq 0$ for the energy eigenvalues.

The general solution of the time-dependent Schrödinger equation is given by a linear combination of waves of different energies. This is tantamount to stating that the various components of different energy $E$ superimposed in the solution change independently of one another with time.

For initial time $t = 0$ the functions $\psi_E$ and $\varphi_E$ coincide. An initial condition prescribed at $t = 0$ determines the coefficients in the linear combination of spectral components of different energies. Therefore the procedure for solving the equation for a given initial condition has three steps. First, we determine the stationary solutions $\varphi_E(x)$ of the time-independent Schrödinger equation. Second, we superimpose them with appropriate coefficients to reproduce the initial condition $\psi(x,0)$ at $t = 0$. Finally, we introduce into every term of this linear combination the time-dependent factor $\exp[-(i/\hbar)Et]$ corresponding to the energy of the stationary solution $\varphi_E$ and sum them up to give $\psi(x,t)$, the solution of the time-dependent Schrödinger equation.

In the next section we study methods of obtaining the stationary solutions.

## 4.2 Stationary Scattering Solutions: Piecewise Constant Potential

As in classical mechanics, the scattering of a particle by a force is called elastic if only its momentum is changed while its energy is conserved. A force is said to be of finite range if it is practically zero for distances from the center of force larger than a finite distance $d$. This distance $d$ is called the range of the force. The elastic scattering of a particle through a force of finite range consists of three stages subsequent in time.

1. The incoming particle moves unaccelerated in a force-free region toward the range of the force.

2. The particle moves under the influence of the force. The action of the force changes the momentum of the particle.

3. After the scattering the outgoing particle moves away from the range of the force. Its motion in the force-free region is again unaccelerated.

In Section 3.3 we have seen that the force-free motion of a particle of mass $m$ can be described by a wave packet of de Broglie waves,

$$\psi_p(x - x_0, t) = \frac{1}{(2\pi\hbar)^{1/2}} \exp\left[-\frac{i}{\hbar}(Et - px + px_0)\right] \quad,$$

$$E = \frac{p^2}{2m} \quad.$$

They can be factored into the time-dependent factor $\exp[-(i/\hbar)Et]$ and the stationary wave $(2\pi\hbar)^{-1/2}\exp[(i/\hbar)p(x - x_0)]$. This stationary wave is a solution of the time-independent Schrödinger equation with a vanishing potential.

If the spectral function $f(p)$ of the wave packet has values different from zero in a range of positive $p$ values, the wave packet

$$\psi(x, t) = \int_{-\infty}^{+\infty} f(p)\psi_p(x - x_0, t)\,\mathrm{d}p$$

$$= \int_{-\infty}^{+\infty} f(p)\exp\left(-\frac{i}{\hbar}Et\right)\frac{1}{(2\pi\hbar)^{1/2}}\exp\left[\frac{i}{\hbar}p(x - x_0)\right]\mathrm{d}p$$

moves along the $x$ axis from left to right, that is, in the direction of increasing $x$ values.

Now we superimpose de Broglie waves of momentum $-p$,

$$\psi_{-p}(x - x_0, t) = \frac{1}{(2\pi\hbar)^{1/2}} \exp\left[-\frac{i}{\hbar}(Et + px - px_0)\right] \quad,$$

$$E = \frac{p^2}{2m} \quad,$$

with the same spectral function $f(p)$. A simple change of variables $p' = -p$ yields

$$\psi_-(x,t) = \int_{-\infty}^{+\infty} f(-p') \exp\left(-\frac{i}{\hbar}Et\right) \frac{1}{(2\pi\hbar)^{1/2}} \exp\left[\frac{i}{\hbar}p'(x-x_0)\right] dp'$$

$$= \int_{-\infty}^{+\infty} f(-p')\psi_p(x-x_0,t) dp \quad .$$

We obtain a wave packet with a spectral function $f(-p)$ having its range of values different from zero at negative values of $p$. The wave packet $\psi_-(x,t)$ moves along the $x$ axis from right to left, that is, in the direction of decreasing $x$ values. Thus we learn that for a given spectral function, wave packets formed with $\psi_p(x-x_0,t)$ and $\psi_{-p}(x-x_0,t)$ move in opposite directions. This says that the sign of the exponent of the stationary wave $(2\pi\hbar)^{-1/2}\exp[\pm(i/\hbar)p(x-x_0)]$ decides the direction of motion. For a spectral function $f(p)$ different from zero at positive values of momentum $p$, a wave packet formed with the stationary wave,

$$\exp\left[\frac{i}{\hbar}p(x-x_0)\right] = \exp[ik(x-x_0)] \quad , \qquad k = p/\hbar \quad ,$$

moves in the direction of increasing $x$. A wave packet formed with the stationary wave

$$\exp\left[-\frac{i}{\hbar}p(x-x_0)\right] = \exp[-ik(x-x_0)]$$

moves in the direction of decreasing $x$.

Let us consider a particle moving from the left in the direction of increasing $x$. The force

$$F(x) = -\frac{d}{dx}V(x)$$

derived from the potential energy $V(x)$ has finite range $d$. This range is assumed to be near the origin $x = 0$. The initial position $x_0$ of the wave packet is assumed to be far to the left of the origin at large negative values of the coordinate. As long as the particle is far to the left of the origin, the particle moves unaccelerated. In this region the solution is a wave packet of de Broglie waves $\psi_p(x-x_0,t)$. Thus the stationary solution $\varphi_E(x)$ of the time-independent Schrödinger equation for the eigenvalue $E$ should contain a term approaching the function $\exp[(i/\hbar)p(x-x_0)]$ for negative $x$ of large absolute value.

Through the scattering process in one dimension, the particle can only be transmitted or reflected. The transmitted particle will move force free at large positive $x$. Here it will be represented by a wave packet of de Broglie waves of the form $\psi_{p'}(x-x_0,t)$. Therefore the solution of the stationary Schrödinger equation must approach the function $\exp[(i/\hbar)p'(x-x_0)]$ for large positive

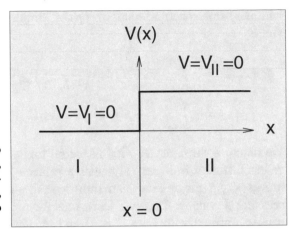

**Fig. 4.1. Space is divided into region I, $x < 0$, and region II, $x > 0$. There is a constant potential in region II, $V = V_0$, whereas in region I there is no potential, $V = 0$.**

$x$. The value $p'$ differs from $p$ if the potential $V(x)$ assumes different values for large negative and large positive $x$. The reflected particle has momentum $-p$ and will leave the range of the potential to the left and thus return to the region of large negative $x$. In this region it will be represented by a wave packet of de Broglie waves $\psi_{-p}(x - x_0, t)$. Therefore the solution of the time-independent Schrödinger equation must also contain a contribution tending toward a function $\exp[-(i/\hbar)p(x - x_0)]$ for large negative $x$. The conditions for large positive and negative $x$ just derived constitute the boundary condition that the stationary solution $\varphi_E(x)$, $E = p^2/2m$ must fulfill if its superpositions forming wave packets are to describe an elastic scattering process. We summarize the boundary conditions for stationary scattering solutions of the time-independent Schrödinger equation in the following statement:

$$\varphi_E(x) \xrightarrow[\text{approaches}]{} \begin{cases} \exp\left[\dfrac{i}{\hbar}p(x - x_0)\right] + B\exp\left[-\dfrac{i}{\hbar}p(x - x_0)\right] \\ \quad \text{for large negative } x \\[2mm] A\exp\left[\dfrac{i}{\hbar}p'(x - x_0)\right] \\ \quad \text{for large positive } x \end{cases} .$$

Since there are no general methods for solving in closed form the Schrödinger equation for an arbitrary potential, we choose for our discussion particularly simple examples. We begin with a *potential step* of height $V = V_0$ at $x = 0$. The potential divides the space into two regions. In region I, that is, to the left of $x = 0$, the potential vanishes. To the right, region II, it has the constant value $V = V_0$ (Figure 4.1).

The time-independent Schrödinger equation has the form

$$-\frac{\hbar^2}{2m}\frac{d^2}{dx^2}\varphi + V_i\varphi = E\varphi$$

in both regions, with $V_i$ assuming different but constant values in the two regions, $V_I = 0$, $V_{II} = V_0$. Thus the stationary solution for given energy $E$ of the incoming wave is

$$\varphi_I = \exp\left[\frac{i}{\hbar}p(x-x_0)\right] + B_I\exp\left[-\frac{i}{\hbar}p(x-x_0)\right] \quad , \quad x < 0 \quad ,$$

$$\varphi_{II} = A_{II}\exp\left[\frac{i}{\hbar}p'(x-x_0)\right] \quad , \quad x > 0 \quad .$$

Obviously, this solution fulfills the boundary conditions we have posed earlier in this section.

In region I the momentum is $p = \sqrt{2mE}$, in region II it is $p' = \sqrt{2m(E-V_0)}$. Since the potential is discontinuous at $x = 0$, the second derivative of $\varphi$ has to reproduce the same discontinuity, reduced by the factor $-\hbar^2/(2m)$. Thus $\varphi$ and $d\varphi/dx$ are continuous at $x = 0$. These conditions determine the complex coefficients $B_I$ and $A_{II}$ which are as yet unknown. The coefficient of the incoming wave has been chosen equal to one, thus fixing the incoming amplitude. The phase of the wave function depends on the initial position parameter $x_0$.

As for light waves (Section 2.2), we denote the three members on the right-hand sides of the two expressions $\varphi_I$ and $\varphi_{II}$ as *constituent waves*. That is, we call

$$\varphi_{1+} = \exp\left[(i/\hbar)p(x-x_0)\right]$$
the incoming constituent wave ,

$$\varphi_{1-} = B_I\exp\left[-(i/\hbar)p(x-x_0)\right]$$
the reflected constituent wave ,

$$\varphi_2 = A_{II}\exp\left[(i/\hbar)p'(x-x_0)\right]$$
the transmitted constituent wave .

As a first example, we choose a repulsive step, $V_0 > 0$, and an incoming wave of energy $E < V_0$, so that in classical mechanics the particle would be reflected by the potential step. The momentum of the transmitted wave in region II,

$$p' = \sqrt{2m(E-V_0)} = i\sqrt{2m(V_0-E)} \quad ,$$

is now imaginary so that the transmitted wave

$$\varphi_{\mathrm{II}} \;=\; A_{\mathrm{II}} \exp\left[\frac{i}{\hbar} p'(x - x_0)\right]$$

$$\;=\; A_{\mathrm{II}} \exp\left[-\frac{1}{\hbar}\sqrt{2m(V_0 - E)}(x - x_0)\right]$$

becomes a real exponential function which falls off with increasing $x$ in region II. We obtain the full solution of the time-dependent Schrödinger equation for a given energy by multiplying the stationary wave by the factor $\exp(-iEt/\hbar)$.

The upper and middle plot of Figure 4.2 show the time developments of the real and imaginary parts of the wave function with fixed energy $E$. The real and imaginary parts behave in region I like standing waves, for they are superpositions of an incoming and a reflected wave of equal frequency and equal amplitude. We are easily convinced of this fact by looking at Figure 4.3, in which the time developments of the incoming and reflected constituent waves in region I are plotted separately. In Figure 4.2, region II, both the real and the imaginary parts are represented by exponentials oscillating in time. The time development of the absolute square of the wave function, which we shall call intensity (Figure 4.2, bottom), shows no variation at all in time. In region I it is periodic in space, but in region II it shows an exponential falloff.

We now examine an incoming wave of energy $E > V_0$. Obviously, the momentum $p' = \sqrt{2m(E - V_0)}$ in region II for $E > V_0$ is real. Therefore the stationary solution in this region, as in region I, is an oscillating function in space.

Figure 4.4 (top) shows the energy dependence of the real parts of the stationary solutions. It includes both energies $E > V_0$ and energies $E < V_0$. For energies $E > V_0$ the wavelength in region II is longer than that in region I. For energies $E < V_0$ the stationary wave function has the exponential falloff just mentioned. The energy dependence of the intensity is given in Figure 4.4 (bottom). For $E > V_0$ the intensity is constant in region II, corresponding to the outgoing wave in this region. The periodic structure of the intensity in region I results from the superposition of the incoming and reflected waves.

For $V_0 < 0$ there is for all energies an oscillating transmitted wave in region II. Figure 4.5 shows the energy dependence of the real part and of the absolute square of the wave function. Since the potential is now attractive, the wavelength of the transmitted wave is decreased in region II.

Since for every energy $E > 0$ there is a stationary solution of the time-independent Schrödinger equation for a potential step, we say that the physical system has a *continuous energy spectrum*. For some types of potential, the Schrödinger equation has solutions only for certain particular values of energy. They form a *discrete energy spectrum*. The most general physical system has an energy spectrum composed of a discrete part and a continuous one (see Section 4.4).

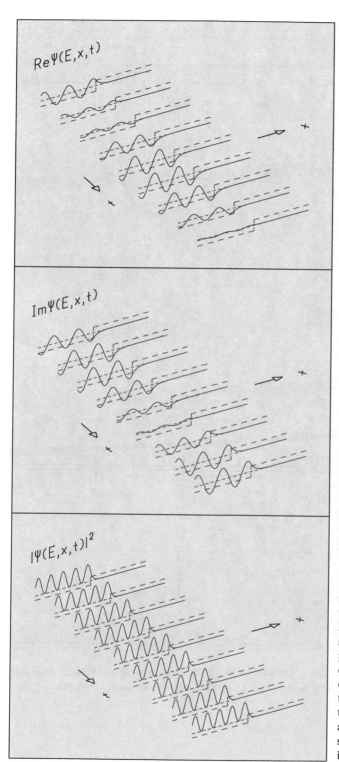

**Fig. 4.2.** Time developments of the real part, the imaginary part, and the intensity of a harmonic wave of energy $E < V_0$ falling onto a potential step of height $V_0$. The form of the potential $V(x)$ is indicated by the line made up of long dashes, the energy of the wave by the short-dash horizontal line, which also serves as zero line for the functions plotted. To the left of the potential step is a standing-wave pattern, as is apparent from the time-independent position of the nodes or zeros of the functions $\mathrm{Re}\,\psi(x,t)$ and $\mathrm{Im}\,\psi(x,t)$. The absolute square $|\psi(x,t)|^2$ is time independent.

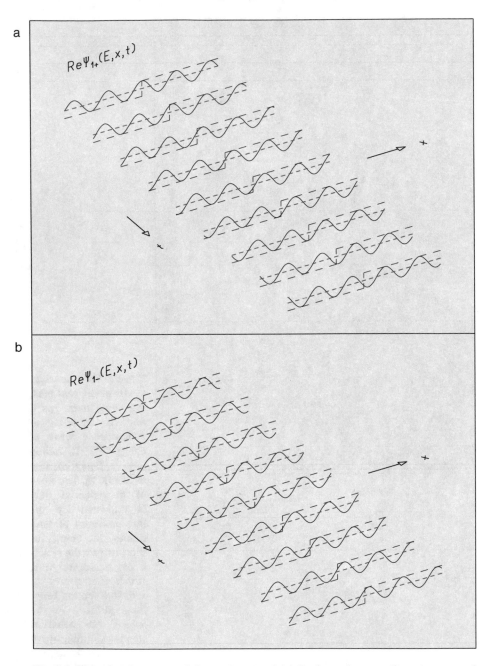

**Fig. 4.3.** Time developments of the real parts of (a) the incoming constituent wave and (b) the reflected constituent wave making up the harmonic wave of Figure 4.2.

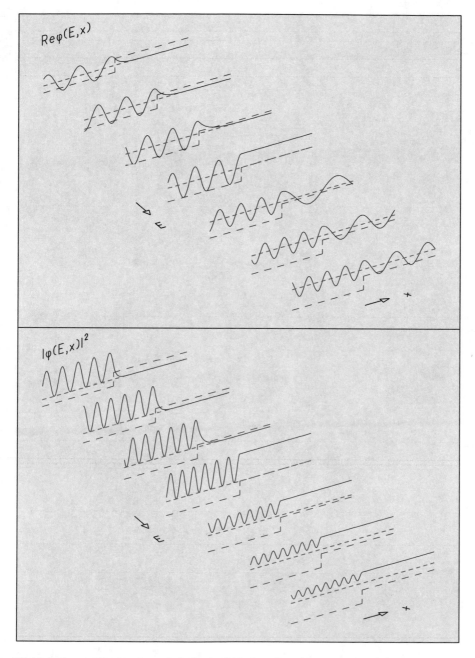

**Fig. 4.4. Energy dependence of stationary solutions for waves incident on a potential step of height $V_0 > 0$. Shown are the real part of the wave function and the intensity. Small energies are in the background.**

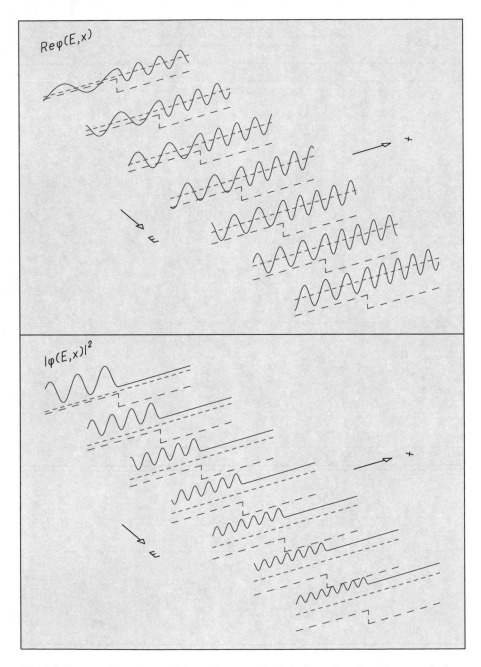

**Fig. 4.5. Energy dependence of the real part and of the intensity of stationary solutions for harmonic waves incident on a potential step of height $V_0 < 0$.**

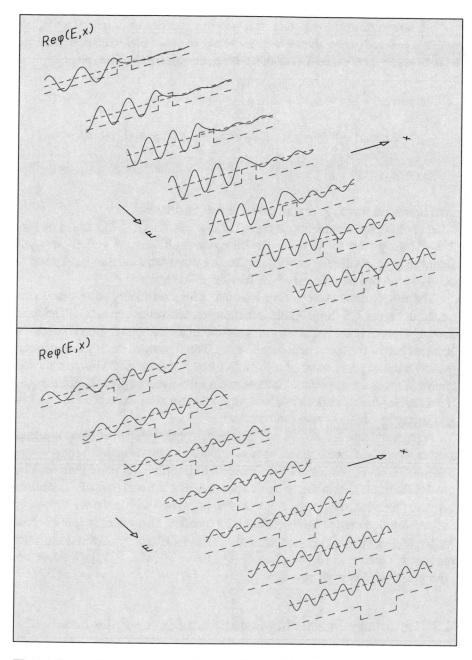

**Fig. 4.6. Energy dependence of stationary solutions for waves incident onto a positive potential barrier (top), $V_0 > 0$, and a negative potential barrier (bottom), $V_0 < 0$, which is also called a square-well potential. The real part of the wave function is shown.**

We now turn to the example of a potential barrier of height $V = V_0$ between $x = 0$ and $x = d$. Outside this interval the potential vanishes. Here we have to study three different regions where the solution is given by

$$\varphi_\mathrm{I} = \exp\left[\frac{i}{\hbar}p(x - x_0)\right] + B_\mathrm{I}\exp\left[-\frac{i}{\hbar}p(x - x_0)\right] , \qquad x < 0 ,$$

$$\varphi_\mathrm{II} = A_\mathrm{II}\exp\left[\frac{i}{\hbar}p'(x - x_0)\right] + B_\mathrm{II}\exp\left[-\frac{i}{\hbar}p'(x - x_0)\right] , \quad 0 < x < d ,$$

$$\varphi_\mathrm{III} = A_\mathrm{III}\exp\left[\frac{i}{\hbar}p(x - x_0)\right] , \qquad d < x$$

for a harmonic wave moving in from the left, that is, $x < 0$.

As before, the momentum in region II is $p' = \sqrt{2m(E - V_0)}$ and is real for $E > V_0$, imaginary for $E < V_0$. The complex coefficients $A$ and $B$ are again determined by continuity conditions for the wave function and its derivative $d\varphi/dx$ at the two boundaries of the barrier, $x = 0$ and $x = d$.

The energy dependence of the real part of the stationary solutions is presented in Figure 4.6 (top). Again transmission and reflection occur. The most striking feature, however, is the transmission of a wave into region III even for energies below the barrier height, $E < V_0$. The transmission of the wave corresponds to the penetration of a particle through the barrier. This remarkable quantum-mechanical phenomenon is called the *tunnel effect*. Within region II, of course, the wave does not have periodic structure since $p'$ is imaginary, giving rise to a real exponential function.

A potential that is constant and negative in region II, that is, $V_0$ is less than zero for $0 < x < d$, and that vanishes in regions I and III is called a *square-well potential*. Here the waves keep their oscillating form in all three regions. Figure 4.6 (bottom) shows the energy dependence of the real part of the stationary wave function. The wavelength is now decreased within the well, through the acceleration caused by the attractive potential. This effect is less obvious for the higher kinetic energies since the relative difference between the wave number $k_\mathrm{I}$ outside and $k_\mathrm{II} = \sqrt{2m(E - V_0)}/\hbar = \sqrt{2m(E + |V_0|)}/\hbar$ decreases with growing kinetic energy $E$.

## 4.3 Stationary Scattering Solutions: Linear Potentials

In Section 4.2 we have investigated the stationary solutions of the Schrödinger equation for piecewise constant potentials. Slightly more complicated is the linear potential

$$V(x) = -mgx + V_0 = -mg(x - x_0) , \qquad x_0 = \frac{V_0}{mg} ,$$

which, for example, governs the free fall of a body under the action of a constant force $F = mg$. The corresponding time-independent, stationary Schrödinger equation reads

$$\left(-\frac{\hbar^2}{2m}\frac{d^2}{dx^2} - mgx + V_0\right)\varphi(x) = E\varphi(x) \quad,$$

or, in normal form,

$$\left[\frac{d^2}{dx^2} + 2\frac{m^2}{\hbar^2}gx + \frac{2m}{\hbar^2}(E - V_0)\right]\varphi(x) = 0 \quad.$$

The position

$$x_T = -\frac{E - V_0}{mg} = -\frac{E}{mg} + x_0$$

is the *classical turning point* of a particle with the total energy $E$. Introducing the dimensionless variable

$$\xi = \frac{1}{\ell_0}(x - x_T) \quad, \qquad \ell_0 = \left(\frac{\hbar^2}{2m^2g}\right)^{1/3} \quad,$$

scaling $x$ in multiples of the length parameter $\ell_0$, we find a differential equation:

$$\left(\frac{d^2}{d\xi^2} + \xi\right)\phi(\xi) = 0$$

with

$$\phi(\xi) = \varphi(\ell_0\xi + x_T) \quad.$$

The differential equation in $\xi$ no longer contains an energy parameter. A solution of this equation is the *Airy function* $\mathrm{Ai}(\xi)$, cf. Appendix F, multiplied with a normalization constant:

$$\phi(\xi) = N\,\mathrm{Ai}(-\xi) \quad, \qquad N = \left(\frac{2m^{1/2}}{g^{1/2}\hbar^2}\right)^{1/3} \quad.$$

The Airy functions $\mathrm{Bi}(x)$ do not yield physical solutions since they diverge for large positive arguments.

The main features of the solution

$$\varphi(x) = N\,\mathrm{Ai}\left(-\frac{x - x_T}{\ell_0}\right)$$

which is shown in Figure 4.7 are easily understood. For $x > x_T$ the wave function oscillates with wavelengths $\lambda$ becoming shorter with growing $x$. This reflects the increase of momentum $p = h/\lambda$ of the particle because of the acceleration in the positive $x$ direction by the force $F = mg$. In fact, for $(x - x_T) \gg \ell_0$ the solution has the asymptotic behavior

$$\varphi_{as}(x) = N \left( \frac{\ell_0}{x - x_T} \right)^{1/4} \sin \left( \frac{2}{3} \left[ \frac{x - x_T}{\ell_0} \right]^{3/2} + \frac{\pi}{4} \right) \quad .$$

This shows that the wavelength $\lambda$ defined by

$$\frac{2}{3} \left( \frac{x + \lambda - x_T}{\ell_0} \right)^{3/2} - \frac{2}{3} \left( \frac{x - x_T}{\ell_0} \right)^{3/2} = 2\pi$$

depends on the position $x$ and is approximately given by

$$\lambda = \frac{2\pi \hbar}{m \sqrt{2g(x - x_T)}} \quad .$$

This corresponds to the momentum

$$p = h/\lambda = m \sqrt{2g(x - x_T)} = mv \quad ,$$

where

$$v = \sqrt{2g(x - x_T)}$$

is the classical velocity of a particle at position $x$ falling from the point $x_T$ under the acceleration $g$. For $x < x_T$ the wave function falls off very quickly to zero. Since $x_T$ is the classical turning point of the particle, this vanishing of the wave function and thus the probability density to the left of $x_T$ is to be expected. For $x - x_T < 0$ and $|x - x_T| \gg \ell_0$ the asymptotic form $\varphi_{as}$ of the wave function is given by

$$\varphi_{as}(x) = N \left( \frac{\ell_0}{|x - x_T|} \right)^{1/4} \exp \left\{ -\frac{2}{3} \left| \frac{x - x_T}{\ell_0} \right|^{3/2} \right\} \quad .$$

As to be expected there is no point to the left of which the wave function is exactly zero. To the far left of the turning point, however, the probability of finding the particle becomes very small.

Let us now consider a piecewise linear potential, i.e., a potential that is either constant or linear in different regions,

$$V(x) = \begin{cases} V_I = c_1 = \text{const}, & x < x_1 = 0 & \text{region I} \\ V_{II,a} + (x - x_1)V_{II}', & x_1 \leq x < x_2 & \text{region II} \\ \vdots & & \\ V_{N-1,a} + (x - x_{N-2})V_{N-1}', & x_{N-2} \leq x < x_{N-1} & \text{region } N-1 \\ V_N = c_2 = \text{const}, & x_{N-1} \leq x & \text{region } N \end{cases},$$

$$V_j' = \frac{V_{j,b} - V_{j,a}}{x_j - x_{j-1}} \quad , \quad j = II, \ldots, N-1 \quad .$$

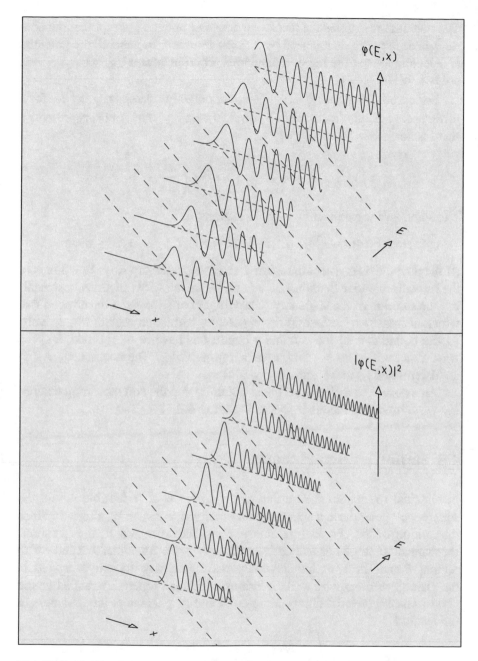

**Fig. 4.7. Stationary-solution wave function** $\varphi(x)$ **(top) and its absolute square (bottom) in a linear potential for various values of the total energy** $E$. **The potential is indicated by the long-dash broken line, the total energy** $E$ **by the short-dash broken line. They intersect at the classical turning point** $x_T$. **The short-dash broken line also serves as the zero line for the functions** $\varphi(x)$ **and** $|\varphi(x)|^2$.

The potential resembles the piecewise constant potential except that the constant in each region is replaced by a linear function. In general, the potential is discontinuous at the region boundaries; continuity at $x_j$, $j = I, \ldots, N-1$, holds only if $V_{j,b} = V_{j+1,a}$.

For a nonzero slope $V_j' = \frac{V_{j,b} - V_{j,a}}{x_j - x_{j-1}}$ of a potential in region $j$, $x_j^{\mathrm{T}}$ is, for a given energy $E$, the (extrapolated) classical turning point, and $\ell_j$ represents a scale factor:

$$x_j^{\mathrm{T}} = x_{j-1} + \frac{E - V_{j,a}}{V_j'} \quad , \quad \ell_j = \left( \frac{\hbar^2}{2m} \frac{1}{V_j'} \right)^{\frac{1}{3}} \quad .$$

The stationary wave function $\varphi_j(x)$ in region $j$ is

$$\varphi_j(x) = A_j \, \mathrm{Ai}((x - x_j^{\mathrm{T}})/\ell_j) + B_j \, \mathrm{Bi}((x - x_j^{\mathrm{T}})/\ell_j) \quad , \quad x_{j-1} \leq x < x_j \quad .$$

(If the potential is constant in region $j$, then the stationary wave function is of the form discussed in Section 4.2.) Ai and Bi are the Airy functions, representing the solution of the stationary Schrödinger equation for a linear potential, which decrease or increase in the classically forbidden region, respectively. At the boundaries $x_j$ the continuity conditions have to be fulfilled, i.e., the wave function has to be *continuously differentiable*. The coefficients $A_j, B_j$ are determined by these boundary conditions.

An example of a piecewise linear potential and the corresponding stationary wave functions therein is shown in Figure 4.8

## 4.4 Stationary Bound States

In classical mechanics the motion of a particle is also possible within the square-well potential at a negative total energy $E = E_{\mathrm{kin}} + V_0$, $E_{\mathrm{kin}} > 0$. Since the motion of the classical particle is restricted to region II, the quantum-mechanical analog is described by stationary wave functions that fall off in regions I and III. Thus the stationary wave functions too are localized in the vicinity of the square well. Solutions with this property are called *bound states*. The exponential falloff in regions I and III guarantees the finiteness of the integral

$$\int_{-\infty}^{+\infty} |\varphi_{\mathrm{unnormalized}}(x)|^2 \, \mathrm{d}x = N^2$$

in contrast to the stationary solutions of positive energy eigenvalues. Dividing the unnormalized solution by $N$ yields the normalized stationary wave function $\varphi(x)$ fulfilling the normalization condition

$$\int_{-\infty}^{+\infty} |\varphi(x)|^2 \, \mathrm{d}x = 1 \quad ,$$

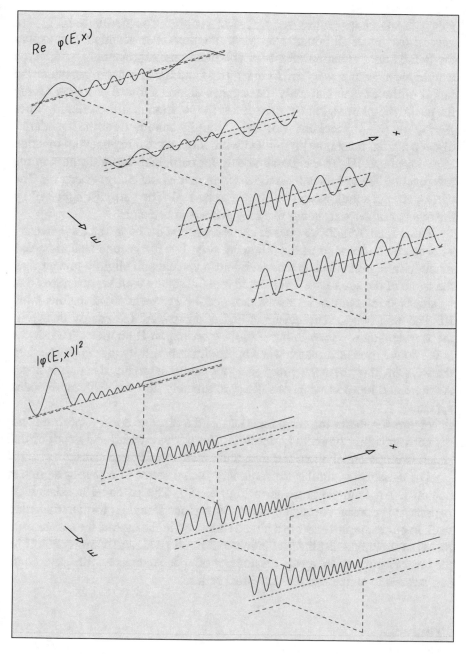

**Fig. 4.8.** Energy dependence of the stationary wave functions $\varphi(x)$ for a piecewise linear potential. In the region where the potential decreases linearly with $x$ the wavelength is observed to fall, since kinetic energy and momentum rise.

which is analogous to the one for wave packets. The normalizability is a general feature of all bound-state wave functions. For negative total energy the Schrödinger equation admits as solutions real exponentials of the type $\exp(\pm i p x / \hbar)$ with $p = i\sqrt{2m|E|}$ in regions I and III. In order to guarantee the falloff of the exponential, only the negative sign is allowed in region I, only the positive sign in region III. In region II the solution is still oscillating since $p' = \sqrt{2m(E - V_0)}$ remains real. The continuity conditions at the boundaries of the potential must now be fulfilled with only one real exponential function in regions I and III. This is possible only for particular discrete values of the total energy. These values form the *discrete part* of the *energy spectrum*. The corresponding solutions can be chosen to be real. They are distinguished by the number of zeros or *nodes* that they possess in region II.

The number of nodes increases as the energy of the bound states increases. This can be understood in the following way. For the ground state the wave function in region II is a cosine with half a wavelength slightly greater than the width of the square well. It is fitted into the square well in such a way that its slopes at the boundaries match those of the exponentials in regions I and III. The next bound state occurs at higher energy. As the energy increases, the wavelength $\lambda = 2\pi\hbar[2m(E - V_0)]^{-1/2}$ in region II shortens. The slopes at the boundaries next match when approximately a full wavelength fits into the well, making the wave function a sine and thus giving rise to one node. As more and more wavelengths fit approximately into the well, more nodes appear.

Figure 4.9 shows the wave functions of the discrete energy spectrum and the corresponding probability densities. For a given width and depth of the square-well potential, there is only a finite number of bound states.

The situation is similar for stationary bound states in a piecewise linear potential. An example is given in Figure 4.10. The potential consists of 4 regions. It is constant in regions I and IV; regions II and III form a triangular well. In our example there are 5 bound states. As in the case of the square well the wave functions display oscillations within the well and a strong falloff in the outer regions. Also, again, the number of nodes increases with energy and can be used to enumerate the individual states.

## Problems

4.1. Solve the stationary Schrödinger equation for energy $E$ with a constant potential $V = V_0$.

4.2. Discuss the behavior of the solutions for energies $E > V_0$, $E < V_0$. Which solutions correspond to the particular energy $E = V_0$? These

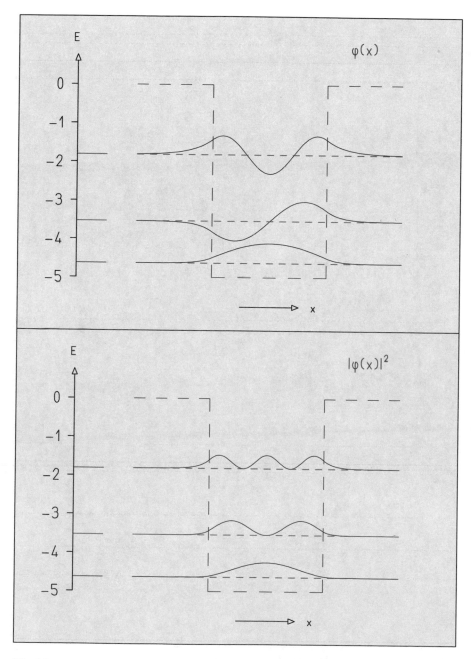

**Fig. 4.9. Wave functions (top) and probability densities (bottom) of the bound states in a square-well potential. On the left side of the picture an energy scale is shown with marks for the bound-state energies ($n = 1, 2, 3$). The form of the potential $V(x)$ is indicated by the long-dash line, the energy $E_n$ of the bound states by the horizontal short-dash lines. The horizontal dashed lines also serve as zero lines for the functions shown.**

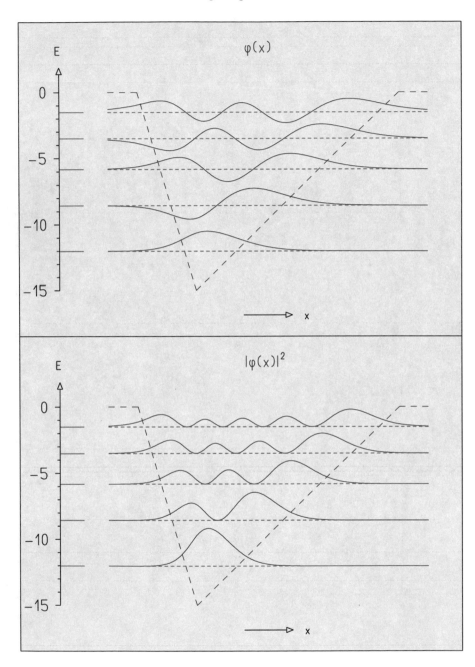

**Fig. 4.10. Wave functions (top) and probability densities (bottom) of the stationary bound states in a piecewise linear potential.**

three cases play a role in the solution of the Schrödinger equation for stepwise constant potentials. Figures 4.4, 4.5, and 4.6 give examples.

4.3. Calculate the intensities $|\varphi_{II}(x)|^2$ of the transmitted stationary wave and of the superposition of the incoming and reflected stationary waves $|\varphi_I(x)|^2$ of Section 4.2.

4.4. Calculate the probability current density

$$j(x) = \frac{\hbar}{2im} \left( \psi^*(x) \frac{\partial}{\partial x} \psi(x) - \psi(x) \frac{\partial}{\partial x} \psi^*(x) \right)$$

for the solution of the stationary Schrödinger equation. Consider a potential step of height $V_0$ as shown in Figure 4.1. Show that the current density is equal in the two spatial regions if the wave function and its derivative fulfill the continuity conditions at the boundaries between the different regions of the potential. Explain the result for $E \geq V_0$.

4.5. Show that the stationary bound-state wave functions can always be chosen to fulfill one of the following two relations:

$$\varphi(-x) = \varphi(x) \qquad \text{or} \qquad \varphi(-x) = -\varphi(x)$$

for an even potential $V(-x) = V(x)$. The function $\varphi(x)$ is said to have positive parity – also called natural or even parity – in the first relation and negative parity – unnatural or odd parity – in the second.

# 5. One-Dimensional Quantum Mechanics: Scattering by a Potential

## 5.1 Sudden Acceleration and Deceleration of a Particle

We now study the motion of a wave packet incident on a potential step. As already discussed at the beginning of Section 4.2, the effect of the potential is the elastic scattering of the particle. In one-dimensional scattering the particle will be transmitted or reflected by the potential.

If we superimpose the stationary solutions of Section 4.2 with the spectral function that was used for the construction of the free wave packet in Section 3.2,

$$f(p) = \frac{1}{(2\pi)^{1/4}\sqrt{\sigma_p}} \exp\left[-\frac{(p-p_0)^2}{4\sigma_p^2}\right] \quad,$$

we obtain an initially Gaussian wave packet which is centered around $x = x_0$ for the values of $x_0$ that are far to the left of the potential step. Its time development is obtained by including the time-dependent factor $\exp(-iEt/\hbar)$, $E = p^2/2m$, in the superposition.

First, we discuss a repulsive potential, that is, a positive step, $V_0 > 0$, and a wave packet with $p_0 > \sqrt{2mV_0}$. Figure 5.1 presents the time developments of the real and imaginary parts of the wave function and of the probability density. Figure 5.1c also shows the position of a classical particle having the same momentum $p_0$ as the expectation value of the quantum-mechanical wave packet. Of course, the classical particle moves to the right in region I with velocity $v = p_0/m$. Entering region II, it is instantaneously decelerated to velocity $v' = p_0'/m = \sqrt{p_0^2 - 2mV_0}/m$. The most striking effect on the behavior of the wave packet is that it is partly reflected at the potential step. For large times we observe a wave packet moving to the right in region II and in addition a wave packet which is reflected at the step and is moving to the left in region I. The wiggly structure in the probability density that occurs close to the step in region I stems from the interference of the incoming and reflected wave packets. The wiggles are caused by the fast variation of the de Broglie wave function. It is interesting to compare the behavior of our quantum-mechanical

S. Brandt and H.D. Dahmen, *The Picture Book of Quantum Mechanics*,
DOI 10.1007/978-1-4614-3951-6_5, © Springer Science+Business Media New York 2012

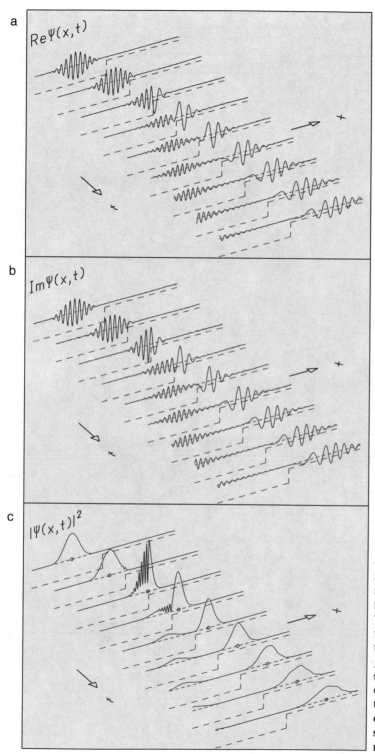

**Fig. 5.1.** Time developments of the real part, of the imaginary part of the wave function, and of the probability density for a wave packet incident from the left on a potential step of height $V_0 > 0$. The form of the potential $V(x)$ is again indicated by the long-dash line, the expectation value of the energy of the wave packet by the short-dash line, which also serves as zero line for the functions plotted. The expectation value of the initial momentum is $p_0 > \sqrt{2mV_0}$. The small circles indicate the positions of a classical particle of the same initial momentum.

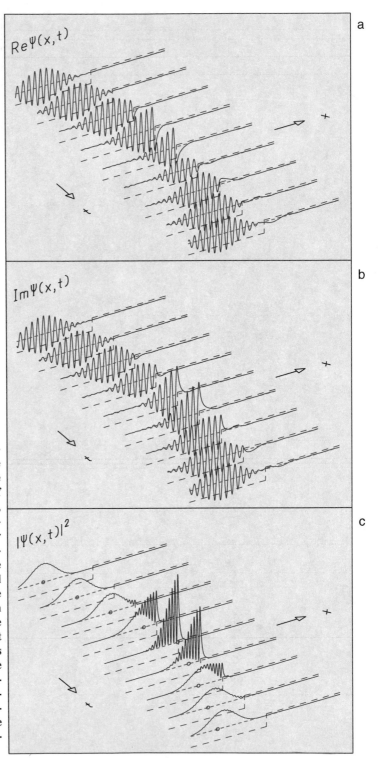

**Fig. 5.2.** Time developments of the real part, of the imaginary part of the wave function, and of the probability density for a wave packet incident from the left on a potential step $V_0 > 0$. The initial momentum expectation value of the incident wave packet is $p_0 < \sqrt{2mV_0}$. The small circles indicate the positions of a classical particle of the same initial momentum.

a

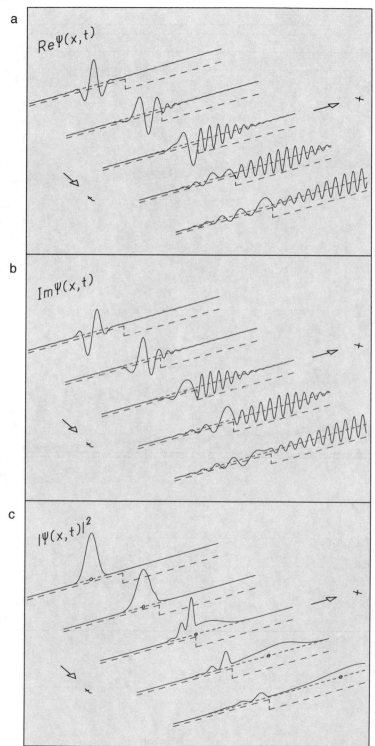

b

c

Fig. 5.3. Time developments of the real part, of the imaginary part, and of the probability density for a wave packet incident from the left on a potential step of height $V_0 < 0$. The small circles in part c indicate the positions of a classical particle incident on the same potential step.

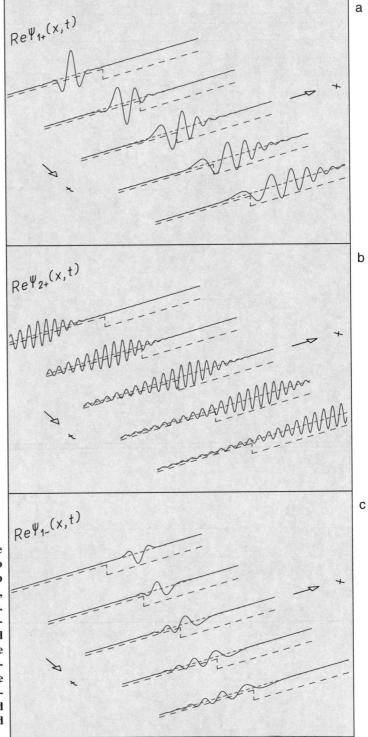

Fig. 5.4. A wave packet falls onto a potential step of height $V_0 < 0$, as in Figure 5.3. The time developments of the real parts of (a) the incident constituent wave, (b) the transmitted constituent wave, and (c) the reflected constituent wave.

wave packet incident on a potential step with that of the packet of light waves incident on a glass surface, which we studied in Section 2.5. The principal difference between the two phenomena is that the optical wave packet shows no dispersion, for its components with different wave numbers all move with the velocity of light.

We now use a wave packet with a lower initial momentum expectation value $p_0$ so that the corresponding classical particle is reflected at the step, that is, $p_0 < \sqrt{2m V_0}$. The time developments of the wave function and the probability density (Figure 5.2) show that part of the wave packet penetrates for a short while, with an exponential falloff into region II that is forbidden for the classical particle. Eventually, the wave packet is also completely reflected. The penetration into region II is analogous to the reflection of light off a metal surface with finite conductivity.

For an attractive potential, that is, $V_0 < 0$, the picture is similar to Figure 5.1. The classical particle is now suddenly accelerated at the potential step and so is the transmitted part of the wave packet. Part of the wave packet is also reflected, however. The time developments of the wave function and the probability density are shown in Figure 5.3. The reflection is not too evident in Figure 5.3 but becomes apparent in Figure 5.4. Here the real parts of the incoming, transmitted, and reflected constituent waves are plotted separately. The constituent waves are shown in their mathematical form for the whole range of $x$ values. The physical significance of $\psi_{1+}$ and $\psi_{1-}$ is restricted to region I, and that of $\psi_2$ to region II. Figure 5.4c shows that there is indeed a sizable reflected constituent wave moving to the left in region I.

## 5.2 Sudden Deceleration of a Classical Phase-Space Distribution

In Section 3.6 we discussed the time development of a classical phase-space probability density describing a particle which at time $t = 0$ is characterized by the position and momentum expectation values and uncertainties $(x_0, p_0)$ and $(\sigma_{x0}, \sigma_p)$, respectively. It was the bivariate Gaussian

$$
\rho_+^{cl}(x, p, t) = \frac{1}{2\pi\sigma_{x0}\sigma_p} \exp\left\{ -\frac{1}{2(1-c^2)} \left[ \frac{(x - [x_0 + p_0 t/m])^2}{\sigma_x^2(t)} \right.\right.
$$
$$
\left.\left. -2c\frac{(x - [x_0 + p_0 t/m])(p - p_0)}{\sigma_x(t)} \cdot \frac{(p - p_0)}{\sigma_p} + \frac{(p - p_0)^2}{\sigma_p^2} \right] \right\}
$$

with

$$
\sigma_x(t) = \sqrt{\sigma_{x0}^2 + \sigma_p^2 t^2/m^2} \quad , \qquad c = \frac{\sigma_p t}{\sigma_x(t) m} \quad ,
$$

shown in Figure 3.7.

We now study the reflection of the particle described by $\rho_+^{cl}(x,p,t)$ by a very high potential step at $x = 0$. It has the effect to reverse the momentum of the particle the moment it hits the point $x = 0$.

For definiteness we assume that $x_0 < 0$ and $|x_0| \gg \sigma_{x0}$. The phase-space density is initially (for $t = 0$) concentrated around the point $(x_0, p_0)$ far away from the step, i.e., it is given by $\rho^{cl} = \rho_+^{cl}$ and moves to the right just as in Figure 3.7. For times $t \gg m|x_0|/p_0$ reflection has taken place. The phase-space density moves towards the left and behaves just as if it would have started at $t = 0$ with the expectation values $(-x_0, -p_0)$, i.e., it is described by

$$\rho_-^{cl}(x,p,t) = \frac{1}{2\pi\sigma_{x0}\sigma_p} \exp\left\{ -\frac{1}{2(1-c^2)} \left[ \frac{(x - [-x_0 - p_0t/m])^2}{\sigma_x^2(t)} \right.\right.$$
$$\left.\left. -2c\frac{(x - [-x_0 - p_0t/m])}{\sigma_x(t)}\frac{(p + p_0)}{\sigma_p} + \frac{(p + p_0)^2}{\sigma_p^2} \right] \right\} \quad .$$

It is now obvious that for all times we can describe the phase-space distribution under the action of a reflecting force at $x = 0$ by the sum

$$\rho^{cl}(x,p,t) = \rho_+^{cl}(x,p,t) + \rho_-^{cl}(x,p,t) \quad , \qquad x < 0 \quad .$$

For positive $x$ values we have, of course,

$$\rho^{cl}(x,p,t) = 0 \quad , \qquad x > 0 \quad .$$

In Figure 5.5 we show the time development of $\rho^{cl}(x,p,t)$. The initial situation is identical to that of Figure 3.7 for the force-free case.

The marginal distribution in $x$,

$$\rho_x^{cl}(x,t) = \rho_{x+}^{cl}(x,t) + \rho_{x-}^{cl}(x,t)$$
$$= \frac{1}{\sqrt{2\pi}\sigma_x(t)} \exp\left\{ -\frac{(x - [x_0 + p_0t/m])^2}{2\sigma_x^2(t)} \right\}$$
$$+ \frac{1}{\sqrt{2\pi}\sigma_x(t)} \exp\left\{ -\frac{(x - [-x_0 - p_0t/m])^2}{2\sigma_x^2(t)} \right\} \quad ,$$

is simply the sum of the marginal distributions of $\rho_+^{cl}$ and $\rho_-^{cl}$ for $x < 0$ and vanishes for $x > 0$. It is the classical spatial probability density and also shown in Figure 5.5. For times long before or long after the reflection process it is identical to the quantum-mechanical probability density, which for a similar case was shown in Figure 5.2. During the period of reflection, however, $t \approx m|x_0|/p_0$, the quantum-mechanical probability density $\rho(x,t) = |\psi(x,t)|^2$ shows the typical interference pattern, whereas the classical density $\rho_x^{cl}(x,t)$ is smooth. This striking difference is due to the fact that in the quantum-mechanical calculation the wave functions $\psi_{1+}(x,t)$ and $\psi_{1-}(x,t)$ are added and the absolute square of the sum is taken to form $\rho(x,t)$ whereas in the classical calculation the marginal densities $\rho_{x+}^{cl}(x,t)$ and $\rho_{x-}^{cl}(x,t)$ of the constituent phase-space distributions are added directly.

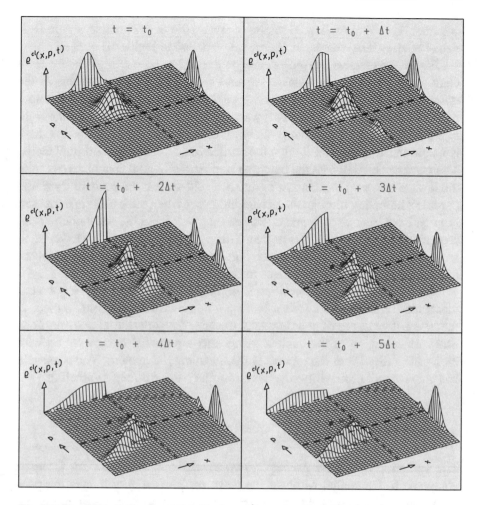

**Fig. 5.5. Classical phase-space distribution** $\rho^{\text{cl}}(x, p, t)$ **reflected at a high potential wall at** $x = 0$ **shown for various times. The marginal distribution** $\rho_x^{\text{cl}}(x, t)$ **is shown over the margin in the background, the marginal distribution** $\rho_p^{\text{cl}}(p, t)$ **over the right-hand margin of the individual plots.**

## 5.3 Tunnel Effect

In Figure 5.2 we studied the behavior of a wave packet that was reflected at a potential step higher than the average energy of the wave packet. We observed that, during the process of reflection, the wave packet penetrated to a certain extent into the region of high potential. It would be interesting now to see what happens if the region of high potential extends only over a distance comparable to the depth of penetration. We therefore study a wave packet which is under the influence of a *potential barrier*. The potential is constant, $V = V_0$, in a limited region of space, $0 < x < d$, called region II. It vanishes elsewhere, that is, in region I, $x < 0$, and in region III, $x > d$.

Figure 5.6a shows the time development of the probability density for a Gaussian wave packet incident in region I on such a potential barrier. At the upstep of the barrier at $x = 0$, we observe the expected pattern of a reflection. At the downstep at $x = d$ we see a wave packet emerge and travel to the right in region III. According to our probability interpretation, this means that there is a nonvanishing probability that the particle described by the original Gaussian wave packet will pass the potential barrier, although it cannot do so under the laws of classical mechanics. Figure 5.6b shows that the probability of the particle's tunneling through the barrier increases when the barrier is narrower. Finally, from Figures 5.6b and c, we see that the probability of tunneling decreases as the barrier becomes higher. These general features have to be taken with caution, for in some potentials there are discrete energy ranges in which the tunneling probability possesses maxima.

The tunnel effect just described is one of the most surprising in quantum mechanics. It is the basis for explaining a number of phenomena, including the radioactive decay of atomic nuclei through the emission of an $\alpha$ particle. The surface region of the nucleus represents a potential barrier which with high probability keeps the $\alpha$ particle from leaving the nucleus. The $\alpha$ particle has only a small probability of penetrating the barrier through tunneling.

## 5.4  Excitation and Decay of Metastable States

The scattering of a wave packet on two repulsive barriers that are far apart compared to the spatial width of the wave packet is a very interesting phenomenon. The width of the two barriers is chosen so that the tunnel effect allows a sizable fraction of the probability to pass through the two barriers. Figure 5.7 shows the time development of the packet entering from the left. We observe that although the major part of the probability is reflected at the first barrier, another part enters the region between the barriers and retains its bell shape at least while distant from the barriers. At a later moment in time the injected packet hits the barrier on the right, and again there is partial reflection and transmission. Later on in the process the particle is with a certain probability confined between the two walls, continuously bouncing back and forth and each time losing part of the probability to the outside region. Except for the continuous broadening of the particle wave packet, this situation is very similar to the analogous process in optics, namely a light wave packet falling onto a glass plate, which was shown in Figure 2.12.

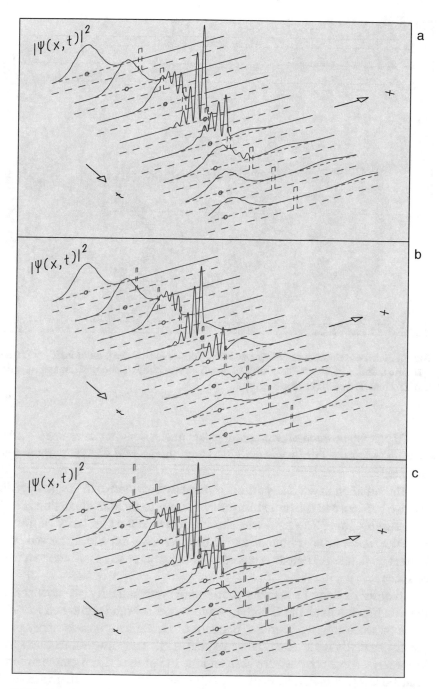

**Fig. 5.6. Tunnel effect. (a)** Time development of the probability density for a wave packet incident from the left onto a potential barrier of height $V_0$. The small circles indicate the positions of a classical particle incident onto the same potential barrier. **(b)** Same as for part a, but for a barrier of half the width. **(c)** Same as for part b, but for a barrier of double the height.

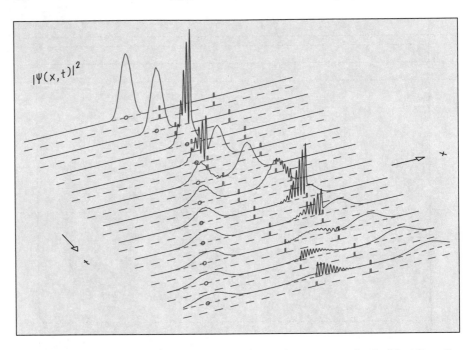

$$|\Psi(x,t)|^2$$

**Fig. 5.7. Time development of the probability density for a wave packet incident from the left onto a double potential barrier. The small circles indicate the positions of a classical particle incident onto the same barrier.**

Of course, no such phenomenon exists in classical physics, since particles transmitted at the left barrier will also pass the second barrier without being reflected.

The situation in which a particle is partially confined to the region between the two barriers and the probability slowly leaks to the outside region is called a *metastable state*. The term was chosen to invoke the similarity of this state with the *stable state* or bound state, which we have already discussed briefly in Section 4.4. A particle in a bound state is permanently confined to a region of space.

In order to study metastable states more systematically, we now consider the situation in which the Gaussian wave packet is broad compared to the distance between the two barriers. Because of Heisenberg's uncertainty principle, the spatially wide Gaussian wave packet has a narrow momentum spread. The energy spectrum between zero and the top of the barrier can therefore be scanned in small intervals. For the two barriers of Figure 5.8, there are various energy, and thus momentum, values for which a fraction of the probability enters the region between the barriers and stays there for quite some time, even though the wave packet has traveled rather far away from the reflecting barrier. Figure 5.8 shows the time developments of the probability densities for wave

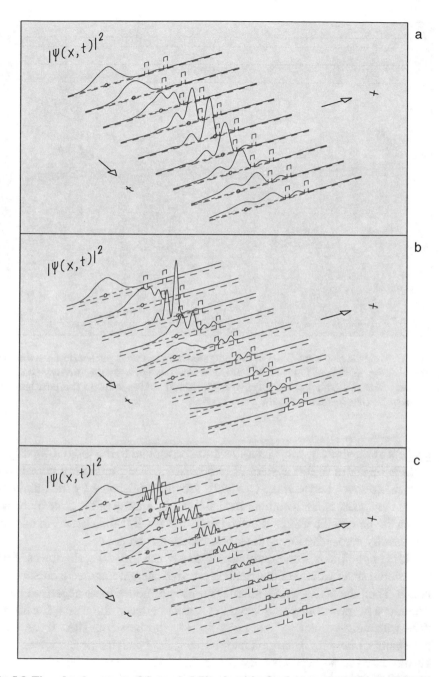

Fig. 5.8. Time developments of the probability densities for wave packets of mean energies corresponding to (a) the first, (b) the second, and (c) the third metastable states in a system of two barriers. The wave packets, which are rather wide in space and thus possess a small momentum width, are incident from the left onto the double potential barrier. The small circles indicate the positions of a classical particle incident on the same barrier.

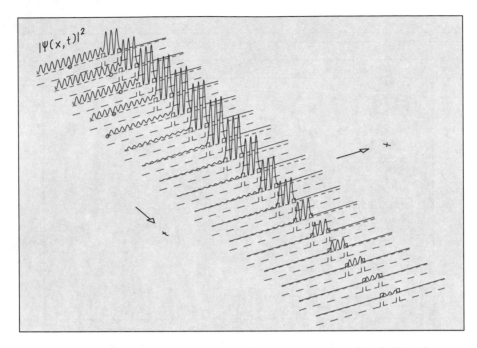

**Fig. 5.9. Time development of the probability density for a wave packet that has the same mean energy as that of Figure 5.8c but is ten times as wide. Again, the wave packet is incident onto a double potential barrier. The small circles indicate the positions of a classical particle incident onto the same barrier.**

packets of the three average energies that correspond to the three lowest-lying metastable states in this system of two barriers. The probability densities of metastable states between the two walls are distinguishable by the number of nodes they possess. This number increases as the energy of the state increases. When the potential between the two barriers is not less than zero – in our case it is exactly zero – the lowest metastable state has no node.

If the potential is sufficiently negative between the walls, the lowest-lying metastable state, which, of course, has positive energy, may have one or more nodes. Then the states with a smaller number of nodes have negative energy. Therefore no probability can leave the region between the walls, for no particle with negative energy can exist outside the barriers. Thus these states are stable or bound. To complicate matters, the behavior of the wave packets discussed so far depends not only on their average energy but also on their spectral function in momentum space, that is, on their spatial form. In order to rid ourselves of this complication, we shall study wave packets with a smaller and smaller momentum spread. They are of course very wide in space.

Figure 5.9 shows the time development of the probability density for a wave packet whose average energy is equal to that of the third metastable state. It corresponds to Figure 5.8c, except that the wave packet now has ten times the spatial width; its width exceeds by far the dimension of the figure. In the region to the left of the barriers, we observe the wiggly pattern typical for the interference between the incoming and the reflected wave packet. Between the two barriers the probability density keeps increasing with time up to the maximum amplitude, which is reached when the bulk of the wave packet has been reflected and has moved to the left. From then on the metastable state with two nodes decays slowly, in fact exponentially. The excitation of the metastable state in Figure 5.9 is much greater than the excitation of that in Figure 5.8c. The greater width of the wave packet in Figure 5.9 implies a narrower spectral function, which therefore contains more probability within the energy range of the metastable state.

To study the lifetime of metastable states, we observe their excitation and decay, as shown in Figure 5.9, over a longer period of time. In Figure 5.10a it is easy to see that the amplitude in the region between the two barriers drops exponentially with time. We can measure the lifetime by determining the time in which the amplitude decreases by a factor of two. This we call the *half-life* of the state. Figure 5.10b shows the excitation and decay of the metastable state with only one node, corresponding to a lower energy, in the same time scale as the metastable state with two nodes. The half-life is now considerably longer. Even longer is the lifetime of the metastable state with no nodes. In Figure 5.10c, which is in the same time scale as Figures 5.10a and b, the amplitude has not decreased yet; the time interval of the figure is still in the excitation phase.

## 5.5 Stationary States of Sharp Momentum

We have just discussed the one-dimensional scattering of wave packets of narrow momentum spread and large extension in space. By reducing the momentum spread further, we obtain as the limiting case a harmonic wave $\psi_E(x,t)$ of fixed energy and momentum. After separating off the energy-dependent phase factor $\exp(-iEt/\hbar)$, we are left with a stationary state $\varphi_E(x)$, which was discussed in Chapter 4. The intimate relation between wide wave packets and stationary states allows a direct physical interpretation, in terms of particle mechanics, of the characteristic features of stationary states. A stationary state can be thought of as a limiting description of a particle with sharp momentum.

We are able to understand important details about metastable states through the study of stationary states in our potential with two barriers. We recall that within the barriers, that is, in regions II and IV, the potential is constant and positive, $V = V_0 > 0$. Outside the barriers, in regions I, III, and V, it van-

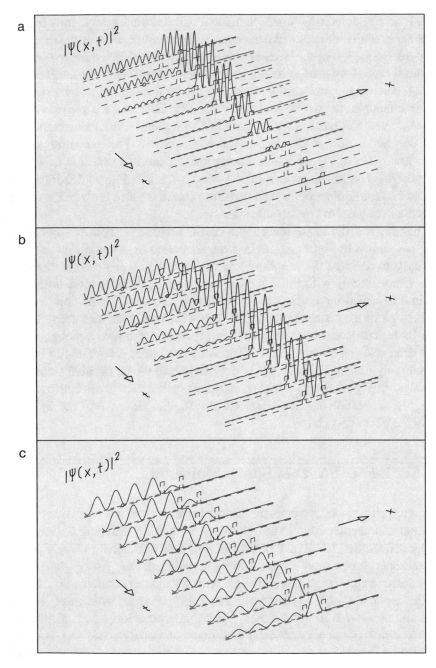

**Fig. 5.10.** (a) Time development of the process shown in Figure 5.9 but observed over a longer period of time. Once the bulk of the wave packet has been reflected, the metastable state decays like an exponential function in time. Parts b and c are the same as part a but for the two metastable states that lie higher in energy. Parts a, b, and c of this figure correspond to parts a, b, and c of Figure 5.8. The wave packets, however, are much broader, and the time interval shown is much longer.

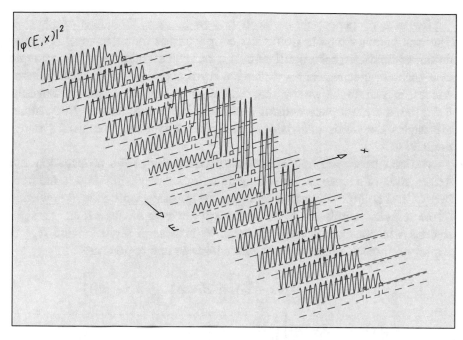

**Fig. 5.11.** Energy dependence, over a small range of energies, of the intensity of a harmonic wave incident onto a double potential barrier. The middle line corresponds to a resonance energy.

ishes, $V = 0$. Figure 5.11 shows the energy dependence of the stationary state $\varphi_E(x)$ in the potential with two barriers. That is, the solution of the time-independent Schrödinger equation for this potential is energy dependent. The quantity plotted in Figure 5.11 is the intensity, introduced in Section 4.2, of this stationary solution. The range of energies shown in the figure comprises the energy of the metastable state with two nodes, which we discussed earlier and showed in Figures 5.8c, 5.9, and 5.10c. When the energy is lower than that of the metastable state – in the background of the picture – only a small fraction of the intensity is transmitted through the barriers into region V. There is a prominent interference pattern in region I from the superposition of the incoming and the reflected wave. As the energy approaches that of the metastable state, the reflection decreases to zero, the interference pattern vanishes, and the full intensity of the incoming wave is transmitted through both barriers into region V. At the energy of the metastable state, the intensity in region III, between the barriers, reaches its maximum and assumes the two-node structure that is characteristic of the metastable state. This phenomenon is called a *resonance* of the system. As the energy increases further, the intensity in region III decreases as does the transmission into region V. The interference pattern in region I reappears as reflection grows.

Resonance phenomena are well known in many branches of physics. The best-known example from classical physics is the resonance of a pendulum excited to forced oscillation of a particular frequency. Our example of a quantum-mechanical resonance has a striking similarity to optical resonances. In Section 2.3 we saw that light at particular frequencies is transmitted through a glass plate without reflection. In the terminology of quantum mechanics, the words *metastable state* and *resonance* are often used synonymously.

As long as we are not interested in the details of the propagation and deformation of a wave packet with definite initial shape, but only in the fraction of probability with which reflection or transmission occurs, knowledge of the complex amplitudes of the stationary waves in the far left and far right regions – in our example regions I and V, in general regions I and $N$ – is entirely sufficient. The stationary waves in these two regions are

$$\varphi_I(x) = \exp\left[\frac{i}{\hbar}p(x - x_0)\right] + B_I \exp\left[-\frac{i}{\hbar}p(x - x_0)\right] \quad ,$$

$$\varphi_N(x) = A_N \exp\left[\frac{i}{\hbar}p(x - x_0)\right] \quad .$$

The fact that we are dealing with a particle that can only be reflected or transmitted obviously requires that

$$|A_N|^2 + |B_I|^2 = 1 \quad .$$

This relation, which expresses the conservation of the total probability of observing the particle, is called the *unitarity relation* for the scattering amplitudes $A_N$ and $B_I$. For vanishing reflection, $B_I = 0$, the unitarity relation allows a circle of radius 1 in the complex plane for $A_N$. Figure 5.12a shows the energy dependence of the complex number $A_N$, again for the problem with two potential barriers. The upper left part of the figure is an *Argand diagram*. A plane is spanned by the axes $\text{Re}\,A_N$ and $\text{Im}\,A_N$. For a fixed energy value the complex number $A_N$ is given by the point $A_N = (\text{Re}\,A_N, \text{Im}\,A_N)$ in

**Fig. 5.12. (a) Energy dependence of the complex amplitude $A_N$ of the part of a harmonic wave that is transmitted through the system with a double potential barrier. The energy ranges from zero to a value twice the barrier height. The energy dependence of $A_N$ is shown as a line, starting from the origin, in the complex plane at upper left. The circle around the origin indicates the maximally allowed region for $A_N$. The energy dependence of the real part, projection onto the real axis, is shown below, that of the imaginary part, projection onto the imaginary axis, to the right. The lower right of the figure shows the energy dependence of $|A_N|^2$. (b) The parts of this figure are the same as those of part a, but they are for the transmission-matrix element $T_T = (A_N - 1)/(2i)$. The line starts at point $i/2$ in the complex plane. The circle around point $i/2$ indicates the maximally allowed region for $T_T$.**

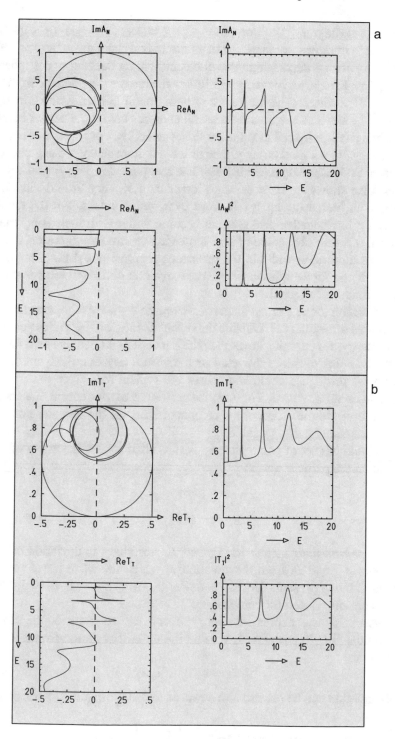

**Fig. 5.12.**

the Argand diagram. The line in this figure shows the variation with energy of $A_N$ as a complex number. The outer circle corresponds to $|A_N| = 1$. Obviously, $A_N$ always stays within this circle, indicating the energy dependence of $\text{Im}\, A_N$ and $\text{Re}\, A_N$, respectively. We follow the energy dependence from $E = 0$ to $E = 2V_0$, where $V_0$ is the height of both potential barriers. The imaginary part of $A_N$ stays near zero for almost all energies below the barrier height $V_0$, slowly deviating from that value for larger energies. For resonance regions of the energy, $\text{Im}\, A_N$ rises quickly, then falls even more steeply to negative values in order to rise quickly again to zero. The real part of $A_N$ displayed below the circle also shows for most energies lower than $V_0$ very little deviation from zero. With increasing energy it drops to negative values. For the resonance regions it has negative peaks which become wider with increasing energies. Finally, the absolute square of $A_N$, shown in the lower right corner, again has peaks of increasing width in the resonance regions. $|A_N|^2$ has a tendency to approach one for energies far above the barrier height. For these energies total transmission is expected.

Returning to the Argand diagram, we are now able to recognize the typical signatures of resonances. Outside the resonance region $A_N$ varies very slowly with energy; for energies lower than the barrier height, it stays near the origin of the complex plane. In the resonance region it passes quickly and counter-clockwise through a circle and causes the typical resonance patterns in the real and imaginary parts. For energies above the barrier height, the circles no longer return to the origin of the complex plane, for the transmission outside the resonance regions is sizable.

Yet another set of parameters is used to characterize the effect of the potential on the particle waves,

$$
\begin{aligned}
A_N &= 1 + 2iT_\mathrm{T} \quad, \\
B_\mathrm{I} &= 2iT_\mathrm{R} \quad.
\end{aligned}
$$

The *transition-matrix elements* $T_\mathrm{T}$ and $T_\mathrm{R}$ describe the deviation of the parameters $A_N$ and $B_\mathrm{I}$ from the situation in which the wave travels without a potential being present, that is, the deviation from $A_N = 1$, $B_\mathrm{I} = 0$. The factors $2i$ are introduced for convenience.

Inserting these expressions into the unitarity relation for the scattering amplitudes, $|A_N|^2 + |B_\mathrm{I}|^2 = 1$, we find the unitarity relation for the $T$-matrix elements:

$$
\text{Im}\, T_\mathrm{T} = T_\mathrm{T} T_\mathrm{T}^* + T_\mathrm{R} T_\mathrm{R}^* \quad.
$$

This equation can be rewritten in terms of real and imaginary parts of $T_\mathrm{T}$:

$$
(\text{Re}\, T_\mathrm{T})^2 + \left(\text{Im}\, T_\mathrm{T} - \frac{1}{2}\right)^2 = \frac{1}{4} - T_\mathrm{R} T_\mathrm{R}^* \quad.
$$

For $T_R T_R^* = 0$ this relation describes complex numbers $T_T$ on a circle of radius $1/2$ centered around the point $i/2$. Because of $|B_I|^2 \leq 1$, we have $|T_R|^2 \leq 1/4$ so that the right-hand side of the equation remains positive or zero. For non-vanishing $T_R$ the complex numbers $T_T$ therefore fall within the circle. Figure 5.12b shows in the upper left the Argand diagram of $T_T$ with the circle of radius $1/2$ centered around $i/2$ limiting its possible values. It also contains the projections $\text{Im}\,T_T(E)$, $\text{Re}\,T_T(E)$ as well as $|T_T(E)|^2$. Because of the simple relation between $T_T$ and $A_N$, the features of these diagrams are in one-to-one correspondence with those of Figure 5.12a.

In elementary particle physics Argand diagrams of the type given in Figure 5.12 are used to study the complex scattering amplitude. This amplitude describes the collision probability of two particles. Detection of characteristic circular features is equivalent to the discovery of metastable states. Such states are considered to be elementary particles with very short lifetimes.

## 5.6 Free Fall of a Body

In Section 4.3 we dealt with a constant force, i.e., a linear potential. The motion of a body of mass $m$ under the action of a constant force $F = mg$ is described by an initially Gaussian wave packet. The initial expectation values are $x_0$, $p_0 = mv_0$, the initial spatial width is $\sigma_{x0}$, or equivalently in momentum space, $\sigma_p = \hbar/2\sigma_{x0}$. The time-dependent wave function of the wave packet is given by

$$
\psi(x,t) = \frac{1}{\sqrt[4]{2\pi}\sqrt{\sigma_x(t)}} \exp\left\{-\left(\frac{x - \langle x(t)\rangle}{2\sigma_x(t)}\right)^2\right\}
$$
$$
\times \exp\left\{\frac{i}{\hbar}\left[\left(c(t)\sigma_p \frac{x - \langle x(t)\rangle}{2\sigma_x(t)} + \langle p(t)\rangle\right)[x - \langle x(t)\rangle] + \hbar\alpha(t)\right]\right\} \quad .
$$

Here

$$
\alpha(t) = \frac{m}{\hbar}\left[\left(gx_0 + \frac{v_0^2}{2}\right)t + gv_0 t^2 + \frac{1}{3}g^2 t^3\right] - \frac{1}{2}\arctan\left(\frac{2\sigma_p^2}{\hbar m}t\right)
$$

is a time-dependent phase,

$$
\langle x(t)\rangle = x_0 + v_0 t + \frac{g}{2}t^2
$$

is the position of a free-falling classical particle of initial position $x_0$ and velocity $v_0$, and

$$
\langle p(t)\rangle = p_0 + mgt
$$

its momentum. The time-dependent width of the wave packet is as in the force-free motion

$$\sigma_x(t) = \left[ \sigma_{x0}^2 + \frac{\sigma_p^2}{m^2} t^2 \right]^{1/2} \quad .$$

The quantity $c(t)$ is given by

$$c(t) = \frac{\sigma_p t}{m \sigma_x(t)} \quad .$$

The phase of the wave packet contains a term proportional to $[x - \langle x(t) \rangle]$,

$$\phi(t) = \frac{1}{\hbar} \left[ \langle p(t) \rangle + c(t) \frac{x - \langle x(t) \rangle}{2 \sigma_x(t)} \sigma_p \right] [x - \langle x(t) \rangle] \quad ,$$

which can again be interpreted as a product of the coordinate and a time- and position-dependent effective wave number:

$$k_{\text{eff}}(x,t) = \frac{1}{\hbar} \left( p_0 + mgt + c(t) \frac{x - \langle x(t) \rangle}{2 \sigma_x(t)} \sigma_p \right) \quad .$$

It reveals the correlation between $x$ and $p = \hbar k_{\text{eff}}$. For fixed $t$ and at $x = \langle x(t) \rangle$ the effective wave number is equal to

$$k_{\text{eff}}(\langle x(t) \rangle, t) = \frac{1}{\hbar} (p_0 + mgt) \quad .$$

This is to say, at the classical position of the particle at time $t$ the momentum has the classical value $p_0 + mgt$. For positions $x > \langle x(t) \rangle$, the effective wave number is larger than $(p_0 + mgt)/\hbar$. For $x < \langle x(t) \rangle$ the opposite holds true.

In Figure 5.13 we show the time development of the wave function and its absolute square for a particle initially at rest, $v_0 = 0$, and being pulled to the right by the constant force $F = mg$. It illustrates the free fall of a particle as described by quantum mechanics.

In Figure 5.14 the situation is slightly more complicated. The particle has an initial velocity $v_0 > 0$ and the constant force now pulls to the left, $g > 0$. In classical mechanics the resulting motion is that of a stone being thrown upwards against the direction of the gravitational force.

The motion of a classical particle described by a Gaussian probability distribution in phase space under the action of a constant force is easily described if one uses arguments analogous to those of Section 3.6. Also in this case the classical phase-space probability density stays a bivariate Gaussian determined by the expectation values $\langle x(t) \rangle$ and $\langle p(t) \rangle$ of position and momentum, the widths $\sigma_x(t)$ and $\sigma_p$ – the latter being constant – and the correlation coefficient $c(t)$. For all five parameters the classical and the quantum-mechanical

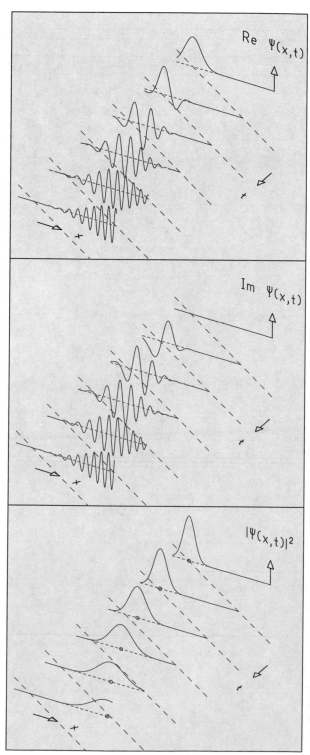

**Fig. 5.13.** Time development of the real part, the imaginary part, and the absolute square of the wave function for a wave packet which is initially at rest and which is pulled to the right by a constant force. The (linear) potential of the force is indicated by the long-dash line, the expectation value of the energy of the wave packet by the short-dash line which also serves as zero line for the function plotted. The small circles indicate the position of a classical particle with initial position and momentum equal to the corresponding expectation values of the wave packet.

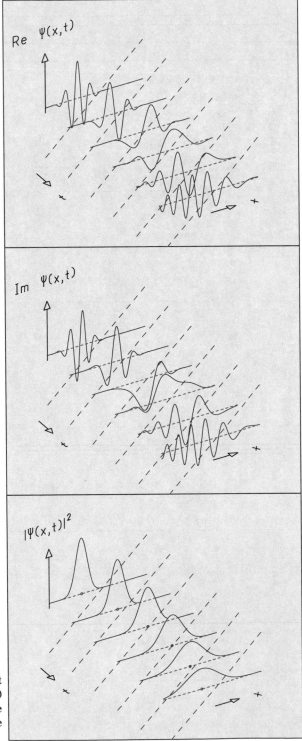

**Fig. 5.14. As Figure 5.13 but for an initial velocity $v_0 > 0$ and for a constant force pulling the particle to the left.**

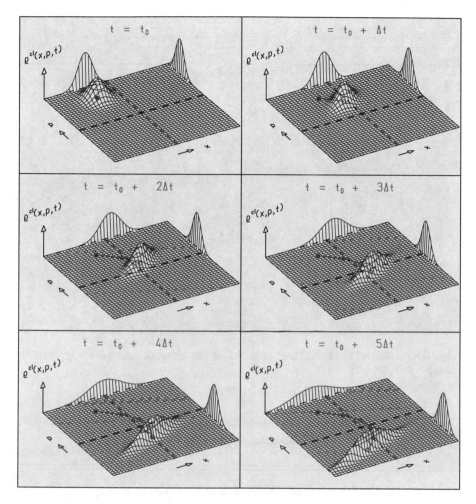

**Fig. 5.15.** Time development of the classical phase-space probability density $\rho^{cl}(x,p,t)$ corresponding to the quantum-mechanical situation of Figure 5.14. The trajectory of the point $(\langle x(t)\rangle, \langle p(t)\rangle)$ defined by the expectation values of position and momentum between the time $t = t_0 = 0$ and the actual time is shown for each plot. Also shown are the marginal distributions $\rho^{cl}_x(x,t)$ in position and $\rho^{cl}_p(p,t)$ in momentum.

calculations yield the same result. Moreover, the widths and the correlation coefficient are the same as in the case of the force-free particle.

The time development of the classical phase-space probability density of a particle with an initial velocity opposite to the direction of the constant force is shown in Figure 5.15. The point $(\langle x(t)\rangle, \langle p(t)\rangle)$, given by the expectation values in position and momentum, moves on a parabola in the $x, p$ plane. The initially uncorrelated distribution develops an increasing positive correlation between position and momentum. While the momentum width stays constant

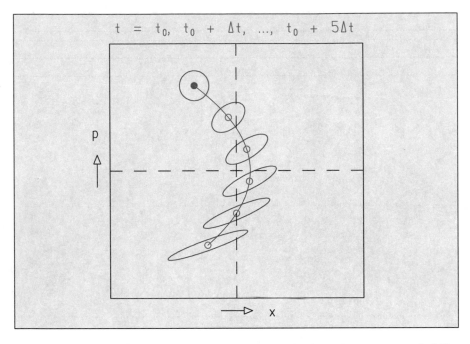

**Fig. 5.16. Motion of the covariance ellipsoid of the classical phase-space probability density of Figure 5.15.**

the spatial width increases. All these features are also apparent in Figure 5.16 which shows the motion of the covariance ellipse in phase space.

## 5.7 Scattering by a Piecewise Linear Potential

The general features of scattering are similar for a piecewise constant potential used for the examples in Sections 5.1 to 5.5 and for a piecewise linear potential. The solution of the Schrödinger equation for the latter was discussed in Section 4.3. Here we present a few examples. The potential in Figures 5.17 to 5.19 can be seen as a triangular well between two triangular barriers with perpendicular outer edges. Figure 5.17 contains the stationary states for various energies of the incoming wave. For the central energy value the wave function and its absolute square display the typical resonance features familiar from Sections 5.4 and 5.5: The absolute square is constant on both sides of the potential structure, indicating that there is no reflection, i.e., no interference between incoming and reflected wave. Morever, the wave function and its absolute square display a prominent shape within the potential structure. This shape – two nodes of the wave function's real part, two zeros of its absolute square – remind us of the third resonance for the double-barrier

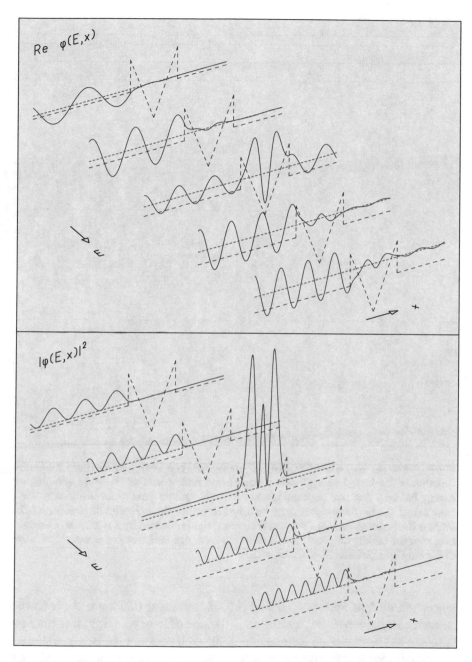

**Fig. 5.17. Stationary scattering states in a piecewise linear potential. Shown is the real part (top) and the absolute square (bottom) of the stationary wave functions for different energy values. The central diagram in each of the two plots corresponds to a resonance energy.**

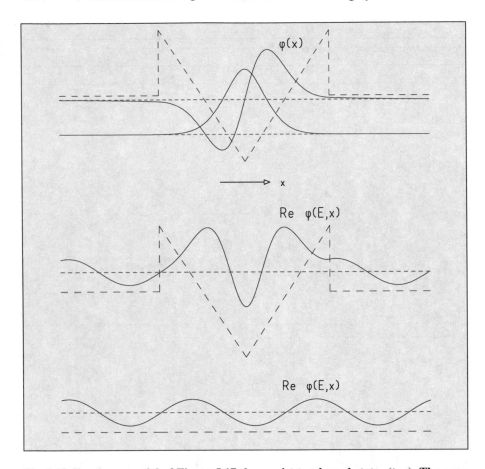

**Fig. 5.18. For the potential of Figure 5.17 there exist two bound states (top). The wave function of the lowest energy state possesses no nodes; that of the state with higher energy has one. The real part (middle) and the imaginary part of the stationary wave function of the first resonance have two nodes within the potential structure. At the bottom the incident wave is shown as free wave, not influenced by a potential. Comparison with the middle diagram reveals that, at resonance, incident and transmitted wave differ only by a phase shift of about $\pi/2$.**

system studied in Sections 5.4 and 5.5, in particular of Figure 5.11. In the present case, however, the resonance is the one of lowest energy; it is the first resonance. The two-node form is due to the well between the two barriers.

A potential of this form can accomodate bound states. In our particular case there are two, possessing zero and one node, respectively. The first resonance displays symmetry features quite similar to those the next bound state would have, if it existed. For comparison the two bound states and the stationary scattering state of the first resonance are shown in Figure 5.18. The figure also demonstrates that, apart from a phase shift of about $\pi/2$, incident and transmitted wave are identical at resonance. We shall come back to this feature in more detail when we discuss scattering in three dimensions in Chapters 12 and 15.

To exemplify the nature of the resonance a little further we display, in Figure 5.19, the time development of both the real and the imaginary part of the wave function at resonance. The two plots span a time interval equal to half the oscillation period of the incoming harmonic wave. In this time both the incoming wave left of the potential structure and the transmitted wave on its right have equal amplitudes and wavelengths and travel along the positive $x$ direction by half a wavelength without changing their forms. Within the structure, on the other hand, one observes the ups and downs of a standing wave pattern. The absolute square of the wave function is independent of time and has the form of the central diagram in the bottom part of Figure 5.17.

We conclude this section by observing the passage of a wave packet through various piecewise linear potentials. Figure 5.20 displays a Gaussian wave packet over a time interval in which it passes a wide triangular potential well. The corresponding classical particle (indicated by the dot in the bottom part of Figure 5.20) travels with constant velocity in the regions before and behind the well; it is accelerated in the first half of the well and decelerated in the second. The probability density is stretched in width while being pulled "downhill" and compressed again when it climbs "uphill". Overall, some widening of the packet results by the displacement of the packet, very similar as in the case of a free wave packet. The real part of the wave function allows us to study the momentum structure of the packet by observing the local variation of its wavelength. It is large (indicating low momentum) where the potential is constant and varies considerably in the region of the well, being smallest at its minimum. There is only very little distortion of the wave packet through interference of reflected and transmitted waves at those points where the derivative of the potential is discontinuous. It shows up in the form of some slight wiggles in the shape of the probability density.

We pursue this phenomenon by comparing the passage of a wave packet through a triangular well with that through two sorts of well with somewhat different shapes. One is similar to the square well potential. However, its edges are skew; in these regions the potential is a linear function of $x$. The other one is composed of two consecutive triangular wells. In Figure 5.21 we observe that the passage through the trapezoidal well which resembles closely that through the triangular well. Also quite similar is the passage through two consecutive triangular wells, although the wave packet gets a little more distorted. In both cases there is only marginally more interference than for the single triangular well, even though there are now more points in which the derivative of the potential is discontinuous.

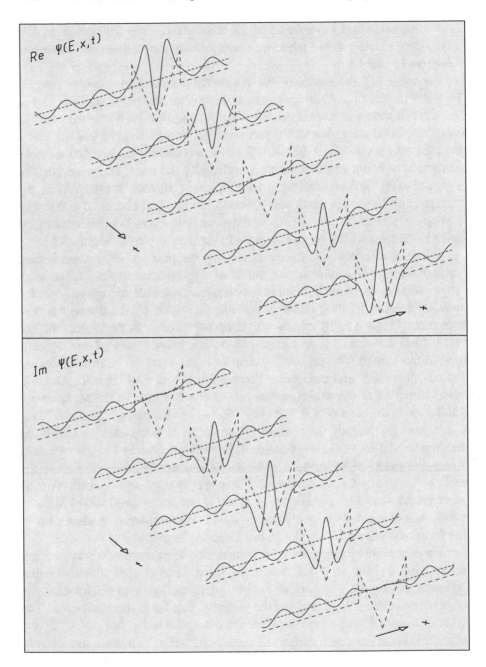

**Fig. 5.19. The piecewise linear potential of Figure 5.17 is traversed by a harmonic wave at resonance energy. The time development of its real part (top) and imaginary part (bottom) is shown over half an oscillation period.**

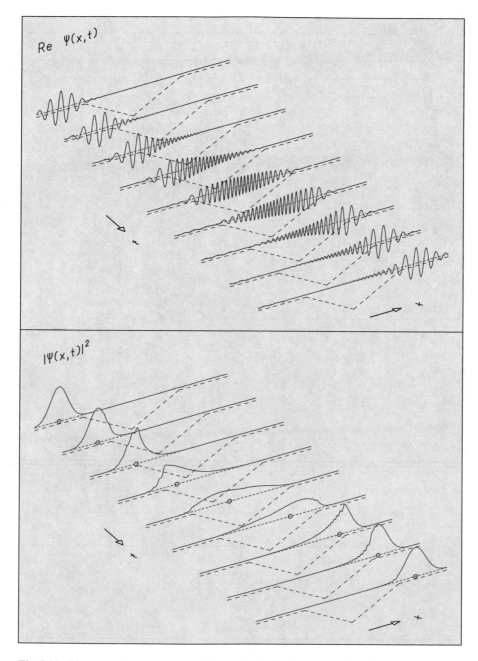

$Re \ \Psi(x,t)$

$|\Psi(x,t)|^2$

**Fig. 5.20. Wave packet traversing a triangular well. Time development of the real part of the wave function (top) and of the probability density (bottom).**

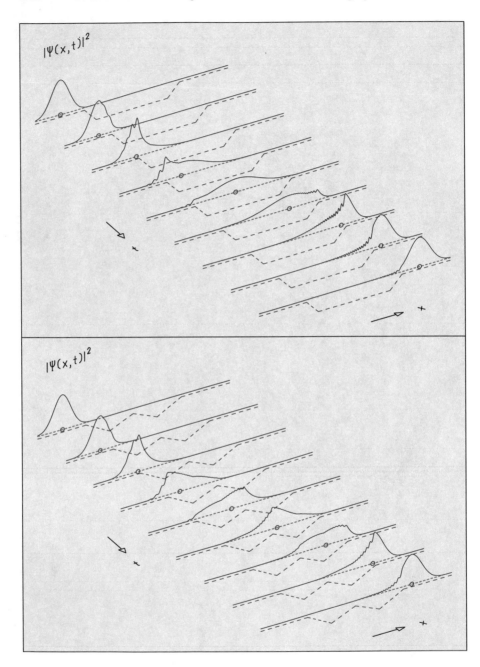

**Fig. 5.21. Wave packet traversing a trapezoidal well, i.e., a "square" well with skew edges (top), and a double triangular well (bottom). For both cases only the probability density is displayed. There is some interference due to the discontinuity of the potential but hardly any reflection.**

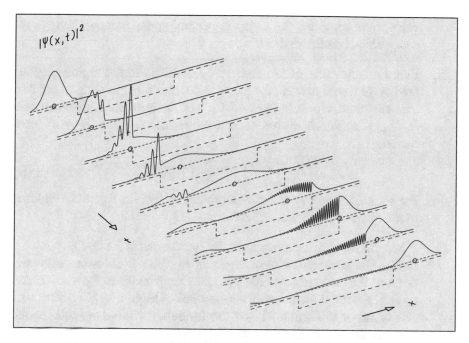

**Fig. 5.22. Wave packet traversing a true square well with sharp edges. The probability density exhibits important interference and reflection.**

In contrast, the passage through a true square well, depicted in Figure 5.22, displays considerable reflection and interference at both of its sharp edges where not only the derivative but the potential itself is discontinuous. From this observation we may conclude that, if we want to model a quantum-mechanical problem with a smoothly varying potential, its approximation by a piecewise linear potential is preferable to that by a piecewise constant one.

# Problems

5.1. Figure 5.1c shows the probability density and the classical position of a particle moving toward and beyond a potential step. Why is the wave packet narrower immediately after passing the positive potential step than before passing it? Predict the behavior of the wave packet at a negative potential step and verify this in Figure 5.3c.

5.2. Determine the ratios of the amplitudes of the metastable state at successive equidistant moments in time. Use a ruler to measure the amplitudes in Figure 5.9. For later moments in time, the ratios tend to a constant value, indicating that the decay is becoming exponential. Why is the decay slower earlier in time?

5.3. Plot the amplitudes of the probability densities in the region between the two potential barriers for the energies $E_i$ corresponding to the thirteen situations shown in Figure 5.11. The energies are equidistant, that is, $E_{i+1} - E_i = \Delta E = $ constant. Fit the result to a *Breit–Wigner distribution*,

$$f(E) = A \frac{\Gamma^2/4}{(E - E_r)^2 + \Gamma^2/4} \quad .$$

For $E_r$, use the energy of the maximum amplitude and give the width $\Gamma$ of the distribution in units of $\Delta E$.

5.4. Figure 5.12 shows the energies of the resonances in the double potential barrier. Calculate the ratios of the energies of the three lowest resonance peaks as they are given in Figure 5.12b. Compare the ratios to the corresponding ones of the bound-state energies of Figure 4.9. Compare both sets of ratios to Figure 6.1 and the formula for the deep square well given at the beginning of Section 6.1.

# 6. One-Dimensional Quantum Mechanics: Motion within a Potential, Stationary Bound States

So far we have dealt with the motion of particles with a total energy $E = E_{kin} + V$ that is positive at least in region I, the region of the incoming particle. Of course, classical motion inside a finite region where the potential is negative is also possible for negative total energies, as long as kinetic energy $E_{kin} = E - V$ is positive. We now study this system from the point of view of quantum mechanics.

## 6.1 Spectrum of a Deep Square Well

As a particularly simple system, let us consider the force-free motion in a region of zero potential between two infinitely high potential walls at $x = -d/2$ and $x = d/2$. Since the potential outside this region is infinite, the solutions of the time-independent Schrödinger equation vanish there. Within this region they have the simple forms

$$\varphi_n(x) = \sqrt{\frac{2}{d}} \cos(n\pi \frac{x}{d}) \quad , \qquad n = 1, 3, 5, \ldots \quad ,$$

or

$$\varphi_n(x) = \sqrt{\frac{2}{d}} \sin(n\pi \frac{x}{d}) \quad , \qquad n = 2, 4, 6, \ldots \quad .$$

The energies of these bound states are

$$E_n = \frac{1}{2m} \left( \frac{\hbar n \pi}{d} \right)^2 \quad , \qquad n = 1, 2, 3, \ldots \quad ,$$

as we easily verify by inserting $\varphi_n$ into the time-independent Schrödinger equation

$$-\frac{\hbar^2}{2m} \frac{d^2}{dx^2} \varphi_n = E_n \varphi_n \quad ,$$

S. Brandt and H.D. Dahmen, *The Picture Book of Quantum Mechanics*,
DOI 10.1007/978-1-4614-3951-6_6, © Springer Science+Business Media New York 2012

which is valid between the two walls. Figure 6.1 presents the wave function, the probability density, and the energy spectrum. The lowest-lying state at $E_1$, called the *ground state*, has a finite energy $E_1 > 0$, which implies a kinetic energy $E_{kin} > 0$ since the potential energy $V$ is zero by construction. Already this situation differs from that in classical mechanics, where the state of least energy is of course the state of rest with $E = E_{kin} = 0$. The higher states increase in energy proportionally to $n^2$. The *quantum number $n$* is equal to one plus the number of nodes of the wave function in the region $-d/2 < x < d/2$; that is, the boundaries $x = \pm d/2$ are excluded. The wave function has even or odd symmetry with respect to the point $x = 0$, depending on whether $n$ is odd or even, respectively. Even wave functions, here the cosine functions, are said to possess *even* or *natural parity*, odd wave functions *odd* or *unnatural parity*. Obviously, wave functions with an even number of nodes have even parity, those with an odd number odd parity. This property also holds for other one-dimensional potentials that are mirror-symmetric.

## 6.2  Particle Motion in a Deep Square Well

In Section 6.1 we found the spectrum of eigenvalues $E_n$ and the wave functions describing the corresponding eigenstates $\varphi_n(x)$ for the deep square well. The solutions of the time-dependent Schrödinger equation are obtained by multiplying $\varphi_n(x)$ with a factor $\exp(-iE_nt/\hbar)$. Through a suitable superposition of such time-dependent solutions, we form a moving wave packet which at the initial time $t = 0$ is bell shaped with a momentum average $p_0$. Its wave function is

$$\psi(x,t) = \sum_{n=1}^{\infty} a_n(p_0,x_0)\varphi_n(x)\exp\left[-\frac{i}{\hbar}E_nt\right] \quad ,$$

where the coefficients $a_n(p_0,x_0)$ have been chosen to ensure a bell shape around location $x_0$ for $t = 0$ and the momentum average $p_0$.

Figure 6.2 shows the time development of the probability density $|\psi(x,t)|^2$ for such a wave packet. We observe that for $t = 0$ the wave packet is well localized about initial position $x_0$ of the classical particle. It moves toward one wall of the well, where it is reflected. Here it shows the pattern typical of interference between incident and reflected waves. The pattern is very similar to that caused by a free wave packet incident on a sharp potential step, shown in Figure 5.2c. It continues to bounce between the two walls and is soon so wide that the packet touches both walls simultaneously, showing interference patterns at both walls.

It is interesting to see how the spatial probability density $\rho^{cl}(x,t)$ derived from a classical phase-space probability density behaves in time. This is

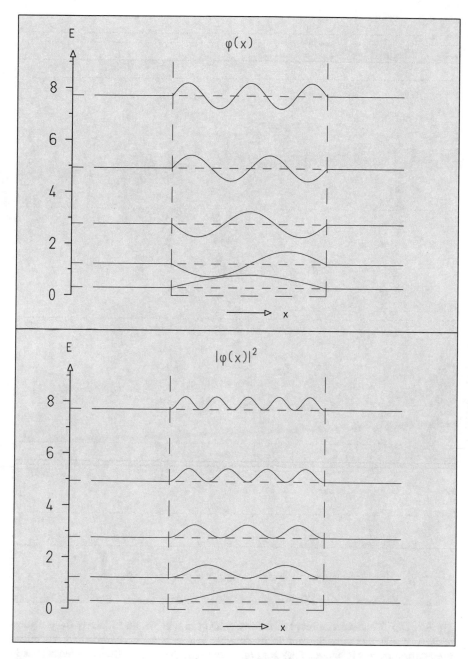

**Fig. 6.1. Bound states in an infinitely deep square well. The long-dash line indicates the potential energy** $V(x)$**. It vanishes for** $-d/2 < x < d/2$ **and is infinite elsewhere. Points** $x = \pm d/2$ **are indicated as vertical walls. On the left side an energy scale is drawn, and to the right of it the energies** $E_n$ **of the lower-lying bound states are indicated by horizontal lines. These lines are repeated as short-dash lines on the left. They serve as zero lines for the wave functions** $\varphi(x)$ **and the probability densities** $|\varphi(x)|^2$ **of the bound states.**

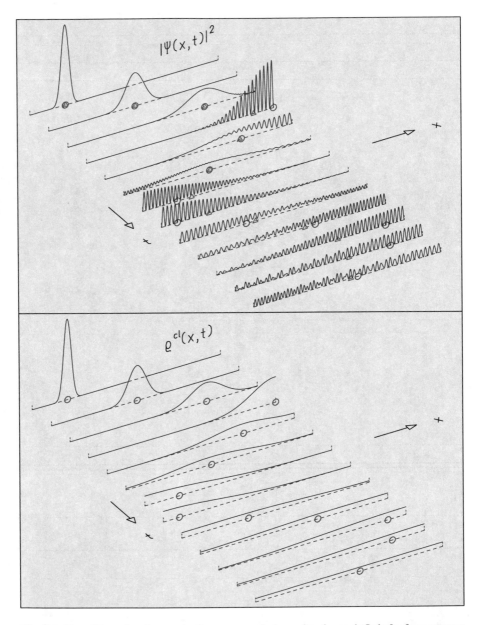

**Fig. 6.2.** Top: Time development of a wave packet moving in an infinitely deep square well. At $t = 0$, in the background, the smooth packet is well concentrated. Its initial momentum makes it bounce back and forth between the two walls. The characteristic interference pattern of the reflection process, as well as the dispersion of the packet with time, is apparent. The small circle indicates the position of the corresponding classical particle. The quantum-mechanical position expectation value is shown by a small triangle. Bottom: Time development of the spatial probability density computed from the classical phase-space distribution corresponding to the quantum-mechanical wave packet.

shown in the bottom part of Figure 6.2. As long as the bulk of the probability density is not close to the walls the quantum-mechanical density $|\psi(x,t)|^2$ and the classical density $\rho^{\mathrm{cl}}(x,t)$ are very similar.

Near the walls, however, the quantum-mechanical wave packet displays the interference pattern typical for the superposition of the two wave functions incident on and reflected by the wall. As the packet disperses the interference pattern fills the whole well. No interference is observed in the time development of the classical phase-space density. It is obtained as the sum

$$\rho_x^{\mathrm{cl}}(x,t) = \frac{1}{\sqrt{2\pi}\,\sigma_x(t)} \sum_{n=-\infty}^{\infty} \left\{ \exp\left[ -\frac{(x - v_0 t - 2nd)^2}{2\sigma_x^2(t)} \right] \right.$$
$$\left. + \exp\left[ -\frac{(x + v_0 t - (2n+1)d)^2}{2\sigma_x^2(t)} \right] \right\}$$

with the time-dependent width of a free wave packet:

$$\sigma_x(t) = \sigma_{x0}\sqrt{1 + \left( \frac{\sigma_p t}{\sigma_{x0} m} \right)^2}$$

by a simple generalization of the sum at the end of Section 5.2 from the reflection at one high potential wall to the repeated reflection between two high walls.

We now want to study the quantum-mechanical wave packet in a deep well over a much longer period of time. At the end of the time interval studied in Figure 6.2 the quantum-mechanical probability density $|\psi(x,t)|^2$ occupies the full width of the well and one might be inclined to think that it continues to do so. It is easy to see, however, that the quantum-mechanical wave function $\psi(x,t)$ must be periodic in time, the period being

$$T_1 = \frac{2\pi}{\omega_1} \quad,$$

where $\omega_1$ is the frequency of the ground-state wave function

$$\omega_1 = \frac{E_1}{\hbar} = \frac{\hbar}{2m} \left( \frac{\pi}{d} \right)^2 \quad.$$

Since all energies $E_n$, $n = 2, 3, \ldots$, are integer multiples of $E_1$, the period $T_1$ of the ground state is also the period of the superposition $\psi(x,t)$ that describes the wave packet. Because of this periodicity in time the original wave packet must be restored after the time $T_1$ has elapsed. In Figure 6.3 we show the time dependence of the same wave packet as in Figure 6.2 over a full period $T_1$ and find our expectation verified.

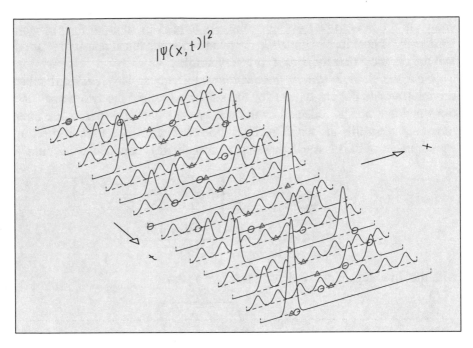

$|\psi(x,t)|^2$

**Fig. 6.3. Time development of the same wave packet as in Figure 6.2 but observed of a full revival period $T_1$. The time interval shown in Figure 6.2 is $T_1/60$.**

The periodicity is called *revival* of the wave packet. As we shall see in Section 13.5, the phenomenon is also encountered in the wave-packet motion in the Coulomb potential, e.g., in the hydrogen atom as an approximate revival. To a larger or lesser degree it exists in all systems with discrete spectra of reasonable spacing. In the case of the deep square well it is, however, an exact revival.

In addition to the revival at $t = T_1$ we can also observe *fractional revivals* at the times $t = (k/\ell)T_1$. Here $k$ and $\ell$ are integer numbers. Since in Figure 6.3 the time $T_1$ is divided into 16 equal intervals it is easy to observe the packet at the times $t = T_1/2$, $T_1/4$, $T_1/8$, and $T_1/16$. For these times the function $|\psi(x,t)|^2$ consists of 1, 2, 4, and 8 well-separated "Gaussian" humps.

## 6.3 Spectrum of the Harmonic-Oscillator Potential

The particle in a deep square well experiences a force only when hitting the wall. A simple, continuously acting force $F(x)$ can be thought of as the force of a spring, which follows Hooke's law,

$$F(x) = -kx \quad , \qquad k > 0 \quad .$$

This force, also called a *harmonic force*, is proportional to the displacement $x$ from equilibrium position $x = 0$. A physical system in which a particle moves under the influence of a harmonic force is called a *harmonic oscillator*. The proportionality constant $k$ gives the stiffness of the spring. The potential energy stored in the spring is

$$V(x) = \frac{k}{2}x^2 \quad .$$

A classical particle of mass $m$ performs *harmonic oscillations* of angular frequency

$$\omega = \sqrt{k/m}$$

so that $V(x)$ can be equivalently expressed by $V(x) = (m/2)\omega^2 x^2$. Introducing this expression into the time-independent Schrödinger equation yields

$$\left(-\frac{\hbar^2}{2m}\frac{d^2}{dx^2} + \frac{m}{2}\omega^2 x^2\right)\varphi(x) = E\varphi(x) \quad .$$

With the help of the dimensionless variable

$$\xi = \frac{x}{\sigma_0} \quad , \qquad \sigma_0 = \sqrt{\hbar/(m\omega)} \quad ,$$

the equation above simplifies to the reduced form

$$\frac{1}{2}\left(-\frac{d^2}{d\xi^2} + \xi^2\right)\phi(\xi) = \varepsilon\phi(\xi) \quad , \qquad \phi(\xi) = \varphi(\sigma_0\xi) \quad .$$

The dimensionless eigenvalue $\varepsilon = E/\hbar\omega$ measures the energy of the oscillator in multiples of Planck's quantum of energy $\hbar\omega$.

The solutions of the Schrödinger equation for the harmonic oscillator can be normalized (see Section 4.4) for the eigenvalues

$$\varepsilon_n = n + \frac{1}{2} \quad , \qquad n = 0, 1, 2, \ldots \quad ,$$

thus determining the energy eigenvalues of the harmonic oscillator,

$$E_n = \varepsilon_n \hbar\omega = (n + \frac{1}{2})\hbar\omega \quad .$$

The state of lowest energy $E_0 = \hbar\omega/2$ is the ground state. The energies $E_n$ of the higher states differ from the ground-state energy by the energy of $n$ quanta, each having the energy $\hbar\omega$ of Planck's quantum (see Chapter 1).

The eigenfunctions, normalized in $\xi$, can be represented in the form

$$\phi_n(\xi) = (\sqrt{\pi}2^n n!)^{-1/2} H_n(\xi) e^{-\xi^2/2} \quad , \qquad n = 0, 1, 2, \ldots \quad ,$$

where the $H_n(\xi)$ are the *Hermite polynomials*. They are given by

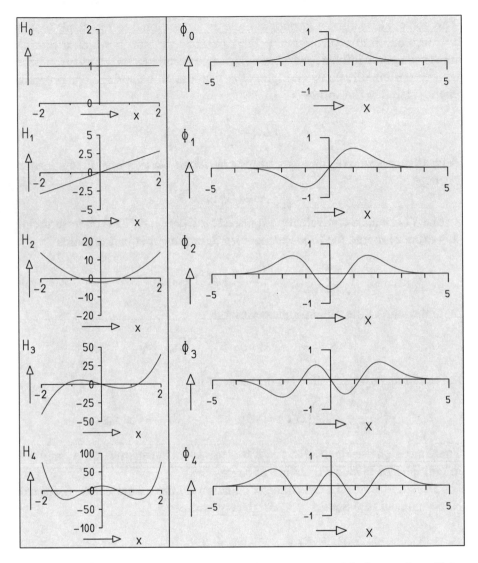

**Fig. 6.4. Hermite polynomials $H_n(x)$ and eigenfunctions $\phi_n(x)$ of the harmonic oscillator for low values of $n$.**

$$H_0(\xi) = 1 \quad , \qquad H_1(\xi) = 2\xi \quad ,$$

and for higher values of $n$ by the recurrence relation

$$H_n(\xi) = 2\xi\, H_{n-1}(\xi) - 2(n-1)H_{n-2}(\xi) \quad , \qquad n = 2, 3,\dots \quad .$$

Figure 6.4 shows the Hermite polynomials $H_n(\xi)$ and the normalized eigenfunctions $\varphi_n(\xi)$ for low values of $n$.

The eigenfunctions $\varphi_n(x)$, normalized in $x$, are

$$\varphi_n(x) = (\sigma_0\sqrt{\pi}\,2^n n!)^{-1/2} H_n\left(\frac{x}{\sigma_0}\right)\exp\left(-\frac{x^2}{2\sigma_0^2}\right) \quad.$$

They are plotted in Figure 6.5 together with the potential energy $V(x)$. The dashed lines indicate the energy eigenvalues in relation to the bottom of the potential energy. They serve as zero lines for the corresponding $\varphi_n$. On the left-hand side the energy spectrum is shown. The exponential factor $\exp(-\xi^2/2)$ in the formula for $\varphi_n$ ensures that

$$\varphi_n(x) \to 0 \qquad \text{for} \quad |x| \to \infty$$

rendering these wave functions normalizable.

Figure 6.5 gives the probability densities $|\varphi_n(x)|^2$, showing that, even in regions where $E$ is smaller than $V$, there is a certain probability of observing the particle. The absolute square of the wave function of the ground state formulated in terms of the position variable $x = \sigma_0\xi$ has the form

$$|\varphi_0(x)|^2 = \frac{1}{\sqrt{\pi}\,\sigma_0}\exp\left(-\frac{x^2}{2\sigma_0^2/2}\right) \quad.$$

The exponent in this equation shows that the width of the probability density of the harmonic oscillator's ground state is $\sigma_0/\sqrt{2}$.

## 6.4 Harmonic Particle Motion

We now consider the quantum-mechanical description of a particle moving under the influence of harmonic force. The particle at initial time $t = 0$ is at rest when placed in a position $x = x_0 \neq 0$, which is not the equilibrium position of the oscillator. In terms of a wave function, the initial state consists of a Gaussian wave packet of width $\sigma$ with zero average momentum and an expectation value at position $x = x_0$ of the corresponding classical particle. This wave packet can be decomposed into a sum over eigenfunctions $\varphi_n(x)$ of the harmonic oscillator,

$$\varphi(x) = \sum_{n=0}^{\infty} a_n\varphi_n(x) \quad.$$

The time-dependent solution of the Schrödinger equation with $\varphi(x)$ as initial wave function at $t = 0$ is then simply

$$\psi(x,t) = \sum_{n=0}^{\infty} a_n\varphi_n(x)\exp\left(-\frac{i}{\hbar}E_n t\right) \quad,$$

where $E_n = (n + \frac{1}{2})\hbar\omega$.

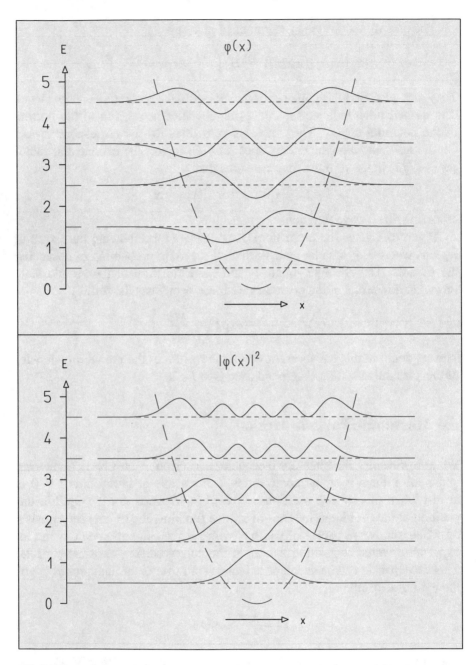

**Fig. 6.5. Bound states in a harmonic-oscillator potential. The potential is drawn as a long-dash line, a parabola. The eigenvalue spectrum of bound states (in units of $\hbar\omega$) is indicated by the horizontal lines on the left side. Repeated on the right as short-dash lines, they serve as zero lines for the wave functions $\varphi(x)$ and the probability densities $|\varphi(x)|^2$ of the bound states.**

The infinite sum can be added up explicitly. For brevity we give here only the result for the absolute square of the wave function,

$$|\psi(x,t)|^2 = \frac{1}{\sqrt{2\pi}} \frac{2\sigma}{\sqrt{\sigma_0^4 s^2 + 4\sigma^4 c^2}} \exp\left[-\frac{2\sigma^2}{\sigma_0^4 s^2 + 4\sigma^4 c^2}(x - cx_0)^2\right] ,$$

where $c$ and $s$ represent $\cos\omega t$ and $\sin\omega t$, respectively, and where $\sigma_0/\sqrt{2}$ is the width of the probability distribution of the harmonic oscillator's ground state, as introduced in Section 6.3. This equation represents a Gaussian distribution with oscillating expectation value $x_0(t) = x_0 \cos\omega t$ and oscillating width $\sigma(t) = \sqrt{\sigma_0^4 \sin^2\omega t + 4\sigma^4 \cos^2\omega t}/(2\sigma)$. Of course, for the initial time $t = 0$ the time-dependent width $\sigma(t)$ reduces to the initial width $\sigma$.

Figure 6.6a shows a time development of a wave packet in the harmonic oscillator with initial width $\sigma < \sigma_0/\sqrt{2}$. As expected, the time dependence of the average position performs the same oscillation as the corresponding classical particle. The width oscillates with twice the frequency of the oscillator, starting with $\sigma$ and increasing for the first quarter period $T/4 = \pi/2\omega$ to its maximum value $\sigma(T/4) = \sigma_0^2/(2\sigma)$. In Figure 6.6b the initial width is $\sigma > \sigma_0/\sqrt{2}$. Here the wave packet is wide initially and becomes narrower in the first quarter period, decreasing to the minimum value $\sigma_0^2/(2\sigma)$. The case $\sigma = \sigma_0/\sqrt{2}$ (Figure 6.6c), in which the width of the packet remains constant in time, represents the border line between the two situations. The particular value $\sigma_0/\sqrt{2}$ is exactly the width of the absolute square $|\varphi_0|^2$ of the ground-state wave function (shown in Figure 6.5). The factor $\sqrt{2}$ appears since $\sigma_0$ was defined conventionally as the width of the wave function $\varphi_0$ itself. In all three situations the behavior of the position expectation value is identical and equal to that of the classical particle.

We now look at half a period of the oscillation in more detail. In Figure 6.7 (top), which depicts this time interval, the time development of the probability density is plotted again for a wave packet with initial width smaller than the ground-state width $\sigma_0/\sqrt{2}$. The real and imaginary parts of the wave function are shown in the lower plots of Figure 6.7. At the turning points, $t = 0$ and $t = T/2$, the wave function is purely Gaussian and either real or imaginary. For other moments in time, the wiggly structure originates from the superposition of the eigenfunctions of the harmonic oscillator. As is true of the eigenfunctions themselves, the distance between two nodes increases in the vicinity of the turning points. For a free harmonic wave the distance between two nodes is half the wavelength; a large wavelength signifies low momentum. We can therefore interpret the increasing distance between nodes in the vicinity of the turning points as the slowing down of the particle.

Finally, we look at the particular situation of a particle "at rest" in the center of the oscillator (Figure 6.8). Initially, the particle is sharply local-

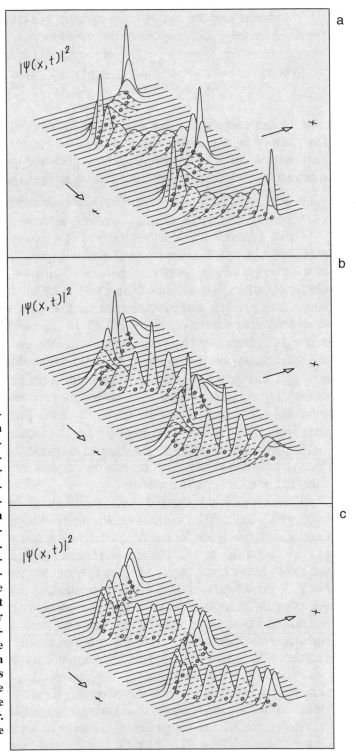

**Fig. 6.6. Time development of a Gaussian wave packet, represented by its probability density, under the influence of a harmonic force. The circles show the motion of the corresponding classical particle. The broken lines extend between its turning points. The wave packet is initially at rest at an off-center position. (a) The initial width of the wave packet is smaller than that of the oscillator's ground state. (b) The initial width of the wave packet is greater. (c) Both widths are equal.**

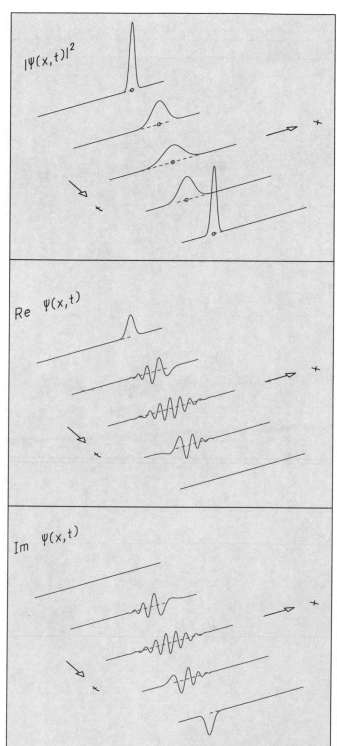

**Fig. 6.7. Time development of a Gaussian wave packet under the influence of a harmonic force, observed over half an oscillation period. Shown are the probability density, the real part of the wave function, and the imaginary part of the wave function.**

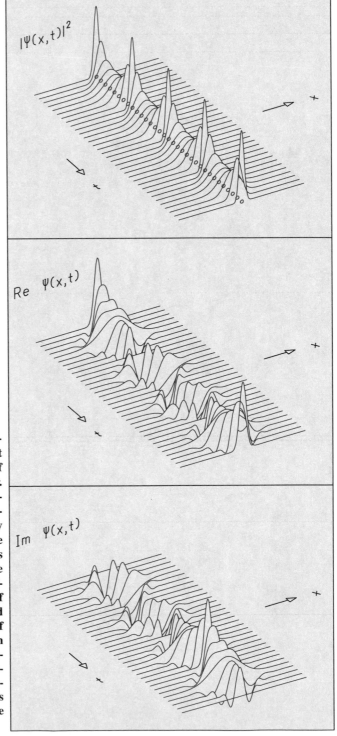

**Fig. 6.8. Time development of a wave packet at rest in the center of a harmonic oscillator. The packet is represented by its probability density, and by the real part, and the imaginary part of its wave function. Since its initial width is different from that of the oscillator's ground state, the width of the packet oscillates in time with twice the oscillator frequency. Except for the initial position, all parameters are identical to those of Figure 6.7.**

ized compared to the ground-state width, that is, $\sigma < \sigma_0/\sqrt{2}$. The expectation value in space remains at $x = 0$, just as the classical particle does. The width of the wave packet, however, oscillates with twice the oscillator frequency between its initial value $\sigma$ and its maximum value $\sigma_0^2/(2\sigma)$. Only for initial width $\sigma = \sigma_0/\sqrt{2}$ does the absolute square of the wave packet keep its position as well as its shape.

The wave packet of Figure 6.6c is called a *coherent state* of the oscillator. While oscillating the wave packet keeps its width equal to the ground-state width of the oscillator. At all times it is a *minimum-uncertainty state*, that is to say, it fulfills Heisenberg's uncertainty principle as an equation $\Delta x \, \Delta p = \hbar/2$.

The ground state of the harmonic oscillator is a particular coherent state because it is also an eigenstate of the Hamilton operator. The other coherent states are not among the eigenstates but are particular superpositions of eigenstates of the harmonic oscillator. Since the various eigenstates differ in energy, a coherent state, except for the ground state, is a superposition of states with different numbers of energy quanta $\hbar\omega$. The weights $p(n)$, with which these states of different numbers $n$ of energy quanta $\hbar\omega$ contribute to the coherent state, follow a Poisson distribution, cf. Appendix G,

$$p(n) = \frac{\langle n \rangle^n}{n!} e^{-\langle n \rangle} \quad .$$

Here $\langle n \rangle$ is the expectation value of the number of quanta given by

$$\left( \langle n \rangle + \frac{1}{2} \right) \hbar\omega = \langle E \rangle \quad ,$$

where $\langle E \rangle$ is the energy expectation value of the coherent state. It therefore has a nonvanishing variance of the number of energy quanta and of the energy. If an external force acts upon a harmonic oscillator in its ground state, the oscillator responds with a transition into another coherent state. If the action of the external force is terminated at some time $t_0$, the state of the oscillator behaves as the coherent state of Figure 6.6c. It performs a harmonic oscillation along a classical trajectory with the frequency $\omega$ of the classical oscillator. Coherent states play an important role in quantum optics and quantum electronics.

The initial packets shown in Figures 6.6a and b are not coherent states. Their initial widths are different from the ground-state width $\sigma_0/\sqrt{2}$. They are called *squeezed states*. Squeezed states are not minimum-uncertainty states at all moments of time. Four times during one period of oscillation, however, they develop into minimum-uncertainty states. As we have seen in Figures 6.6a and b, wave packets representing squeezed states also oscillate so that their expectation values follow the classical trajectories. Their widths, however, vary with time. They oscillate back and forth between a minimum and a maximum value. The distribution of the numbers of the energy quanta contributing to a squeezed state deviates from a Poisson distribution. Not being minimum-uncertainty states, squeezed states allow one observable quantity of an oscillator to be less uncertain than it is in the ground state, at the cost of the other observables occurring in Heisenberg's uncertainty principle. For this reason squeezed states are of great interest in the theory of measurement of weak signals.

## 6.5 Harmonic Motion of a Classical Phase-Space Distribution

We will show later in this section that the classical phase-space distribution of Section 3.6, i.e., a phase-space distribution which at the initial time $t = 0$ fulfills the uncertainty relation $\sigma_{x0}\sigma_{p0} = \hbar/2$, behaves in the harmonic-oscillator potential in the very same way as a quantum-mechanical wave packet. Before we do so we will present a qualitative argument showing that a classical Gaussian phase-space density indeed oscillates as does the quantum-mechanical probability density in Figure 6.7.

A classical particle described by a phase-space distribution of large initial spatial width $\sigma_{x0}$ possesses a rather well-defined momentum. For a classical particle initially at rest at $x = x_0$ the period $T$ of oscillation is independent of $x_0$. Thus, particles at rest at different initial positions $x_0$ all reach the point $x = 0$ at the same time, $t = T/4$. Since the initial momentum spread is small but not zero the spatial distribution at $t = T/4$ will have a finite spread $\sigma_x(T/4) < \sigma_{x0}$.

For a small initial spatial width, on the other hand, the spatial definition of the particle is rather well-defined, but the particle may start at this position with rather different momenta. Consequently the distribution spreads in space and, at $t = T/4$, has a rather large width, $\sigma_x(T/4) > \sigma_{x0}$.

There is a particular intermediate initial width, which will turn out to be $\sigma_{x0} = \sigma_0/\sqrt{2}$, for which the classical phase-space distribution keeps its shape while oscillating as a whole. This is the classical analog of the coherent state of quantum mechanics.

We mentioned that for constant forces or for forces that depend linearly on the coordinates the temporal evolution of the Wigner distribution (cf. Appendix D) of a quantum-mechanical wave packet is identical to that of a classical phase-space density. The phase-space probability density corresponding to a Gaussian wave packet without correlation between momentum and position at the initial time $t = 0$ is

$$\rho_i^{cl}(x_i, p_i) = \frac{1}{2\pi \sigma_{x0}\sigma_{p0}} \exp\left\{ -\frac{1}{2}\left[ \frac{(x_i - x_{0i})^2}{\sigma_{x0}^2} + \frac{(p_i - p_{0i})^2}{\sigma_{p0}^2} \right] \right\} .$$

Here $x_{0i}$, $p_{0i}$ are the initial expectation values and $\sigma_{x0}$, $\sigma_{p0}$ are the initial widths of position and momentum, respectively.

The covariance ellipse of the bivariate Gaussian is characterized by the exponential being equal to $-1/2$,

$$\frac{(x_i - x_{0i})^2}{\sigma_{x0}^2} + \frac{(p_i - p_{0i})^2}{\sigma_{p0}^2} = 1 .$$

The classical motion of a particle in phase space under the action of a harmonic force is simply

$$\begin{aligned} x &= x_i \cos\omega t + q_i \sin\omega t , \\ q &= -x_i \sin\omega t + q_i \cos\omega t . \end{aligned}$$

Here we have introduced the variables

$$q(t) = \frac{p(t)}{m\omega} , \qquad q_i = \frac{p_i}{m\omega} .$$

A classical particle rotates with angular velocity $\omega$ on a circle around the origin in the $x,q$ plane. For a given time $t$ and given values $x(t)$, $q(t)$ the initial conditions of a particle are then

$$\begin{aligned} x_i &= x \cos\omega t - q \sin\omega t , \\ q_i &= x \sin\omega t + q \cos\omega t . \end{aligned}$$

Introducing this result into the equation for the initial covariance ellipse,

$$\frac{(x_i - x_{0i})^2}{\sigma_{x0}^2} + \frac{(q_i - q_{0i})^2}{\sigma_{q0}^2} = 1 ,$$

which describes an ellipse with the center $(x_{0i}, q_{0i})$ and the semi-axes $\sigma_{x0}$ and $\sigma_{q0}$ which are parallel to the $x$ axis and the $q$ axis, respectively, we get

$$\frac{([x - x_0]\cos\omega t - [q - q_0]\sin\omega t)^2}{\sigma_{x0}^2}$$
$$+ \frac{([x - x_0]\sin\omega t + [q - q_0]\cos\omega t)^2}{\sigma_{q0}^2} = 1 .$$

This is again an equation of an ellipse with principal semi-axes of length $\sigma_{x0}$ and $\sigma_{q0}$. They are, however, no longer parallel to the coordinate directions but rotated by an angle $\omega t$ with respect to these. The center of the ellipse is the point $(x_0, q_0)$ to which the set of initial expectation values $(x_{0i}, q_{0i})$ has moved at the time $t$.

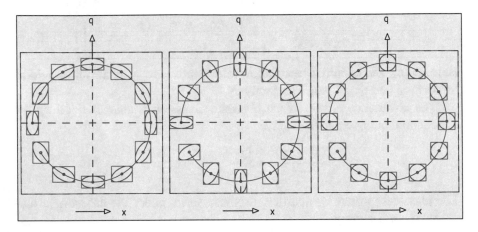

**Fig. 6.9. Motion of the covariance ellipse of a classical phase-space density under the influence of a harmonic force. The large circle is the trajectory of the center of the ellipse. The ellipse is shown for equidistant moments in time. The rectangle circumscribing it has sides $\sigma_x, \sigma_q$. The small circles indicate the centers of the ellipses. For the initial time it is drawn as a full dot. The relation between the initial widths is (left) $\sigma_{x0} < \sigma_{q0}$, (middle) $\sigma_{x0} > \sigma_{q0}$, and (right) $\sigma_{x0} = \sigma_{q0}$.**

We summarize our discussion by the following simple statements:

1. A classical phase-space distribution described by a bivariate Gaussian keeps its Gaussian shape.

2. Its center, which is the center of the covariance ellipse, moves on a circle around the center of the $x, q$ plane with angular velocity $\omega$.

3. The covariance ellipse keeps its shape but rotates around its center with the same angular velocity $\omega$.

In Figure 6.9 we illustrate the motion of the covariance ellipse for the three cases $\sigma_{x0} < \sigma_{q0}$, $\sigma_{x0} > \sigma_{q0}$, $\sigma_{x0} = \sigma_{q0}$.

Rotation of the covariance implies a time dependence of the widths $\sigma_x(t)$ and $\sigma_q(t)$ in $x$ and $q$ as well as a nonvanishing correlation coefficient $c(t)$ which also depends on time. We can rewrite the equation of the covariance ellipse in the form known from Section 3.5,

$$\frac{1}{1 - c^2(t)} \left\{ \frac{(x - x_0)^2}{\sigma_x^2(t)} - 2c(t) \frac{(x - x_0)(q - q_0)}{\sigma_x(t)\sigma_q(t)} + \frac{(q - q_0)^2}{\sigma_q^2(t)} \right\} = 1$$

with

$$\sigma_x(t) = \sqrt{\sigma_{x0}^2 \cos^2 \omega t + \sigma_{q0}^2 \sin^2 \omega t} \quad ,$$

$$\sigma_q(t) = \sqrt{\sigma_{x0}^2 \sin^2 \omega t + \sigma_{q0}^2 \cos^2 \omega t} \quad,$$

$$c(t) = \frac{(\sigma_{q0}^2 - \sigma_{x0}^2) \sin 2\omega t}{\sqrt{4\sigma_{x0}^2 \sigma_{q0}^2 + (\sigma_{x0}^2 - \sigma_{q0}^2)^2 \sin^2 2\omega t}} \quad.$$

The time dependence both of the position expectation value $x_0(t)$ and of the width $\sigma_x(t)$ is exactly equal to what we have found in the quantum-mechanical calculation.

In the particular case

$$\sigma_{x0} = \sigma_{q0}$$

the covariance matrix is a circle, $\sigma_x$ and $\sigma_q$ are independent of time, and the correlation vanishes for all times. If we require the minimum-uncertainty relation of quantum mechanics,

$$\sigma_{x0}\sigma_{p0} = \frac{\hbar}{2} \quad,$$

to be fulfilled for our classical phase-space probability density as we have done in Section 3.6 we have

$$\sigma_{q0} = \frac{\sigma_{p0}}{m\omega} = \frac{\hbar}{2m\omega\sigma_{x0}} \quad.$$

Together with the requirement $\sigma_{x0} = \sigma_{q0}$ we get

$$\sigma_{x0} = \frac{1}{\sqrt{2}}\sqrt{\frac{\hbar}{m\omega}} = \frac{\sigma_0}{\sqrt{2}} \quad, \qquad \sigma_0 = \sqrt{\frac{\hbar}{m\omega}} \quad.$$

For this particular value of the initial width, the width stays constant. For $\sigma_{x0} \neq \sigma_0/\sqrt{2}$ the spatial width of the classical phase-space density oscillates exactly as the quantum-mechanical probability density does as shown in Figure 6.6.

## 6.6 Spectra of Square-Well Potentials of Finite Depths

In Section 4.4 we studied the stationary bound states in a square-well potential. We found that these states exist only for discrete negative energy eigenvalues, which form the discrete spectrum of bound-state energies. The probability densities of these states are concentrated for the most part in the square well. We now discuss the bound-state spectra for different shapes of the square well.

Figure 6.10 shows the wave functions and the energy spectra for several square-well potentials of equal widths but different depths. For a well of finite

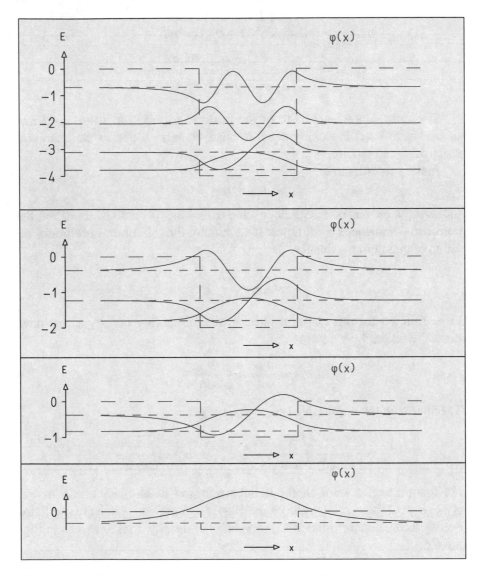

**Fig. 6.10. Bound-state wave functions and energy spectra for square-well potentials of different finite depths but identical widths. The number of bound states increases with the depth of the potential.**

depth, there is only a finite number of bound states. Their number increases with depth. In contrast to the wave functions of an infinitely deep well, the wave functions of a finite square well are different from zero outside the well but drop there exponentially to zero. The exponential falloff is fastest for the ground state. Figure 6.11 indicates that for a fixed depth the number of bound states increases as the well becomes wider.

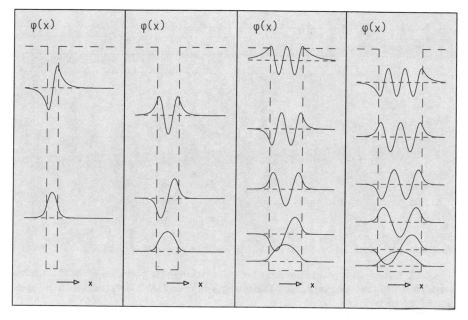

**Fig. 6.11. Bound-state wave functions for square-well potentials of identical depth but different widths. The number of bound states increases with the width of the well.**

## 6.7 Stationary Bound States in Piecewise Linear Potentials

A potential accomodating bound states can also be constructed from piece-wise linear sections; the solution of the Schrödinger equation in this case was discussed in Section 4.4. In its simplest form such a potential is a triangular well. For three such wells of equal depth but different opening angles the stationary bound states are shown in Figure 6.12. As for the square wells in Figure 6.11 the number of bound states rises with the opening of the well. Also the general features of the wave function, characterized by number and ordering of maxima, minima, and nodes, are similar to those of the square well. However, for the example of Figure 6.12, there is no left–right symmetry because the potential itself does not display this symmetry.

In Figure 6.13 we can compare the bound states in a skew triangular well with a vertical edge with those in a symmetric well. Both wells have the same depth and are equally wide at the top. At the sharp edge of the skew well the wave functions extend somewhat into the classically forbidden region left of the well where the difference $V - E$ of potential and total energy is negative. If we call $D(x) = |V(x) - E|$ the thickness of the "roof" over a state in a classically forbidden region, then the state extends the farther into that region, the thinner the roof is. Left of the sharp edge the extension grows as the energy rises; at the same time the average curvature of the wave function

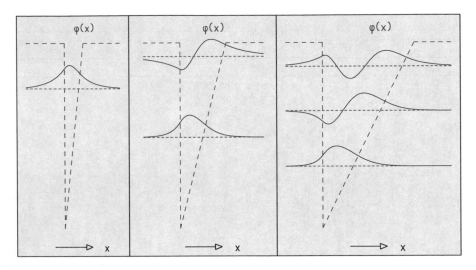

**Fig. 6.12. Bound-state wave functions for skew triangular potential wells of identical depth but different opening angles. The number of bound states increases with the opening of the well.**

falls. To the right of the inclined edge, however, the wave function appears to be quite independent of the energy; in the region where the wave function is not essentially zero the roof thickness is practically independent of energy, except for the highest energy value. The symmetric potential has two inclined edges which facilitate extension of the states. It is interesting to note that here

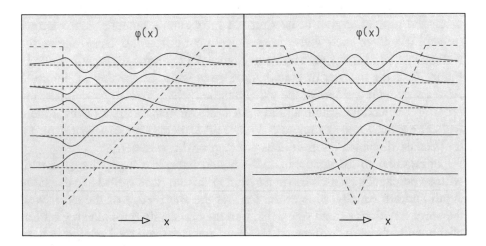

**Fig. 6.13. Bound-state wave functions for a skew and for a symmetric triangular potential well. Both are equally wide at the top. In the symmetric potential the lowest state is lower and the highest state is higher than in the skew potential.**

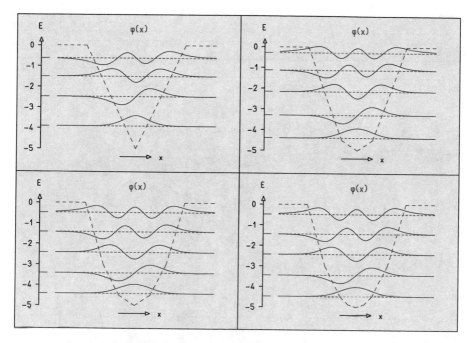

Fig. 6.14. **Successive approximation of the harmonic-oscillator potential of Fig. 6.5 by a piecewise linear potential.**

the lowest state has lower and the highest state has higher energy than in the skew potential. This can now be understood if the connection between kinetic energy and average curvature on page 137 is invoked.

It can be useful in calculations to approximate a more complicated potential by a piecewise linear one. As an example we show in Figure 6.14 the approximation of the harmonic-oscillator potential by a system of linear potentials with more and more regions.

## 6.8 Periodic Potentials, Band Spectra

As a first step in discussing periodic potentials as they occur in *crystals*, let us look at two potential wells more or less distant from each other. Figure 6.15 shows such potentials as well as the spectra of eigenvalues and eigenfunctions. When the two wells have some distance between them, we observe pairs of energy eigenvalues group closely together. Of the eigenfunctions belonging to each pair, one is always symmetric, the other antisymmetric. Comparing the eigenfunctions of two single wells with those of a single well, we observe that in corresponding regions they strongly resemble one another. The symmetric wave function of the double well is a smooth symmetric match with the two

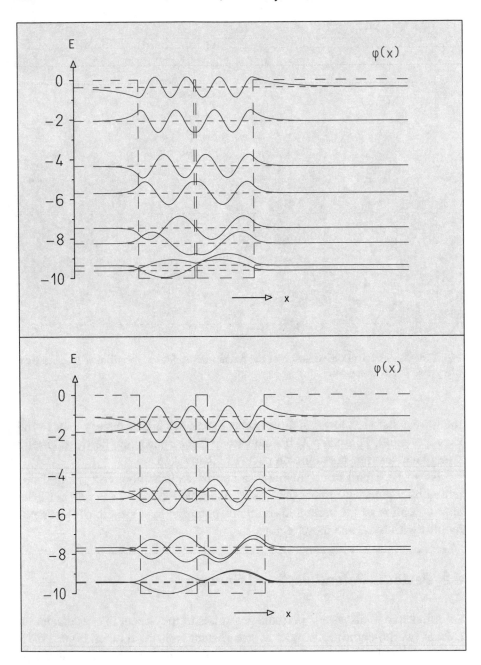

**Fig. 6.15. Bound-state wave functions and energy spectra for systems of two square wells. In one system the wells are very close together, in the other some distance apart.**

wave functions of the two single wells. The antisymmetric wave function of the double well is an antisymmetric match. In the limiting case when the distance between the two wells becomes zero, that is, when the wall vanishes, the eigenfunctions and the spectra become those of a single well of double width.

We now need to study the structure of the pairs of wave functions in two wells in more detail. The relation of their structure to that of the wave functions in a single well is easily explained, using the same reasoning given in Section 4.4. To this end we divide the $x$ axis into five regions,

$$
\begin{array}{llll}
\text{I} & -\infty \ < x < \ -d_2 \ , & V(x) = 0 \ , \\
\text{II} & -d_2 \ \leq x < \ -d_1 \ , & V(x) = -V_0 \ , \\
\text{III} & -d_1 \ \leq x < \ d_1 \ , & V(x) = 0 \ , \\
\text{IV} & d_1 \ \leq x < \ d_2 \ , & V(x) = -V_0 \ , \\
\text{V} & d_2 \ \leq x < \ \infty \ , & V(x) = 0 \ ,
\end{array}
$$

where the potential has a constant value. Notice that the potential is completely symmetric with respect to the point $x = 0$. That is, it does not change if $x$ is replaced by $-x$. In regions I and V the wave function must show exponential falloff for large values of $|x|$. In regions II and IV it oscillates as a superposition of two complex exponentials.

The behavior of the wave function is determined in particular by its structure in region III, which encompasses the origin. In this domain the wave function is a linear combination of real exponentials which, because of the symmetry of the problem, are either symmetric (s) or antisymmetric (a):

$$
\varphi_{\text{III}}^{\text{s}} = A_s \frac{1}{2}(e^{\kappa_s x} + e^{-\kappa_s x}) = A_s \cosh(\kappa_s x)
$$

and

$$
\varphi_{\text{III}}^{\text{a}} = A_a \frac{1}{2}(e^{\kappa_a x} - e^{-\kappa_a x}) = A_a \sinh(\kappa_a x) \quad .
$$

The parameters $\kappa_a$, $\kappa_s$ are given by

$$
\kappa_s = -\frac{i}{\hbar} p_s' = \frac{1}{\hbar}\sqrt{-2m E_s} \ ,
$$

$$
\kappa_a = -\frac{i}{\hbar} p_a' = \frac{1}{\hbar}\sqrt{-2m E_a} \ ,
$$

where $E_s$ and $E_a$ are the negative bound-state energies of the symmetric and antisymmetric solutions, respectively. The wave function in region III connects the wave functions of regions II and IV. It therefore determines the overall symmetry. The total wave function is symmetric if in region III it is of the symmetric type, $\varphi_{\text{III}}^{\text{s}} = A_s \cosh(\kappa_s x)$. Since the antisymmetric solution has the larger average curvature, it possesses the greater kinetic energy (classically $E_{\text{kin}} = p^2/(2m)$)

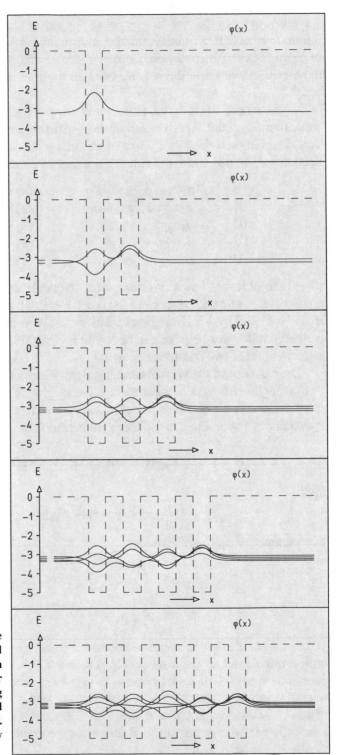

**Fig. 6.16. Bound-state wave functions and energy spectra for a potential well and for potentials consisting of two, three, four, and five neighboring wells. The states have very similar energies.**

$$E_{kin} = -\int_{-\infty}^{+\infty} \varphi(x)\frac{\hbar^2}{2m}\frac{d^2}{dx^2}\varphi(x)\,dx$$

compared to the symmetric solution. This explains why the splitting of the two energy eigenvalues of the bound states increases when the two wells approach each other. When the separating wall in region III has disappeared, the symmetric solution no longer has a dent in the middle.

It is now plausible that for a potential consisting of a periodic repetition of $N$ neighboring wells, each single-well eigenvalue reflects itself in a set of $N$ bound states of the periodic system of square-well potentials. The spacing of the energy eigenvalues of these states may be very narrow. They are said to form an *energy band*. A crystal consists of a large number ($N \approx 10^{23}$) of regularly spaced atoms. They form a periodic electric potential pattern in three dimensions giving rise to analogous band structures.

Figure 6.16 shows how the band structure, starting with the ground state of a single square well, takes form when two, three, four, and five potential wells are placed at equal distances next to it and to one another. The number of states forming the band is equal to the number of potential wells. Their spacing in energy becomes narrower as the number of wells increases. Certainly, for large numbers of potential wells forming a periodic structure, each individual band contains a large number of states represented by periodic wave functions. The wave functions of a single band can be linearly combined to form wave packets describing localized particles. If the time dependence of the eigenstates is included in the superposition (see Section 6.2), the wave packets describe particles moving freely in the periodic potential structure. In this way the free motion of electrons in the *conduction band* of the lattice of a metal or a semiconductor can be explained.

In Figure 6.17 we illustrate some details of band formation. From the bottom of a square well a zigzag-shaped potential rises in several steps, forming four separate wells of increasing depth. The four lowest states are gradually pulled together in a band. Although their wave functions are changed, they still retain some of the original symmetry properties. The upper states, at least at first, are much less affected.

## Problems

6.1. Calculate the integrals over the products of the eigenfunctions $\varphi_n(x)$ as given in Section 6.1 for the bound states of the deep square well,

$$\int_{-d/2}^{d/2} \varphi_n(x)\varphi_m(x)\,dx \quad .$$

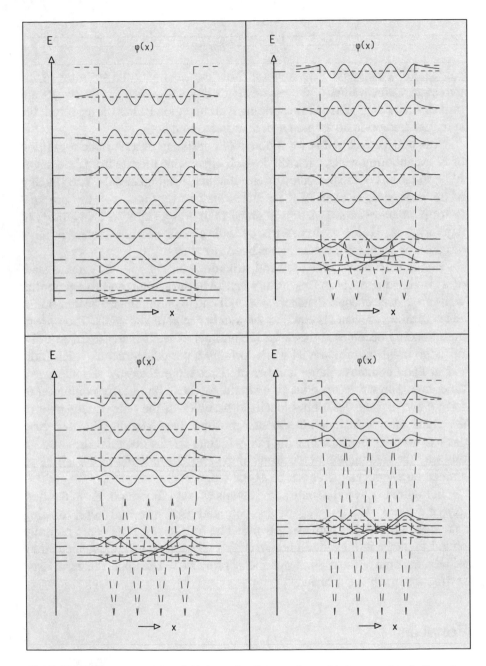

**Fig. 6.17. The square well (top left) is gradually transformed into a quasiperiodic potential with four wells. As a result the lowest four bound states form a band with closely space energies.**

6.2. What determines the frequency of the oscillation in the deep square well shown in Figure 6.2? What determines the wavelength of the interference wiggles in Figure 6.2?

6.3. Show that for $n = 0, 1$ the functions $\phi_n(\xi)$, $\xi = x/\sigma_0$ given in Section 6.3, are solutions of the stationary Schrödinger equation for the harmonic oscillator.

6.4. In terms of the momentum operator $\hat{p} = (\hbar/i)(d/dx)$, the operator of total energy of a harmonic oscillator is

$$H = \frac{1}{2m}\hat{p}^2 + \frac{m}{2}\omega^2 x^2 \quad .$$

In a bound state of the harmonic oscillator, the expectation values $\langle p \rangle$ and $\langle x \rangle$ of momentum and position vanish. Thus the expectation values

$$\begin{aligned} \langle p^2 \rangle &= (\Delta p)^2 + \langle p \rangle^2 = (\Delta p)^2 \quad , \\ \langle x^2 \rangle &= (\Delta x)^2 + \langle x \rangle^2 = (\Delta x)^2 \end{aligned}$$

are equal to the squares of the uncertainties of momentum and position. Use the uncertainty principle, from Section 3.3, to calculate the minimum energy of a bound state in a harmonic-oscillator potential.

6.5. Give an argument why the real and imaginary parts of the wave functions of Figure 6.7 have long wavelengths to the left or right when they are close to their left or right classical turning points, but not when the wave packet is in the center of the oscillator potential.

6.6. Compare the ratio $R = E_2/E_1$ of the energies $E_2$, $E_1$ of the two lowest levels in the different parts of Figure 6.11 with the corresponding ratio in the infinitely deep potential well; they are given in Section 6.1. Explain your result.

6.7. A rough approximation of the wave functions of the multiple square-well potentials in Figure 6.16 is given by

$$\varphi_n(x) = \sqrt{\frac{B_N}{N}} \varphi_n(x, B_N) \sum_{\ell=1}^{N} \varphi_1(x - x_\ell, d) \quad .$$

Here $\varphi_1(x - x_\ell, d)$ is the ground-state wave function of a single potential well of width $d$ and depth $V_0$ symmetric around $x = x_\ell$. With $B_N$ the width of the whole arrangement of all $N$ square-well potentials, including their $N - 1$ separating walls, $\varphi_n(x, B_N)$ is the eigenfunction of quantum number $n$ of the square-well potential with depth $V_0$ and width $B_N$.

Using Figure 6.11 for $\varphi_n(x, B_N)$ and Figure 6.16 (top) for $\varphi_1(x - x_\ell, d)$, sketch the wave functions $\varphi_n(x)$ for $n = 1, 2, \ldots, N$ and $N = 2, 3, 4, 5$. Compare their appearance with that of the wave functions in Figure 6.16. Discuss their symmetry properties.

6.8. What is the parity of the ground state with respect to reflection about the symmetry point of the potential for all examples given in this chapter? Explain the result, using the square well and the harmonic-oscillator potential as examples.

# 7. Quantile Motion in One Dimension

In classical mechanics the position $x(t)$ of a point particle and its velocity $v(t) = \mathrm{d}x(t)/\mathrm{d}t$ are well defined. This is not the case in quantum mechanics. For a free wave packet one can use the expectation value $\langle x(t) \rangle$ and its time derivative $\mathrm{d}\langle x(t) \rangle/\mathrm{d}t$ to characterize the position and the velocity of a particle. But for a particle under the influence of a force this description is not adequate. In the case of the tunnel effect, for instance, the expectation value $\langle x(t) \rangle$ may never pass through the barrier. In the following we shall see that mathematical statistics allows us to define a quantile position $x_P(t)$ and a quantile velocity $\mathrm{d}x_P(t)/\mathrm{d}t$ in all cases where we deal with a probability distribution $\varrho(x,t)$ and that this velocity can be related to experiment. (This chapter and Section 10.2 are based on the following publication: S. Brandt, H.D. Dahmen, E. Gjonaj, T. Stroh, Physics Letters A 249, 265 (1998).)

## 7.1 Quantile Motion and Tunneling

For a probability density $\varrho(x)$ the *quantile* $x_Q$ associated with the probability $Q$ is defined by

$$Q = \int_{-\infty}^{x_Q} \varrho(x)\,\mathrm{d}x \quad .$$

For the time-dependent probability density $\varrho(x,t)$ and time-independent probability $P, 0 \leq P \leq 1$, we define the time-dependent *quantile position* $x_P(t)$ by

$$\int_{x_P(t)}^{\infty} \varrho(x,t)\,\mathrm{d}x = P \quad .$$

The function $x = x_P(t)$ describes the *quantile trajectory* (in the $x,t$ plane) of a point moving along the $x$ axis. Its time derivative

$$v_P(t) = \frac{\mathrm{d}x_P(t)}{\mathrm{d}t}$$

defines the *quantile velocity* $v_P(t)$ of the point $x_P(t)$.

The upper plot of Figure 7.1 exhibits the time development of the scattering of an initially Gaussian wave packet by a repulsive barrier of height $V_0$.

S. Brandt and H.D. Dahmen, *The Picture Book of Quantum Mechanics*,
DOI 10.1007/978-1-4614-3951-6_7, © Springer Science+Business Media New York 2012

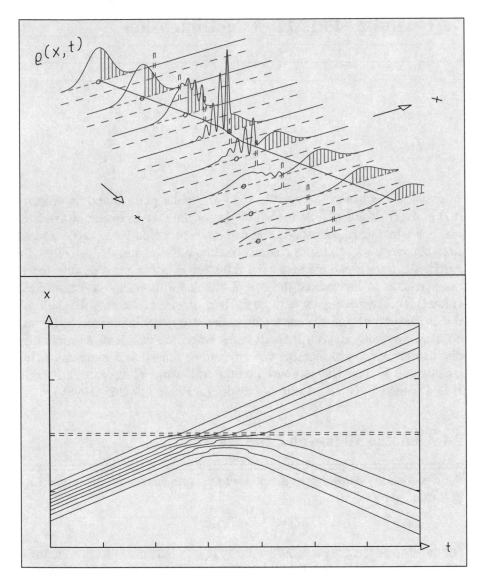

**Fig. 7.1. Quantile trajectories of the tunnel effect. The upper plot represents the time development of the scattering of an initially Gaussian wave packet by a repulsive potential barrier of height $V_0$. The expectation value of the kinetic energy is smaller than $V_0$. The small circles indicate the position of the classical particle. The shaded areas under the curves correspond to the probability $P = 0.4$ in the interval $x_P(t) \leq x < \infty$. The line cutting through the plot from the upper left to the lower right is the quantile trajectory for $P = 0.4$. The lower plot presents the quantile trajectories for the value $P = 0.1$ for the top curve and in steps of $\Delta P = 0.1$ for the lower curves up to $P = 0.9$. The thick curve is the same quantile trajectory as the one in the upper plot.**

The expectation value of the kinetic energy $E$ of the wave packet is smaller than $V_0$. The wave packet is partly reflected by the barrier. The other part tunnels through the barrier moving more or less like a force-free wave packet to the right. The shaded areas under the curves cover the probability $P = 0.4$. The line cutting through the plot from the upper left to the lower right is the quantile trajectory $x_P(t)$ corresponding to $P = 0.4$. The lower plot of Figure 7.1 presents the quantile trajectories for $P$ in the range between 0.1 and 0.9 in steps of $\Delta P = 0.1$. In the region of the repulsive potential barrier the quantile velocities are smaller than in regions far from the barrier.

In Figure 7.2 the quantile trajectories for the scattering of an initially Gaussian wave packet by a double barrier are shown. The upper plot exhibits the time development of the wave packet incident on the double barrier from the left. We observe the partially reflected and transmitted parts of the wave packet and the resonance behavior due to the oscillation of part of the probability between the two barriers. The resonance decay is due to the tunneling through the left or right barrier which occurs whenever the wave packet moving between the two barriers interacts with one of them. This leads to repeated reflected and transmitted pulses following the first one's with some time delay. The shaded areas under the curves representing the probability density correspond to a probability of $P = 0.4$. It has been chosen to be larger than the probability contained in the earliest transmitted pulse. Therefore the quantile trajectory does not leave the double-barrier region together with the first transmitted pulse. In fact, for the probability $P = 0.4$ the quantile trajectory oscillates once between the two barriers and leaves this region together with the second transmitted pulse.

The lower part of Figure 7.2 exhibits the set of quantile trajectories starting with the probability $P = 0.1$ for the top curve passing through steps of $\Delta P = 0.1$ ending with the value $P = 0.9$ for the bottom curve. The thick line is the same quantile trajectory as in the top picture. A fine-tuning of the probability $P$ to values slightly smaller than the one for the thick line would produce quantile trajectories showing more and more oscillations between the barriers. The quantile trajectory for the total transmission probability $P_T$ never leaves the region between the two barriers. All trajectories for values $P > P_T$ are eventually reflected.

## 7.2 Probability Current, Continuity Equation

In Section 3.3 we discussed the probability interpretation of quantum mechanics introduced by Max Born. Following his reasoning, the absolute square $|\psi(x,t)|^2 = \varrho(x,t)$ of the wave function represents a probability density in the following sense: The probability $dP$ of finding the particle in the interval

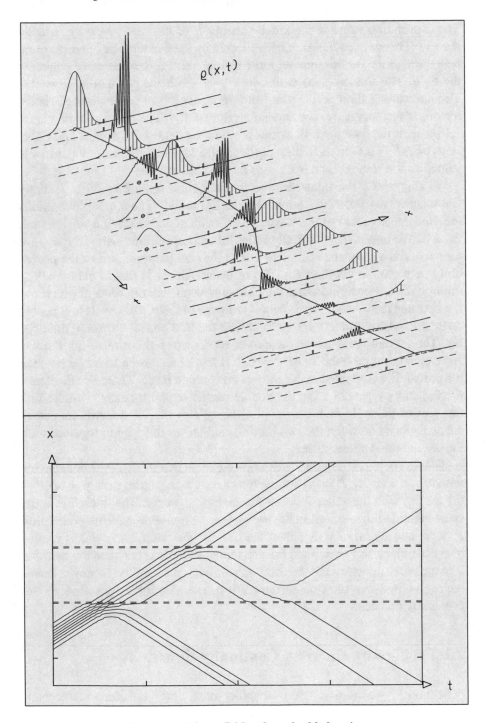

**Fig. 7.2. As Figure 7.1 but for a double barrier.**

$(x, x + dx)$ is given by

$$dP = \varrho(x,t)\,dx = |\psi(x,t)|^2\,dx \quad .$$

Since the probability is conserved the time variation of the probability

$$P_{12} = \int_{x_1}^{x_2} dP = \int_{x_1}^{x_2} \varrho(x,t)\,dx$$

in one interval $(x_1, x_2)$ must result in a flow entering or leaving this interval. This flow can be described by a *probability current density* $j(x,t)$ defined by the requirement

$$j(x,t) - j(x_1,t) = -\frac{d}{dt}\int_{x_1}^{x} \varrho(x',t)\,dx' = \int_{x_1}^{x}\left(-\frac{\partial\varrho(x',t)}{\partial t}\right)dx' \quad .$$

It states that the resulting flow at the borders $x_1$ and $x$ is equal to the rate of decrease of the probability inside the interval.

For a very narrow interval $(x_1, x)$ we may approximate the integral on the right-hand side by $[-\partial\varrho(x,t)/\partial t]\,\Delta x$, $\Delta x = x - x_1$. In the limit $\Delta x \to 0$ we obtain the expression

$$\left.\frac{\partial j(x,t)}{\partial x}\right|_{x=x_1}\Delta x = \lim_{\Delta x \to 0}\left(\frac{j(x_1 + \Delta x, t) - j(x_1, t)}{\Delta x}\right)\Delta x$$

for the left-hand side. Altogether we get the *continuity equation*

$$-\frac{\partial\varrho(x,t)}{\partial t} = \frac{\partial j(x,t)}{\partial x} \quad .$$

For current densities $j(x,t)$ vanishing for $x \to \pm\infty$ we derive

$$-\frac{d}{dt}\int_{-\infty}^{\infty}\varrho(x,t)\,dx = \int_{-\infty}^{\infty}\frac{\partial j(x,t)}{\partial x}\,dx = 0 \quad ,$$

the conservation of the total probability. The explicit form of the current density can be derived by observing

$$\frac{\partial\varrho(x,t)}{\partial t} = \psi^*(x,t)\frac{\partial\psi(x,t)}{\partial t} + \frac{\partial\psi^*(x,t)}{\partial t}\psi(x,t) \quad .$$

The time derivatives of the wave functions $\psi$ and $\psi^*$ are determined by the Schrödinger equations for $\psi$ and its complex conjugate $\psi^*$,

$$i\hbar\frac{\partial\psi}{\partial t} = -\frac{\hbar^2}{2m}\frac{\partial^2\psi}{\partial x^2} + V(x)\psi \quad , \qquad -i\hbar\frac{\partial\psi^*}{\partial t} = -\frac{\hbar^2}{2m}\frac{\partial^2\psi^*}{\partial x^2} + V(x)\psi^* \quad .$$

Inserting the expressions for $\partial\psi/\partial t$ and $\partial\psi^*/\partial t$, we obtain

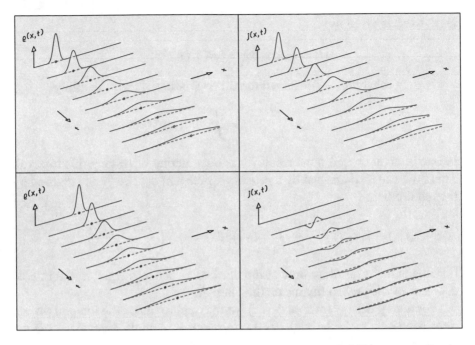

**Fig. 7.3. Time development of the density** $\varrho(x,t)$ **and the probability current density** $j(x,t)$ **of the force-free motion of a Gaussian wave packet. The graphs in the top row refer to a moving wave packet, the bottom row to a wave packet at rest,** $\langle x(t) \rangle = \text{const} = 0$. **The small circles indicate the position expectation value** $\langle x(t) \rangle$ **of the wave packet. In the bottom row the change with time of the wave packet is entirely due to its broadening because of dispersion. The probability density remains even with respect to** $x = 0$, **the current density stays odd; thus, the integral over the current density vanishes.**

$$\frac{\partial \varrho}{\partial t} = -\frac{\hbar}{2mi} \left[ \psi^* \frac{\partial^2}{\partial x^2} \psi - \psi \frac{\partial^2 \psi^*}{\partial x^2} \right] \quad .$$

This can be turned into the form of the continuity equation with the expression

$$j(x,t) = \frac{\hbar}{2mi} \left( \psi^* \frac{\partial \psi}{\partial x} - \psi \frac{\partial \psi^*}{\partial x} \right)$$

for the probability current density. In Figure 7.3 the time development of the probability density and the probability current density is shown for a free Gaussian wave packet. The plots in the top row show a moving wave packet with positive momentum expectation value $\langle p \rangle = p_0$, those in the bottom row a wave packet at rest, i.e., $p_0 = 0$. With growing time we observe a broadening of both wave packets, due to dispersion. For the wave packet at rest the dispersion is the only reason for the change in time of the plots. Probability flows to the right for $x > 0$ and to the left for $x < 0$. Thus, the current density is positive for $x > 0$ and negative for $x < 0$. Its integral over the whole $x$ axis vanishes in agreement with $p_0 = 0$.

In an experiment the quantile trajectory can be determined on a statistical basis by a series of time-of-flight measurements: One prepares by the same procedure $N$ single-particle wave packets and sets a clock to zero at the time instant at which the spatial expectation value of a wave packet leaves the source. With a detector placed at position $x_1$ one registers the arrival times $t_{1m}$ of particles for $m = 1, 2, \ldots, N$ and orders them such that $t_{11} < t_{12} < \ldots < t_{1N}$. One picks the time $t_{1n}$ which is the largest of the smallest times measured and chooses $n/N = P$. The time $t_{1n}$ is the arrival time of the quantile $x_P$ at the position $x_1$, i.e., $x_P(t_{1n}) = x_1$. By repeating the experiment with a detector at $x_2$ one obtains $t_{2n}$, etc. The points $x_P(t_{in}), i = 1, 2, \ldots$, are discrete points on the quantile trajectory $x_P(t)$. If $x_1$ and $x_2$ mark the front and rear end of the barrier, resp., then $t_{2n} - t_{1n}$ is the quantile traversal time of the barrier. In the quantum-electronic components of electrical circuits signals are propagated by pulses of a certain number $N$ of electrons. The time needed for a signal to pass a quantum-electronic component is the quantile traversal time defined above.

## 7.3 Probability Current Densities of Simple Examples

We have already found that for the free motion (Section 3.2), the motion under a constant force (Section 5.6), and the harmonic motion (Section 6.4), the probability density of an initially Gaussian wave packet has the form

$$\varrho(x,t) = \frac{1}{\sqrt{2\pi}\sigma_x(t)} \exp\left\{ -\frac{(x - \langle x(t) \rangle)^2}{2\sigma_x^2(t)} \right\} \quad ,$$

i.e., the shape of the packet stays Gaussian. However, its position expectation value $\langle x(t) \rangle$ as well as its width $\sigma_x(t)$ changes with time and that time variation differs for the three examples as indicated in Table 7.1, which was compiled from the sections quoted. The probability current density is

**Table 7.1.** Time dependence of the position expectation value and variance for a Gaussian wave packet

|  | Free Motion | Constant Force | Harmonic Oscillator |
|---|---|---|---|
| $V(x)$ | $0$ | $mgx$ | $\frac{1}{2}m\omega^2 x^2$ |
| $\langle x(t) \rangle$ | $x_0 + v_0 t$ | $x_0 + v_0 t + gt^2/2$ | $x_0 \cos\omega t$ |
| $\sigma_x^2(t)$ | $\sigma_{x0}^2 + \left(\dfrac{\sigma_p}{m}t\right)^2$ | $\sigma_{x0}^2 + \left(\dfrac{\sigma_p}{m}t\right)^2$ | $\dfrac{4\sigma^4 \cos^2\omega t + \sigma_0^2 \sin^2\omega t}{4\sigma^2}$ |

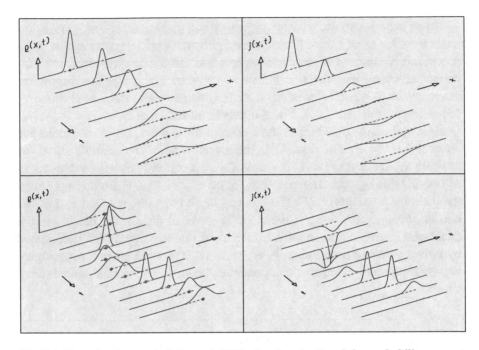

**Fig. 7.4. Time development of the probability density** $\varrho(x,t)$ **and the probability current density** $j(x,t)$ **of a Gaussian wave packet moving under the action of a spatially constant force (top row) and under the action of a harmonic force (bottom row). The small circles indicate the position expectation values of the wave packets. The current density possesses regions of positive or negative values. For the case of a constant force (top row) the wave packet moves to the right for early times; accordingly, the current density is mainly positive. At the turning point (middle of the seven time instants) the current density exhibits regions of positive as well as of negative values. Since the velocity expectation value vanishes at the turning point, the integral of the current density over the whole** $x$ **axis vanishes. The wave packet in the harmonic oscillator is shown over one time period. Since the initial position expectation value** $x_0$ **is positive, the initial velocity expectation value** $p_0/m$ **vanishes. The current density is mainly negative before it reaches the turning point; thereafter its values are mainly positive.**

$$j(x,t) = \left[ \langle v(t) \rangle + \frac{1}{\sigma_x(t)} \frac{\mathrm{d}\sigma_x(t)}{\mathrm{d}t} (x - \langle x(t) \rangle) \right] \varrho(x,t) \quad ,$$

where the velocity expectation value is $\langle v(t) \rangle = \langle p(t) \rangle / m = \mathrm{d}\langle x(t) \rangle / \mathrm{d}t$.

In Figure 7.4 we show the time development of the probability density and the probability current density for wave packets under the influence of a constant and a harmonic force.

## 7.4 Differential Equation of the Quantile Trajectory

By definition the quantile trajectory $x = x_P(t)$ is obtained by solving the equation

$$\int_{x_P(t)}^{\infty} \varrho(x,t)\,dx = P$$

for $x_P(t)$. Since $P = \text{const}$ we have

$$-\frac{dx_P(t)}{dt}\varrho(x_P(t),t) + \int_{x_P(t)}^{\infty} \frac{\partial \varrho(x',t)}{\partial t}\,dx' = \frac{dP}{dt} = 0 \quad .$$

The continuity equation derived in Section 7.2 for the probability density allows us to replace $\partial\varrho/\partial t$ by $-\partial j/\partial x$ in the integral. The integration can then be performed explicitly to yield

$$\varrho(x_P(t),t)\frac{dx_P(t)}{dt} = j(x_P(t),t)$$

as differential equation for the trajectory $x_P(t)$. In terms of the velocity field $v(x,t) = j(x,t)/\varrho(x,t)$ of the probability flow we have

$$\frac{dx_P(t)}{dt} = \frac{j(x_P(t),t)}{\varrho(x_P(t),t)} = v(x_P(t),t) \quad .$$

The initial position $x_P(t_0) = x_0$ needed for solving this differential equation is the quantile position at the initial time $t_0$.

## 7.5 Error Function

For later use we introduce the *(complementary) error function*

$$\text{erfc}\, x = \frac{2}{\sqrt{\pi}} \int_x^{\infty} e^{-u^2}\,du \quad .$$

Since the function $e^{-u^2}$ is positive everywhere and the integration interval shrinks with growing lower boundary, the error function $\text{erfc}\, x$ is a monotonically decreasing function of $x$. For $x \to \infty$ the integration interval shrinks to zero, the value of the integrand tapers off to zero; thus,

$$\lim_{x \to \infty} \text{erfc}\, x = 0 \quad .$$

Making use of the normalization of the Gaussian distribution (cf. Section 2.4), we get

$$\lim_{x \to -\infty} \text{erfc}\, x = \frac{2}{\sqrt{\pi}} \int_{-\infty}^{\infty} e^{-u^2}\,du = 2\frac{1}{\sqrt{2\pi}} \int_{-\infty}^{\infty} e^{-u'^2/2}\,du' = 2 \quad .$$

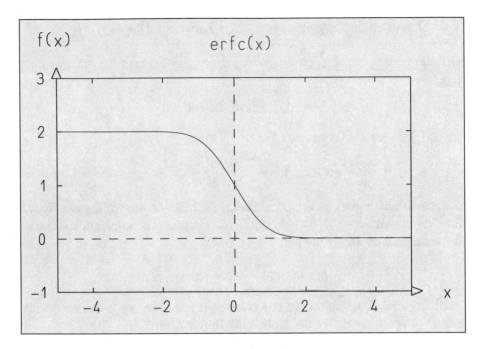

**Fig. 7.5. Plot of the (complementary) error function** erfc $x$.

Since the integrand is an even function, we have for $x = 0$

$$\operatorname{erfc} 0 = \frac{2}{\sqrt{\pi}} \int_0^\infty e^{-u^2}\, du = \frac{1}{2} \frac{2}{\sqrt{\pi}} \int_{-\infty}^\infty e^{-u^2}\, du = 1 \quad .$$

The graph of the error function erfc $x$ is presented in Figure 7.5.

## 7.6 Quantile Trajectories for Simple Examples

Using the error function we have for any Gaussian probability density with mean $\langle x(t) \rangle$ and variance $\sigma_x^2(t)$

$$P = \int_{x_P(t)}^\infty \varrho(x,t)\, dx = \frac{1}{2} \operatorname{erfc} \left( \frac{x_P(t) - \langle x(t) \rangle}{\sqrt{2}\sigma_x(t)} \right) \quad .$$

This equation determines $x_P(t_0)$ for a given value of $P$ and for a given initial time $t_0$. For the three examples of Section 7.3 we then have for the quantile position at time $t$

$$x_P(t) = \langle x(t) \rangle + \frac{\sigma_x(t)}{\sigma_x(t_0)}(x_P(t_0) - \langle x(t_0) \rangle) \quad .$$

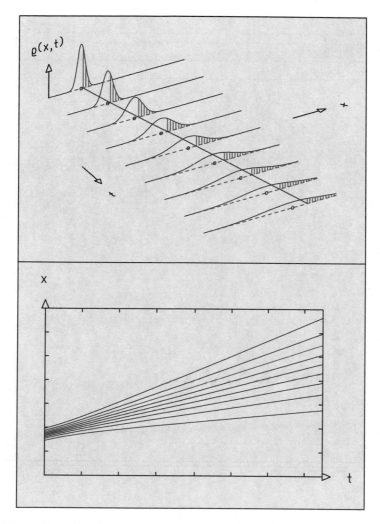

**Fig. 7.6. Quantile trajectories of a force-free Gaussian wave packet. The upper plot presents the time development of the probability density. The small circles on the $x$ axis indicate the position expectation values. The hatched areas correspond to the region $x > x_P(t)$ for $P = 0.3$. The thick line is the corresponding quantile trajectory. The lower plot exhibits the quantile trajectories for this wave packet for different values of $P$. The trajectories correspond to $P = 0.1$ (top line) and $P = 0.9$ (bottom line) in steps of $\Delta P = 0.1$. The thick line is the trajectory shown in the upper plot.**

For the particular value of $P$ for which the initial quantile position $x_P(t_0)$ is equal to the initial expectation value $\langle x(t_0) \rangle$, the quantile trajectory $x_P(t)$ is identical to the trajectory $\langle x(t) \rangle$ of the spatial expectation value in these three examples. In this case the argument of the error function vanishes. Thus, the fraction of probability associated to this particular quantile trajectory is $P = \frac{1}{2}\mathrm{erfc}(0) = 0.5$. For all other values of $P$ the quantile trajectory differs

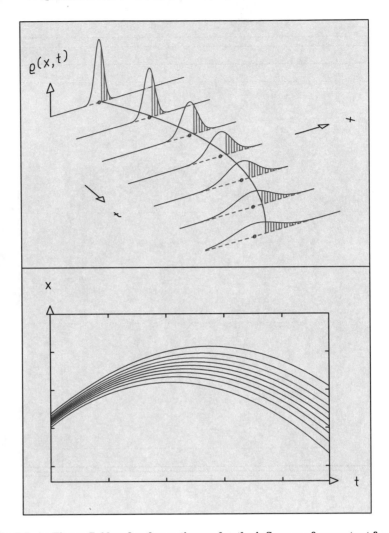

**Fig. 7.7. As Figure 7.6 but for the motion under the influence of a constant force.**

from the trajectory of the expectation value of position. In all three examples the trajectory of the position expectation value is the same as the classical trajectory for the same initial values. Thus, in our examples the quantile trajectory $x_{0.5}(t)$ is the classical trajectory.

Inserting $\langle x(t) \rangle$ and $\sigma_x(t)$ from Table 7.1 into the equation above we can obtain explicitly the quantile trajectories $x_P(t)$ for our three examples of Section 7.3. They are shown in Figures 7.6 to 7.8.

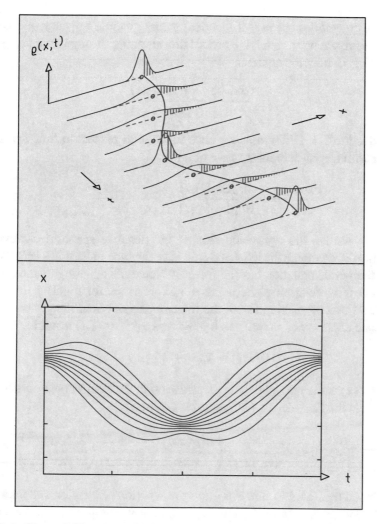

**Fig. 7.8.** As Figure 7.6 but for motion in a harmonic-oscillator potential. The line $x_P(t)$ for $P = 0.5$ is identical to the trajectory $\langle x(t) \rangle$ of the position expectation value. Only this curve is a cosine function. For all other values $P \neq 0.5$ the quantile trajectories deviate from the trigonometric functions. This deviation is due to the time dependence of the width $\sigma_x(t)$, i.e., due to the time-dependent broadening and shrinking of the squeezed state.

## 7.7 Relation to Bohm's Equation of Motion

In this chapter we have introduced quantile trajectories which are strictly based on the probability concept and are therefore quite natural in the framework of "conventional" quantum mechanics and its probability interpretation.

David Bohm in 1952 has given an "unconventional" formalism of quantum mechanics in which particle trajectories are possible. One can show that

Bohm's trajectories are in fact identical to the quantile trajectories discussed above. Here we only sketch the proof without going through all its steps.

We begin with the equation

$$\frac{dx_P(t)}{dt} = \frac{j(x_P(t),t)}{\varrho(x_P(t),t)}$$

from Section 7.4. Differentiating once more with respect to time and multiplying with the particle mass $m$ we obtain

$$m\frac{d^2x_P(t)}{dt^2} = m\frac{d}{dt}\frac{j(x_P(t),t)}{\varrho(x_P(t),t)} = -\frac{\partial U(x,t)}{\partial x}\bigg|_{x=x_P(t)} .$$

We have written the right-hand side as the negative spatial derivative of a potential $U(x,t)$ since the left-hand side is of the type mass times acceleration. Then the whole equation has the form of Newton's equation of motion. The potential $U$ is determined by using the expressions for $\varrho$ and $j$ in terms of $\psi$ and $\psi^*$ and by making use of the time-dependent Schrödinger equation to eliminate expressions of the type $\partial\psi/\partial t$ and $\partial\psi^*/\partial t$. The result is

$$U(x,t) = V(x) + V_Q(x,t) ,$$

where $V(x)$ is the potential energy appearing in the Schrödinger equation and $V_Q(x,t)$ is the time-dependent *quantum potential*

$$V_Q(x,t) = -\frac{\hbar^2}{4m\varrho(x,t)}\left(\frac{\partial^2\varrho(x,t)}{\partial x^2} - \frac{1}{2\varrho(x,t)}\left(\frac{\partial\varrho(x,t)}{\partial x}\right)^2\right)$$

introduced by David Bohm. For a unique solution of Newton's equation two initial conditions $x_P(t_0) = x_0$ and $dx_P(t_0)/dt = v_0$ are necessary. The solution of the differential equation for the quantile trajectory is uniquely determined by the initial condition $x_P(t_0) = x_0$, for a given probability $P$ fixed by the last equation of Section 7.4. The quantile trajectories $x_P(t)$ are identical to those solutions that satisfy the particular initial condition

$$v_0 = \frac{dx_P(t_0)}{dt} = v(x_P(t_0),t_0) .$$

The quantum potential of the free motion, the case of constant force, and the harmonic oscillator has the explicit form

$$V_Q(x,t) = -\frac{\hbar^2}{2m}\frac{1}{2\sigma_x^2(t)}\left[\frac{(x-\langle x(t)\rangle)^2}{2\sigma_x^2(t)} - 1\right] .$$

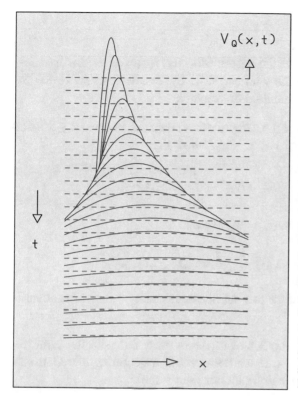

Fig. 7.9. Time development of the quantum potential $V_Q(x,t)$ of a force-free Gaussian wave packet. At any time it is a repulsive, parabolic potential. The force $F_Q = -\partial V_Q/\partial x$ produces the dispersion of the Gaussian wave packet in Bohm's description of quantum mechanics. At $t = 0$ the maximum at $x = \langle x(0)\rangle$ of the potential $V_Q(x,t)$ as well as its curvature are largest; both values decrease with increasing time. The decrease of the quantum potential reflects the fact that the quantile trajectories of the force-free Gaussian wave packets are hyperbolas as functions of time approaching straight lines as asymptotes for large times.

It is a repulsive parabolic potential having its maximum at the position $\langle x(t)\rangle$ of the expectation value and a curvature fixed in terms of the width $\sigma_x(t)$ of the wave packet. In the language of classical mechanics the quantum force $F_Q = -\partial V_Q/\partial x$ is the origin of the dispersion of the Gaussian wave packet. Figure 7.9 presents the time development of the quantum potential $V_Q(x,t)$ for the force-free motion of a Gaussian wave packet. For $t = 0$ the potential has the curvature $-\hbar^2/(4m\sigma_{x0}^4)$, its maximum value is $\hbar^2/(4m\sigma_{x0}^2)$. With growing time the absolute value of the potential at a given location and the curvature decrease.

The price one pays to get a Newtonian equation of motion is to accept the existence of an additional time-dependent potential $V_Q$ and the fixing of the initial condition for the velocity by the initially chosen wave function, which is by no means plausible. We would like to repeat and stress here that Bohm's particle trajectories (which cannot be defined as solutions of Newton's equation without the quantum potential) are identical to the quantile trajectories (which are defined in conventional quantum mechanics).

## Problems

7.1. Calculate the integral of the current density $j(x,t) = \hbar/(2mi)[\psi^* \partial \psi/ \partial x - \psi \partial \psi^*/\partial x]$ over the whole $x$ axis. Express it in terms of the expectation values of momentum or velocity.

7.2. Instead of the integral of the density $\varrho(x,t)$ over the interval $x \leq x' < \infty$ we consider the interval $x \leq x' \leq x_1$. Show that

$$\int_{x_P(t)}^{x_1} \varrho(x',t)\,dx' = \int_t^{t_1} j(x_1,t')\,dt$$

with $t_1$ determined by $x_P(t_1) = x_1$ leads to

$$\varrho(x_P(t),t)\frac{dx_P(t)}{dt} = j(x_P(t),t) \quad .$$

Explain the meaning of the last but one equation in terms of the quantile condition.

7.3. In classical mechanics bodies of different mass fall with the same velocity. Is this statement true for the quantile trajectories of a Gaussian wave packet under the action of a constant force?

7.4. Explain the obviously non-harmonic features of the lowest quantile trajectory of the lower plot of Figure 7.8. What is the condition to be satisfied by $x_P(0)$ for the dent in the top quantile trajectory to appear?

# 8. Coupled Harmonic Oscillators: Distinguishable Particles

So far we have always studied the motion of a single particle under the influence of an external potential. This potential is, however, often caused by another particle. The hydrogen atom, for example, consists of a nucleus, the proton, carrying a positive electric charge, and a negatively charged electron. The electric force between proton and electron is described by the Coulomb potential. The proton exerts a force on the electron, and – according to Newton's third law – the electron exerts a force on the proton. The proton has a mass about 2000 times the electron mass. Therefore the motion of the proton relative to the center of mass of the atom can usually be ignored. In this approximation the electron can be regarded as moving under the influence of an external potential. Generally, however, we have to describe the motion of both particles in a two-particle system. For simplicity we shall consider one-dimensional motion only; that is, both particles move only in the $x$ direction.

## 8.1 The Two-Particle Wave Function

We have seen that the basic entity of quantum mechanics is the wave function describing a system, and we have discussed its interpretation as a probability amplitude. A system consisting of two particles is described by a complex wave function $\psi = \psi(x_1, x_2, t)$ depending on time $t$ and on two spatial coordinates $x_1$ and $x_2$. Its absolute square $|\psi(x_1, x_2, t)|^2$ is the *joint probability density* for finding at time $t$ the two particles at locations $x_1$ and $x_2$. Of course, the wave function is assumed to be normalized, since the probability $\int_{-\infty}^{+\infty} \int_{-\infty}^{+\infty} |\psi(x_1, x_2, t)|^2 \, dx_1 \, dx_2$ of observing the particles anywhere in space has to be one. If the two particles differ in kind, such as a proton and an electron forming the hydrogen atom, they are said to be *distinguishable*. Two particles of the same kind, having the same masses, charges, and so on, as two electrons do, are said to be *indistinguishable*. For distinguishable particles the absolute square $|\psi(x_1, x_2, t)|^2 \, dx_1 \, dx_2$ describes the probability of finding at

S. Brandt and H.D. Dahmen, *The Picture Book of Quantum Mechanics*, 157
DOI 10.1007/978-1-4614-3951-6_8, © Springer Science+Business Media New York 2012

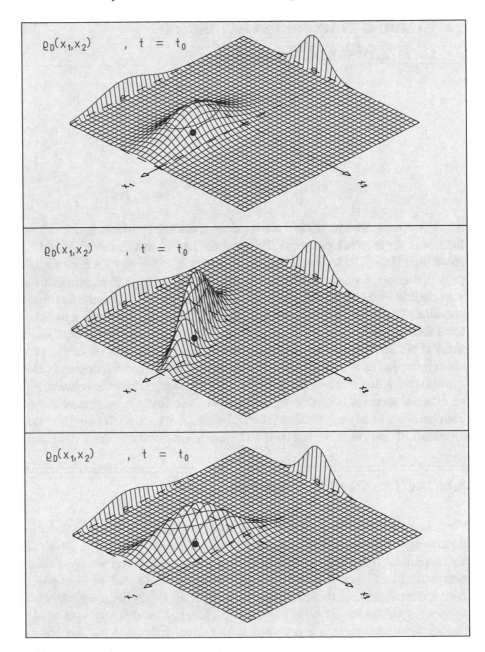

**Fig. 8.1. Joint probability density** $\rho_D(x_1, x_2)$ **for a system of two particles. It forms a surface over the** $x_1, x_2$ **plane. The marginal distributions** $\rho_{D1}(x_1)$ **and** $\rho_{D2}(x_2)$ **are plotted as curves over the margins parallel to the** $x_1$ **axis and the** $x_2$ **axis, respectively. In each plot the classical position** $x_{10}, x_{20}$ **is indicated by a black dot in the** $x_1, x_2$ **plane as well as by its projections on the margins. Also shown is the covariance ellipse. The three plots apply to the cases of (a) uncorrelated variables and (b) positive and (c) negative correlation between** $x_1$ **and** $x_2$.

time $t$ particle 1 in an interval $dx_1$ around position $x_1$ and simultaneously particle 2 in an interval $dx_2$ around $x_2$.

Figure 8.1a illustrates the joint probability density

$$\rho_D(x_1, x_2, t) = |\psi(x_1, x_2, t)|^2$$

for a fixed time $t$. Here a Cartesian coordinate system is spanned by the position variables $x_1$, $x_2$, and $\rho_D$ is plotted in the direction perpendicular to the $x_1, x_2$ plane. In this way the function $\rho_D(x_1, x_2)$ appears as a surface. On two margins of the coordinate plane, functions of only one variable, $x_1$, or the other, $x_2$, are also shown. They are defined by

$$\rho_{D1}(x_1, t) = \int_{-\infty}^{+\infty} \rho_D(x_1, x_2, t) \, dx_2$$

and

$$\rho_{D2}(x_2, t) = \int_{-\infty}^{+\infty} \rho_D(x_1, x_2, t) \, dx_1 \quad .$$

These *marginal distributions* describe the probability of observing one particle at a certain location, irrespective of the position of the second particle.

The black dot under the hump over the $x_1, x_2$ plane marks the expectation values $\langle x_1 \rangle$ and $\langle x_2 \rangle$ of the positions of particles 1 and 2, respectively. From the shape of the surface as well as from the marginal distributions, it is clear that in our example particle 2 is localized more sharply than particle 1.

The function shown in the three parts of Figure 8.1 is a Gaussian distribution of the two variables $x_1, x_2$. The mathematical form of such a bivariate Gaussian probability density and of its marginal distributions was already discussed in Section 3.5.

## 8.2 Coupled Harmonic Oscillators

As a particularly simple and instructive dynamical system, let us investigate the motion of two distinguishable particles of equal mass in external oscillator potentials. Both particles are coupled by another harmonic force. The external potentials are assumed to have the same form,

$$V(x_1) = \frac{k}{2}x_1^2 \quad , \qquad V(x_2) = \frac{k}{2}x_2^2 \quad , \qquad k > 0 \quad .$$

The potential energy of the coupling is

$$V_c(x_1, x_2) = \frac{\kappa}{2}(x_1 - x_2)^2 \quad , \qquad \kappa > 0 \quad .$$

The Schrödinger equation for the wave function $\psi(x_1, x_2, t)$ is then

$$i\hbar \frac{\partial}{\partial t} \psi(x_1, x_2, t) = H \psi(x_1, x_2, t) \quad ,$$

where $H$ is the Hamilton operator of the form

$$H = -\frac{\hbar^2}{2m}\frac{\partial^2}{\partial x_1^2} + V(x_1) - \frac{\hbar^2}{2m}\frac{\partial^2}{\partial x_2^2} + V(x_2) + V_c(x_1, x_2) \quad .$$

This equation is written, in analogy to the single-particle equation, so that its right-hand side is the sum of the kinetic and potential energies of the two particles.

The Schrödinger equation is solved with an initial condition that places the expectation values of the two particles into positions $x_{10} = \langle x_1(t_0)\rangle$ and $x_{20} = \langle x_2(t_0)\rangle$ at time $t = t_0$. We consider the particular situation in which the expectation values of the initial momenta of the two particles are zero. In quantum mechanics there is an infinite variety of wave functions with the expectation values $\langle x_1(t_0)\rangle = x_{10}$, $\langle p_1(t_0)\rangle = 0$, and $\langle x_2(t_0)\rangle = x_{20}$, $\langle p_2(t_0)\rangle = 0$ at initial time $t_0$ describing the particles. Even if we restrict ourselves to the bell-shaped form of a Gaussian wave packet at $t_0$, we still have to specify its widths and correlation. For later moments of time $t > t_0$, the time-dependent solution evolving out of the initial wave packet according to the Schrödinger equation for two coupled harmonic oscillators maintains the Gaussian form. Its parameters, however, become time dependent.

In Figure 8.2 the joint probability distribution $\rho_D(x_1, x_2, t)$ is shown for several times $t = t_0, t_1, \ldots, t_N$ together with its marginal distributions $\rho_{D1}(x_1, t)$ and $\rho_{D2}(x_2, t)$. We observe rather complex behavior. The hump where the probability density is large moves in the $x_1, x_2$ plane and at the same time changes its form; that is, the widths $\sigma_1, \sigma_2$ as well as the correlation coefficient $c$ are time dependent. The motion of the position expectation values $\langle x_1\rangle$, $\langle x_2\rangle$ is shown as a trajectory in the $x_1, x_2$ plane, and the initial positions $x_{10}, x_{20}$ at $t = t_0$ are marked as a black dot at the beginning of the trajectory. The last dot on the trajectory corresponds to the time for which the probability density is plotted. A crude survey can be made by looking only at the marginal distributions.

Figure 8.3a shows the time developments of the marginal distributions of the system in Figure 8.2. The left-hand part contains the marginal distribution $\rho_{D1}(x_1, t)$, the right-hand part $\rho_{D2}(x_2, t)$. The symbols on the $x_1$ and $x_2$ axes indicate position expectation values of the particles that are identical to the classical positions. The initial momenta were chosen so that the particles are, classically speaking, initially at rest. Particle 1 is initially in an off-center position, particle 2 in the center. It is obvious from Figure 8.3a that the position expectation values have the well-known energy exchange pattern of coupled oscillators. The oscillation amplitude of particle 1 decreases with time, whereas that of particle 2 increases until it has reached the initial amplitude of particle 1. At this moment the two particles have interchanged their roles, and the energy is now transferred from particle 2 to particle 1. The time developments of the widths in Figure 8.3a are much less clear.

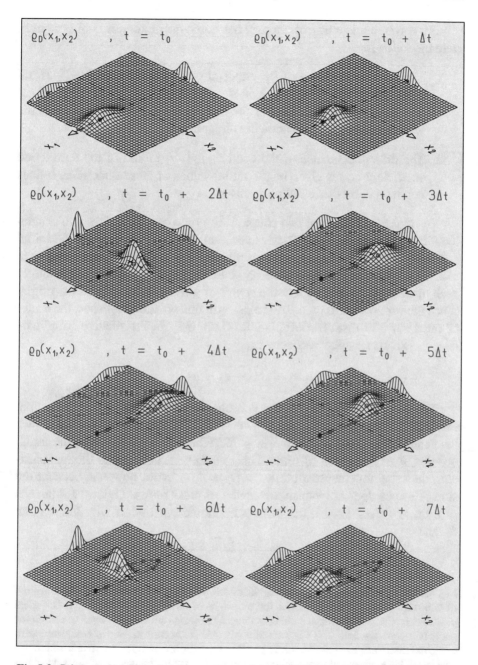

**Fig. 8.2.** Joint probability density $\rho_D(x_1, x_2, t)$ and marginal distributions $\rho_{D1}(x_1, t)$, $\rho_{D2}(x_2, t)$ for two distinguishable particles forming a system of coupled harmonic oscillators. The different plots apply to various times $t_j = t_0, t_1, \ldots, t_N$. The classical position of the two particles at the various moments in time is marked by a dot in the $x_1, x_2$ plane and by two dots on the margins. The initial dot for $t_j = t_0$ is black. The classical motion between $t_0$ and $t_j$ is represented by the trajectory drawn in the $x_1, x_2$ plane.

For a systematic study of coupled harmonic oscillators, it is important to note the following.

1. The time dependence of the expectation values $\langle x_1(t) \rangle$, $\langle x_2(t) \rangle$ is determined by their initial values and it is identical to that of classical particles. It is independent of the initial values $\sigma_{10}$, $\sigma_{20}$, and $c_0$ of the widths and of the correlation coefficient.

2. The time dependence of the widths $\sigma_1(t)$, $\sigma_2(t)$ and of the correlation coefficient $c(t)$ is given by the initial values of these quantities. It does not depend on the initial positions $x_{10}$, $x_{20}$.

The classical system of two coupled harmonic oscillators has two characteristic *normal oscillations*. They can be excited by choosing particular initial conditions. For one of the normal oscillations the center of mass remains at rest. This situation can be realized by choosing initial positions opposite to each other, $x_{10} = -x_{20}$, so that the center of gravity is initially at the origin. Since the sum of forces on the two masses in this position vanishes, the center of mass stays at rest. The oscillation occurs only in the relative coordinate $r = x_2 - x_1$. Its angular frequency is

$$\omega_r = \sqrt{(k + 2\kappa)/m} \quad .$$

The second normal oscillation is brought about by initial conditions that make the force between the two masses vanish. That is, the two particles have the same initial position $x_{10} = x_{20} = R_0$, which is therefore also the initial position $R$ of the center of mass. Since no force acts between the two particles, they stay together at all times, $x_1(t) = x_2(t)$. Now, however, because the sum of forces does not vanish, the center of mass moves under the influence of a linear force. Thus it performs a harmonic oscillation with the angular frequency

$$\omega_R = \sqrt{k/m} \quad .$$

**Fig. 8.3. Time development of the marginal distribution $\rho_{D1}(x_1, t)$ on the left and marginal distribution $\rho_{D2}(x_2, t)$ on the right for a system of coupled oscillators. The classical positions of the two distinguishable particles are plotted on the two axes as circles for particle 1 and particle 2. They coincide with the expectation values computed with marginal distributions. (a) The initial position expectation value of particle 2 is zero. (b) The particles are excited in a normal oscillation in which the center of mass oscillates and there is no relative motion. (c) The particles are excited in a normal oscillation in which there is relative motion and the center of mass is at the rest. In all three cases the initial momentum expectation values of the two particles are zero.**

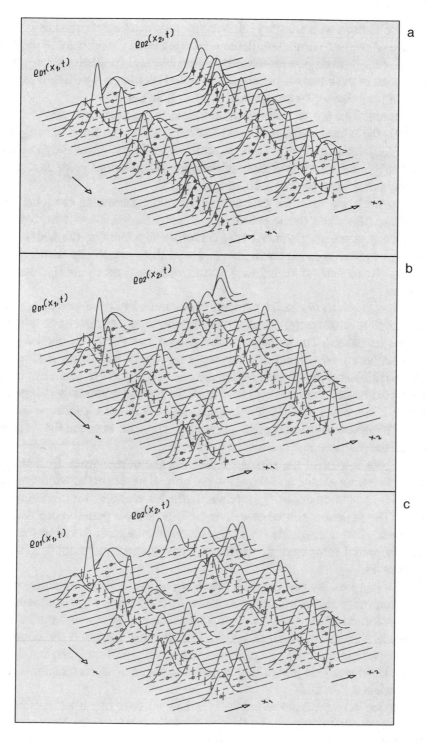

**Fig. 8.3.**

Oscillations with arbitrary initial conditions can be described as superpositions of the two normal oscillations, causing such phenomena as the transfer of energy from one mass to the other. Normal oscillations can also be produced in the quantum-mechanical coupled oscillators by exactly the same prescription. Examples are given in Figures 8.3b and c.

Figure 8.4a presents the oscillations of the expectation values $\langle x_1(t) \rangle$, $\langle x_2(t) \rangle$, the widths $\sigma_1(t)$, $\sigma_2(t)$, and the correlation $c(t)$ for a rather general set of initial conditions. All these quantities have beats. We already know that the beats in the time dependence of the expectation values come from superposition of the two normal oscillations.

As we know from the example of the single harmonic oscillator (Section 6.4), the width of the probability distribution oscillates with twice the frequency of the oscillator. We may therefore stipulate that the widths $\sigma_1(t)$, $\sigma_2(t)$ and the correlation coefficient $c(t)$ will show periodicity with twice the normal frequencies if their initial values $\sigma_{10}$, $\sigma_{20}$, and $c_0$ are appropriately chosen.

Figure 8.4b shows such a particular situation. Here the dependences of the expectation values $\langle x_1(t) \rangle$, $\langle x_2(t) \rangle$ and of the widths and the correlation coefficient are plotted. The initial position expectation values were chosen so that the oscillators have the normal frequency $\omega_R$. The initial widths and correlation coefficient were selected so that the frequency of these quantities is $2\omega_R$. As stated earlier, the time dependence of $\sigma_1$, $\sigma_2$, and $c$ is totally independent of the initial positions. In our example the positions were chosen to oscillate with frequency $\omega_R$ to allow for a simple comparison between the frequency $\omega_R$ of the positions and $2\omega_R$ of the widths.

Figure 8.4c gives the analogous plots for the other normal frequency $\omega_r$. It is interesting to note that preparing the normal modes in the widths requires an initial condition $\sigma_{10} = \sigma_{20}$, a relation which then holds for all moments in time. The variation in time of $\sigma_1$ and $\sigma_2$ is actually a periodic oscillation of frequency $2\omega_R$ or $2\omega_r$ added to a constant. Furthermore, it should be remarked that the initial value $c_0$ of the correlation coefficient is different from zero in both cases.

For one particular set of initial values $\sigma_1$, $\sigma_2$, and $c$, these quantities remain constant independent of time, as shown in Figure 8.5. In this situation the correlation coefficient is always positive, which is easily understood if we remember the attractive force between the two oscillators. If the coordinate of one particle is known, the other one is probably in its neighborhood rather than elsewhere. This probability constitutes the positive correlation between the variables $x_1$ and $x_2$.

In Section 6.5 we discussed the classical behavior of a bivariate Gaussian phase-space distribution under the action of a harmonic force. We pointed out that there is no difference between the temporal evolution of the uncorrelated

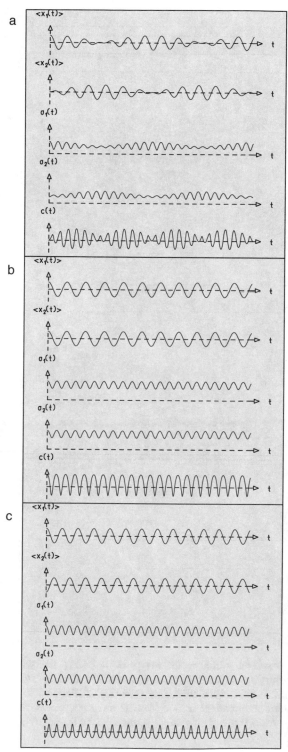

Fig. 8.4. Time dependences of the expectation values $\langle x_1(t) \rangle$, $\langle x_2(t) \rangle$, the widths $\sigma_1(t)$, $\sigma_2(t)$, and the correlation $c(t)$ for a system of coupled harmonic oscillators. (a) Rather general initial conditions were chosen. (b) The oscillation of the expectation values corresponds to an oscillation of the center of mass with frequency $\omega_R$. The initial values $\sigma_1(t_0)$, $\sigma_2(t_0)$, and $c(t_0)$ were chosen so that the two widths and the correlation oscillate with frequency $2\omega_R$. (c) The oscillation of the expectation values corresponds to an oscillation in the relative motion with frequency $\omega_r$; the widths and the correlation oscillate with frequency $2\omega_r$.

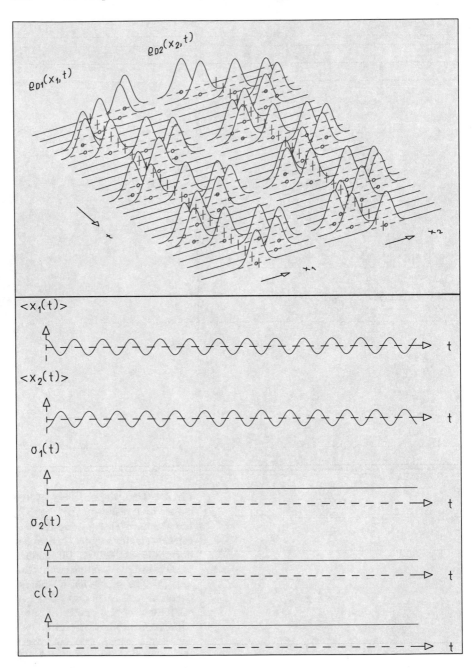

**Fig. 8.5. Coupled harmonic oscillators. The initial conditions** $\langle x_1(t_0)\rangle$, $\langle x_2(t_0)\rangle$ **are the same as in Figure 8.4c, corresponding to an oscillation in the relative motion. The parameters** $\sigma_1(t_0)$, $\sigma_2(t_0)$, **and** $c(t_0)$, **however, were chosen so that the widths and the correlation coefficient remain constant independent of time. Top: Time developments of the marginal distributions. Bottom: Time dependences of the quantities** $\langle x_1(t)\rangle$, $\langle x_2(t)\rangle$, $\sigma_1(t)$, $\sigma_2(t)$, $c(t)$.

classical phase-space distribution of initial spatial width $\sigma_{x0}$ and momentum width $\sigma_p = \hbar/(2\sigma_{x0})$ and of the Wigner distribution (cf. Appendix D) of an uncorrelated Gaussian wave packet of initial spatial width $\sigma_{x0}$. A corresponding statement holds true also for the case of coupled harmonic oscillators of distinguishable particles. This is to say that the figures of this section are identical to the ones derived from the temporal evolution of a Gaussian classical phase-space distribution with the same initial position and momentum data and initially uncorrelated in position and momentum of each particle with widths fulfilling Heisenberg's uncertainty equation $\sigma_{x1}\sigma_{p1} = \hbar/2$, $\sigma_{x2}\sigma_{p2} = \hbar/2$.

## 8.3 Stationary States

The stationary wave functions $\varphi_E$ are solutions of the time-independent Schrödinger equation

$$H\varphi_E(x_1, x_2) = E\varphi_E(x_1, x_2) \quad .$$

The Hamiltonian is that given at the beginning of Section 8.2. As in classical mechanics, the Hamiltonian can be separated into two terms,

$$H = H_R + H_r \quad ,$$

where

$$H_R = -\frac{\hbar^2}{2M}\frac{\mathrm{d}^2}{\mathrm{d}R^2} + kR^2$$

governs the motion of the center of mass

$$R = \tfrac{1}{2}(x_1 + x_2)$$

and

$$H_r = -\frac{\hbar^2}{2\mu}\frac{\mathrm{d}^2}{\mathrm{d}r^2} + \frac{1}{2}\left(\frac{k}{2} + \kappa\right)r^2$$

determines the dynamics of the relative motion in the relative coordinate

$$r = x_2 - x_1 \quad .$$

Here $M = 2m$ denotes the total mass, $\mu = m/2$ the reduced mass of the system.

The separation of the Hamiltonian permits a factored *ansatz* for the stationary wave functions,

$$\varphi_E(x_1, x_2) = U_N(R)u_n(r) \quad ,$$

with the factors fulfilling the equations

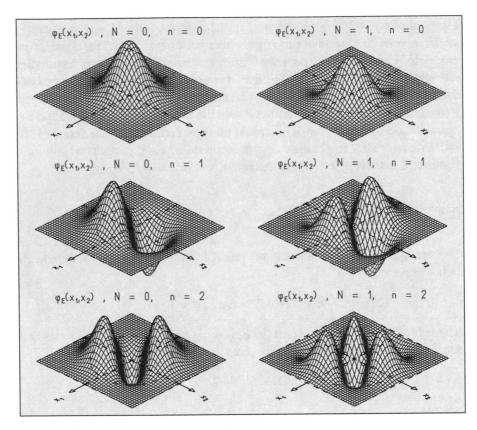

**Fig. 8.6.** Wave function $\varphi_E(x_1, x_2)$ for stationary states of a system of two coupled harmonic oscillators for low values of the quantum numbers $N$ and $n$. Note that $\varphi_E(x_1, x_2)$ is symmetric with respect to the permutation $(x_1, x_2) \rightarrow (x_2, x_1)$ for $n$ even and antisymmetric for $n$ odd. The dashed ellipse in the $x_1, x_2$ plane corresponds to the energetically allowed region for classical particles.

$$H_R U_N(R) = (N + \tfrac{1}{2})\hbar\omega_R U_N(R) \quad,$$
$$H_r u_n(r) = (n + \tfrac{1}{2})\hbar\omega_r u_n(r)$$

for the center-of-mass and relative motions, respectively. The functions $U_N(R)$ and $u_n(r)$ are thus the eigenfunctions for harmonic oscillators of single particles, as discussed in Section 6.3. The total energy $E$ is simply the sum of the center-of-mass and relative energies:

$$E = (N + \tfrac{1}{2})\hbar\omega_R + (n + \tfrac{1}{2})\hbar\omega_r \quad.$$

The energy spectrum now has two independent quantum numbers, $N$ for center-of-mass excitations, and $n$ for relative excitations. Figure 8.6 shows the stationary states $\varphi_E(x_1, x_2)$ for the lowest values of quantum numbers $N$ and $n$.

# Problems

8.1. Determine the coordinate transformation and thus the new coordinates $\xi_1$, $\xi_2$ that transform the exponent of the Gaussian function for $\rho_D(x_1,x_2)$, as given in Section 8.1, to normal form so that we have

$$\rho_D(x_1,x_2) = A\exp\left\{-\frac{1}{2}\left[\frac{(\xi_1-\langle\xi_1\rangle)^2}{\sigma'^2_1} + \frac{(\xi_2-\langle\xi_2\rangle)^2}{\sigma'^2_2}\right]\right\} \quad .$$

8.2. Give an argument why the shape of the wave packet $\rho_D(x_1,x_2,t)$ in Figure 8.2 changes with time as it does. It may help you to look carefully at Figure 6.6 for the single harmonic oscillator.

8.3. Derive the relations given in Section 8.2 for the two normal frequencies $\omega_r$ and $\omega_R$ for a classical system of coupled harmonic oscillators.

8.4. Verify that the Hamiltonian for a system of two coupled oscillators can be decomposed into the Hamiltonian $H_R$ for the center-of-mass motion and the Hamiltonian $H_r$ for relative motion, as given at the beginning of Section 8.3.

8.5. In Section 8.2 the oscillators decouple for $\kappa = 0$. The stationary Schrödinger equation can be solved by a product *ansatz* in the variables $x_1, x_2$,

$$\varphi_E(x_1,x_2) = \varphi_{E_1}(x)\varphi_{E_2}(x_2) \quad , \qquad E = E_1 + E_2 \quad .$$

Show that $\varphi_{E_1}(x_1)$, $\varphi_{E_2}(x_2)$ are then solutions of the stationary Schrödinger equation for the one-dimensional harmonic oscillator.

# 9. Coupled Harmonic Oscillators: Indistinguishable Particles

## 9.1 The Two-Particle Wave Function for Indistinguishable Particles

The probability density $\rho_D(x_1,x_2,t) = |\psi(x_1,x_2,t)|^2$ used in the last chapter described the joint probability of observing particle 1 at position $x_1$ and particle 2 at $x_2$. There is no difficulty with this notion as long as particle 1 can be unambiguously attributed to position $x_1$ and particle 2 to $x_2$. To so attribute them, however, presupposes that particles 1 and 2 have different identities, that they can be distinguished by properties other than being at different locations or having different momenta. They must have different intrinsic properties, for instance, different masses or different electric charges. A system consisting of an electron and a proton is one in which the two particles have different intrinsic properties. A system consisting of two electrons is not. For such a system it is impossible in principle to distinguish the two particles if they are close to each other.

To be more precise, we call two particles close to each other if their position expectation values $\langle x_1 \rangle$, $\langle x_2 \rangle$ differ by no more than the uncertainty to which these positions are known. As usual, we denote the uncertainties in the two positions by $\sigma_1$ and $\sigma_2$. Then the two particles are close if

$$(\langle x_1 \rangle - \langle x_2 \rangle)^2 \leq \sigma_1^2 + \sigma_2^2 \quad .$$

For a system of two indistinguishable particles close to each other, the two situations:

1. Particle 1 is at $x_1$, particle 2 at $x_2$

2. Particle 2 is at $x_1$, particle 1 at $x_2$

cannot be distinguished, and we can only assert that one of the two particles is at $x_1$ and the other at $x_2$.

S. Brandt and H.D. Dahmen, *The Picture Book of Quantum Mechanics*, 170
DOI 10.1007/978-1-4614-3951-6_9, © Springer Science+Business Media New York 2012

Thus, in general, the probability density for such a situation does not allow us to differentiate between the two particles. We therefore have to require that the probability density $|\psi(x_1,x_2,t)|^2$ remain unaltered if the two particles 1 and 2 are interchanged, that is, if their coordinates $x_1$ and $x_2$ are permuted in the argument of $\psi$,

$$|\psi(x_1,x_2,t)|^2 = |\psi(x_2,x_1,t)|^2 \quad .$$

Nor can any of the measurable quantities distinguish the two particles. This means that the potential energy of the two particles must be a symmetric function in the two position variables,

$$V(x_1,x_2) = V(x_2,x_1) \quad ,$$

which, in turn, implies that the Hamiltonian of the two particles is also symmetric not only in the momenta $p_1 = -i\hbar\partial/\partial x_1$, $p_2 = -i\hbar\partial/\partial x_2$ but also in the two position variables $x_1$, $x_2$:

$$\begin{aligned} H(p_1,p_2,x_1,x_2) &= -\frac{\hbar^2}{2m}\frac{\partial^2}{\partial x_1^2} - \frac{\hbar^2}{2m}\frac{\partial^2}{\partial x_2^2} + V(x_1,x_2) \\ &= H(p_2,p_1,x_2,x_1) \quad . \end{aligned}$$

Therefore, together with the solution $\psi'(x_1,x_2,t)$ of the Schrödinger equation

$$i\hbar\frac{\partial}{\partial t}\psi'(x_1,x_2,t) = H\psi'(x_1,x_2,t)$$

the function $\psi'(x_2,x_1,t)$ obtained by exchanging the arguments $(x_1,x_2)$ is also a solution of the Schrödinger equation. Thus any superposition

$$\psi(x_1,x_2,t) = a\psi'(x_1,x_2,t) + b\psi'(x_2,x_1,t) \quad ,$$

where $a$ and $b$ are complex numbers, solves the Schrödinger equation

$$i\hbar\frac{\partial}{\partial t}\psi(x_1,x_2,t) = H\psi(x_1,x_2,t) \quad .$$

The symmetry of the probability density $|\psi(x_1,x_2,t)|^2$ under the permutation of $x_1$ and $x_2$ puts constraints on the coefficients $a$ and $b$. We have

$$\begin{aligned} |\psi(x_1,x_2,t)|^2 &= a^*a|\psi'(x_1,x_2,t)|^2 + b^*b|\psi'(x_2,x_1,t)|^2 \\ &\quad + a^*b\psi'^*(x_1,x_2,t)\psi'(x_2,x_1,t) \\ &\quad + b^*a\psi'^*(x_2,x_1,t)\psi'(x_1,x_2,t) \quad . \end{aligned}$$

Comparing this equation with the corresponding formula for $|\psi(x_2,x_1,t)|^2$, we conclude that the equations for the coefficients are

$$a^*a = b^*b \quad , \qquad a^*b = b^*a \quad .$$

With factoring into absolute value and phase factor,

$$a = |a|e^{i\alpha} \quad , \qquad b = |b|e^{i\beta} \quad ,$$

we find

$$|a| = |b| \quad , \qquad e^{2i\alpha} = e^{2i\beta} \quad .$$

The periodicity of the exponential function fixes phase $2\beta$ relative to $2\alpha$ modulus $2\pi$ only, that is,

$$2\beta = 2\alpha + 2n\pi \quad , \qquad n = 0, \pm 1, \pm 2, \ldots \quad .$$

Thus only two values for the phase factor $e^{i\beta}$ remain,

$$e^{i\beta} = e^{i(\alpha + n\pi)} = \pm e^{i\alpha} \quad ,$$

and therefore

$$b = \pm a \quad .$$

For the superposition we find

$$\psi(x_1, x_2, t) = a[\psi'(x_1, x_2, t) \pm \psi'(x_2, x_1, t)] \quad .$$

The overall phase $e^{i\alpha}$ is arbitrary for any wave function, and the absolute value $|a|$ is fixed by the normalization condition for the wave function $\psi(x_1, x_2, t)$. Putting everything together, we conclude that the wave function for two indistinguishable particles is either symmetric

$$\psi(x_1, x_2, t) = \psi(x_2, x_1, t)$$

or antisymmetric

$$\psi(x_1, x_2, t) = -\psi(x_2, x_1, t)$$

under permutation of the two coordinates $x_1$ and $x_2$.

The behavior of these two types of wave function is characteristically different. The particles having a symmetric two-particle wave function are called Bose–Einstein particles or *bosons*, those with an antisymmetric two-particle wave function Fermi–Dirac particles or *fermions*. The distinction between bosons and fermions becomes clear if we look at the values of their wave functions for the particular locations $x_1 = x_2$. The symmetric wave function is not restricted for these locations, whereas the antisymmetric solution must vanish for them:

$$\psi(x, x, t) = 0 \quad .$$

Thus, in particular, the probability density for two fermions at the same position vanishes. Furthermore, if the two-particle wave function $\psi(x_1, x_2, t)$

is a product of two identical single-particle wave functions, the antisymmetric two-particle wave function vanishes:

$$\psi(x_1,x_2,t) = \varphi(x_1,t)\varphi(x_2,t) - \varphi(x_2,t)\varphi(x_1,t) = 0 \quad.$$

This result must be interpreted as saying that two fermions cannot populate the same state, or that fermions must always populate different states. This phenomenon was discovered in 1925 by Wolfgang Pauli when he was trying to explain the fact that $N$ electrons always populate the $N$ lowest-lying states in atomic shells. The postulate of antisymmetric wave functions for fermions is called the *Pauli exclusion principle*.

## 9.2 Stationary States

As a first example, we look at the wave functions $\varphi_E(x_1,x_2)$ for the stationary states of two bosons or two fermions. They are obtained from solutions of the time-dependent Schrödinger equation factored in time and space dependence in the form

$$\psi(x_1,x_2,t) = \exp\left(-\frac{i}{\hbar}Et\right)\varphi_E(x_1,x_2) \quad.$$

For the stationary wave function the result of the last section requires symmetry for bosons,

$$\varphi_E^B(x_1,x_2) = \varphi_E^B(x_2,x_1) \quad,$$

or antisymmetry for fermions,

$$\varphi_E^F(x_1,x_2) = -\varphi_E^F(x_2,x_1) \quad.$$

For the motion of two indistinguishable particles in a system of coupled harmonic oscillators, we start with the solutions obtained in Section 8.3 for distinguishable particles. The function $u_n(r)$, being a solution of the one-particle Schrödinger equation for harmonic motion in the relative coordinate, is itself either symmetric, $u_n(-r) = u_n(r)$ for even $n$, or antisymmetric, $u_n(-r) = -u_n(r)$ for odd $n$. Therefore the wave functions for two bosons are simply

$$\varphi_E^B(x_1,x_2) = U_N(R)u_n(r) \quad, \qquad n \text{ even} \quad,$$

and correspondingly the wave functions for two fermions are

$$\varphi_E^F(x_1,x_2) = U_N(R)u_n(r) \quad, \qquad n \text{ odd} \quad.$$

The two sets of wave functions together constitute the complete set that we found for distinguishable particles. The symmetry or antisymmetry is apparent in Figure 8.6. The spectrum of energy eigenvalues of coupled harmonic oscillators made up of distinguishable particles splits in two, one describing the bosons,

$$E = (N + \tfrac{1}{2})\hbar\omega_R + (n + \tfrac{1}{2})\hbar\omega_r \quad , \qquad n \text{ even} \quad ,$$

the other one the fermions,

$$E = (N + \tfrac{1}{2})\hbar\omega_R + (n + \tfrac{1}{2})\hbar\omega_r \quad , \qquad n \text{ odd} \quad .$$

## 9.3 Motion of Wave Packets

In order to describe motions in our system of coupled harmonic oscillators, we have to solve the time-dependent Schrödinger equation

$$i\hbar\frac{\partial\psi}{\partial t} = H\psi \quad .$$

If $\psi(x_1, x_2, t)$ is a solution with the initial condition $\psi(x_1, x_2, t_0)$, then $\psi(x_2, x_1, t)$ is also a solution corresponding to the initial condition $\psi(x_2, x_1, t_0)$. This is guaranteed by the symmetry of the Hamiltonian in coordinates and momenta of indistinguishable particles, as discussed in Section 9.1.

Again, by symmetrization or antisymmetrization, we obtain still other solutions of the time-dependent Schrödinger equation. They are

$$\psi_B(x_1, x_2, t) = a_B[\psi(x_1, x_2, t) + \psi(x_2, x_1, t)] \quad ,$$
$$\psi_F(x_1, x_2, t) = a_F[\psi(x_1, x_2, t) - \psi(x_2, x_1, t)] \quad ,$$

and correspond, of course, to symmetric or antisymmetric initial conditions. The numerical factors $a_B$, $a_F$ ensure normalization of the corresponding wave packets.

As a first example, let us consider the motion of two bosons forming a system of coupled harmonic oscillators. In Figure 9.1 the joint probability density

$$\rho_B(x_1, x_2, t) = |\psi_B(x_1, x_2, t)|^2$$

and the marginal distributions $\rho_{B1}(x_1, t)$ and $\rho_{B2}(x_2, t)$ are shown for several times $t = t_0, t_1, \ldots, t_N$. Except for the symmetrization of the wave function, all parameters are the same as those for distinguishable particles, whose motion was illustrated in Figure 8.2. In particular, the trajectory of the classical particles in the $x_1, x_2$ plane is identical in both figures. Since the position expectation values $x_{10}$, $x_{20}$ at initial time $t = t_0$ are farther apart than the width of the unsymmetrized wave packet in Figure 8.2, we observe for $t = t_0$ two well-separated humps corresponding to points $x_1 = x_{10}$, $x_2 = x_{20}$, and $x_1 = x_{20}$,

$x_2 = x_{10}$, respectively. The marginal distribution $\rho_{B1}(x_1, t_0)$, which describes the probability that one particle of the two will be observed at $x_1$, irrespective of the position of the other one, also has two humps. The two humps again reflect the fact that the two particles cannot be distinguished. Then, of course, the marginal distribution $\rho_{B2}(x_2, t)$ has to be identical to the marginal distribution $\rho_{B1}(x_1, t)$. In pursuit of their motion, the particles attain a distance smaller than the width of the unsymmetrized wave packet. In this situation the two humps are no longer separated but merge into one. For a later moment in time they are again separated, and so on.

Figure 9.2 shows the corresponding motion of two fermions. For $t = t_0$, when the two particles are well separated, the situation looks qualitatively similar, but it becomes strikingly different when the particles move close to each other. The hump splits along the direction $x_1 = x_2$, where the probability density is exactly zero as a consequence of the Pauli exclusion principle. In fact, for fermions the probability density vanishes for locations $x_1 = x_2$ at all moments in time. At no time can two fermions be at the same place.

Figures 9.3b and c show the time developments of the marginal distributions $\rho_{B,F}(x, t)$ for two bosons and for two fermions forming a system of coupled harmonic oscillators. The difference between the two is much less striking than that between the corresponding probability distributions of Figures 9.1 and 9.2. But still a trace of the Pauli exclusion principle is visible in the marginal distributions. Near the center of Figures 9.3b and c, where the particles are close to each other, the two humps are farther apart for fermions than for bosons. For purposes of comparison, the time developments of the two marginal distributions for the corresponding system of two distinguishable particles are given in Figure 9.3a.

## 9.4 Indistinguishable Particles from a Classical Point of View

The quantum-mechanical description of the motion of indistinguishable particles poses the question whether the classical concept of the trajectory of a particle can still be upheld or whether it has to be given up. Looking at the joint probability distributions for indistinguishable particles in Figures 9.1 and 9.2, we observe two distinct humps as long as the classical positions are far apart. The center of either of them moves along its classical trajectory with the initial positions

$$x_1 = x_{10} \quad , \qquad x_2 = x_{20} \quad ,$$

or

$$x_1 = x_{20} \quad , \qquad x_2 = x_{10} \quad ,$$

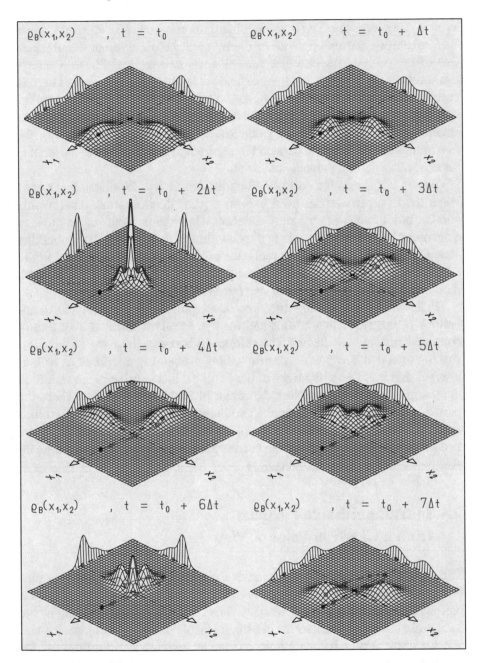

**Fig. 9.1. Joint probability density and marginal distributions for two bosons forming a system of coupled harmonic oscillators. The joint probability density** $\rho_B(x_1, x_2, t)$ **is shown as a surface over the** $x_1, x_2$ **plane, the marginal distribution** $\rho_{B1}(x_1, t)$ **as a curve over the margin parallel to the** $x_1$ **axis, and the marginal distribution** $\rho_{B2}(x_2, t)$ **as a curve over the other margin. The distributions are shown for various times** $t_j = t_0, t_1, \ldots, t_N$. **The positions of the classical particles are indicated by dots in the plane and on the margins; their motion is represented by the trajectory in the** $x_1, x_2$ **plane.**

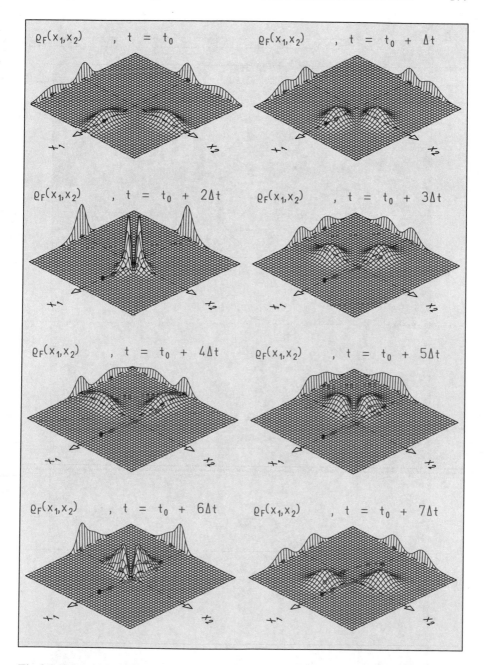

**Fig. 9.2. Joint probability density and marginal distributions for two fermions. All initial conditions are the same as those for Figure 9.1.**

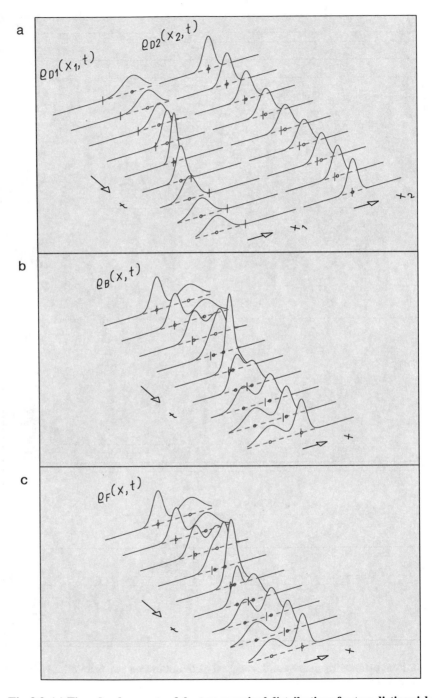

**Fig. 9.3. (a)** Time developments of the two marginal distributions for two distinguishable particles forming a system of coupled harmonic oscillators. Time developments of the marginal distributions $\rho_{B,F}(x,t)$ for the corresponding systems of **(b)** two bosons and **(c)** two fermions.

where $x_{10}$, $x_{20}$ are the position expectation values of the probability distribution for distinguishable particles. This observation suggests that although the particles are indistinguishable in their intrinsic properties, they can under the given circumstances be distinguished by their position. Thus if we call the particle that at $t = t_0$ is in the neighborhood of $x_{10}$ particle 1 and the particle that is close to $x_{20}$ particle 2, it is perfectly consistent to say that particle 1 stays in the neighborhood of the trajectory $\langle x_1(t) \rangle$ and particle 2 in that of $\langle x_2(t) \rangle$, as long as the two humps are well separated. Here, $\langle x_1(t) \rangle$ is the expectation value of the coordinate $x_1$ for the wave packet of distinguishable particles and also the classical position of particle 1 at time $t$. As soon as the particles come closer to each other than the widths of the humps, there is no longer a clear correspondence between the classical trajectory and the structure of the probability density. Once the positions are separated again, a new correspondence can be established.

A look at the relevant formulae justifies this reasoning. The wave functions $\psi_B$, for bosons, and $\psi_F$, for fermions, were obtained from that for distinguishable particles, $\psi$, by symmetrization and antisymmetrization,

$$\psi_{B,F}(x_1, x_2, t) = a_{B,F}[\psi(x_1, x_2, t) \pm \psi(x_2, x_1, t)] \quad .$$

The probability density is found by taking the absolute square,

$$\begin{aligned} \rho_{B,F}(x_1, x_2, t) &= |\psi_{B,F}(x_1, x_2, t)|^2 \\ &= |a_{B,F}|^2 [\rho_D(x_1, x_2, t) + \rho_D(x_2, x_1, t) \pm \tau(x_1, x_2, t)] \quad . \end{aligned}$$

Here

$$\rho_D(x_1, x_2, t) = |\psi(x_1, x_2, t)|^2$$

is the joint probability distribution for distinguishable particles with coordinate $x_1$ corresponding to particle 1 and coordinate $x_2$ to particle 2. The density

$$\rho_D(x_2, x_1, t) = |\psi(x_2, x_1, t)|^2$$

describes the situation in which particles 1 and 2 are interchanged.

The term

$$\tau(x_1, x_2, t) = \psi^*(x_1, x_2, t)\psi(x_2, x_1, t) + \psi^*(x_2, x_1, t)\psi(x_1, x_2, t)$$

is called the *interference term*. This term is practically zero unless the two particles are closer to each other than the width of the single hump. To show this, we consider the particular point $x_1 = x_{10}$, $x_2 = x_{20}$ in the top left-hand corner of Figure 8.2. Clearly here $\psi(x_{10}, x_{20}, t)$ and its complex conjugate have large amplitudes, whereas $\psi(x_{20}, x_{10}, t)$ and its complex conjugate practically vanish. Figure 9.5, which shows the interference term $\tau(x_1, x_2, t)$ for various times $t = t_0, t_1, \ldots, t_N$, verifies the nature of the interference term. The figure

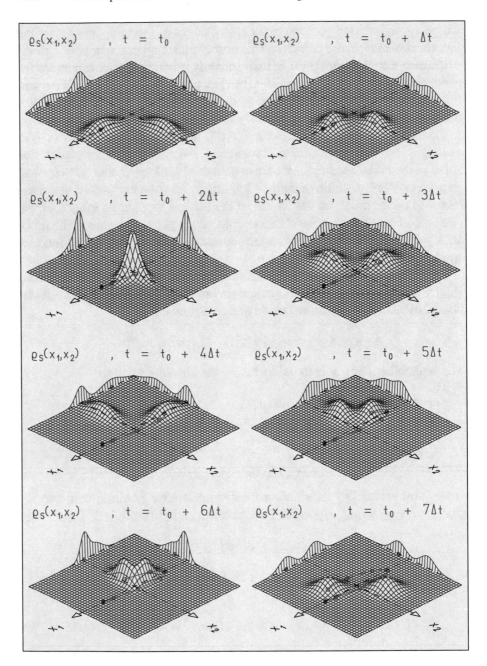

**Fig. 9.4. Symmetrized probability density for two distinguishable particles forming a system of coupled harmonic oscillators. All initial conditions are the same as those for Figure 9.1.**

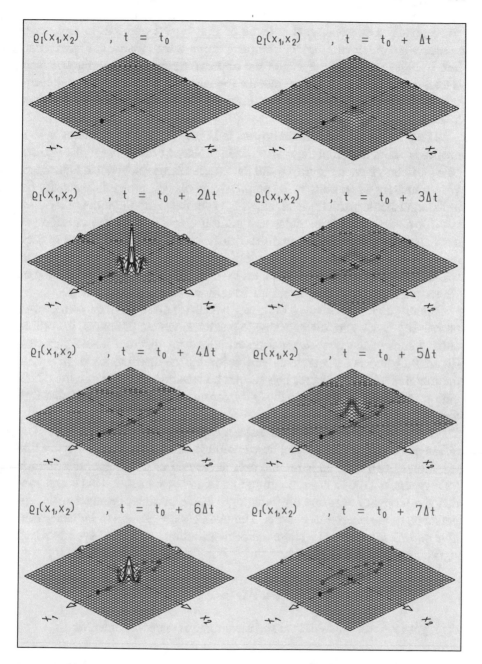

**Fig. 9.5. The interference term for two indistinguishable particles forming a system of coupled harmonic oscillators. The distribution is shown for various times** $t = t_0, t_1, \ldots, t_N$. **All initial conditions are the same as those for Figure 9.1.**

corresponds in all conditions to those of Figures 9.1 and 9.2. In fact, these figures were obtained using the complex formula for $\rho_{B,F}(x_1, x_2, t)$ given earlier. In Figure 9.4 only the sum of the first two terms – the interference term is excluded – is plotted. We see that the interference is comparable to the sum of the other two only when these overlap, that is, when the particles are close to each other.

The probability densities for bosons in Figure 9.1 and for fermions in Figure 9.2 are obtained from the symmetrized probability density for distinguishable particles given in Figure 9.4 and the interference term given in Figure 9.5. We summarize this discussion by emphasizing that the probability density for indistinguishable particles is obtained by symmetrizing the probability density for distinguishable particles and adding or subtracting the interference term. This term contributes only if the particles are sufficiently close to each other. Thus the concept of classical trajectories can be maintained as long as we are able to distinguish the particles by their initial positions and as long as we refrain from localizing them individually in the overlap region.

Finally, Figure 9.6a gives the marginal distribution for the symmetrized probability density for distinguishable particles, which, of course, is nothing but the sum of the two marginal distributions for distinguishable particles. Figure 9.6b shows the marginal distribution for the interference term. Again, the marginal distributions for bosons can be constructed by adding the distributions of Figures 9.6a and b, those for fermions by subtracting the distribution of Figure 9.6b from that of Figure 9.6a.

In Section 8.3 we pointed out that for distinguishable particles there is no difference between the classical time evolution of a Gaussian phase-space distribution of two coupled harmonic oscillators and of the Wigner distribution (cf. Appendix D) of a corresponding Gaussian wave packet. This correspondence no longer holds true for indistinguishable particles because of the appearance of the interference terms. The classical description of indistinguishable particles in terms of a phase-space distribution amounts to symmetrizing $\rho_D(x_1, x_2, t)$ and is thus given by

$$\rho_S(x_1, x_2, t) = \frac{1}{2}[\rho_D(x_1, x_2, t) + \rho_D(x_2, x_1, t)] \quad ,$$

cf. Figure 9.4, for the initial data of the situation under consideration.

## Problems

9.1. Which eigenstates of the system of two coupled harmonic oscillators, as plotted in Figure 8.6, can be occupied by bosons, which by fermions?

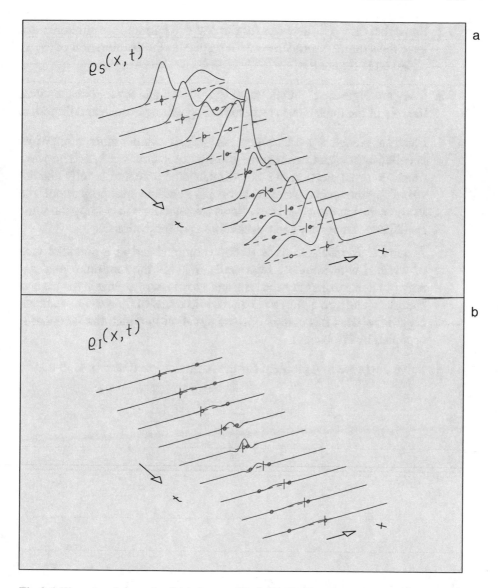

**Fig. 9.6.** Time developments of (a) the marginal distribution for the symmetrized probability density for two distinguishable particles and (b) the marginal distribution for the interference term for two indistinguishable particles. The particles form a system of coupled harmonic oscillators. All initial conditions are the same as those for Figure 9.3.

9.2. Show that the eigenfunctions for the coupled harmonic oscillators must have the symmetry properties with respect to the permutation of $x_1$, $x_2$ observed in Figure 8.6.

9.3. Compare Figures 9.1 and 9.2 with Figures 9.4 and 9.5 and characterize the role of the interference term in distinguishing bosons and fermions.

9.4. Electrons are fermions. They possess intrinsic angular momentum which is called spin $s$ and can assume the two projections $\pm \hbar/2$. The wave function for an electron in a one-dimensional potential is fully characterized by the spatial wave function $\varphi(x)$ and the spin projection. The Pauli exclusion principle then allows two electrons to occupy the same spatial state since they can assume two spin projections.

A number $N$ of electrons is to be accommodated in a potential well of width $d$ with infinitely high walls. What is the minimum total energy of all electrons? For the minimum total energy, what is the highest energy an electron assumes? Express it in terms of the ground-state energy! How does this compare to the situation in which the potential is occupied by $N$ bosons?

9.5. Solve the preceding problem for the harmonic-oscillator potential.

# 10. Wave Packet in Three Dimensions

## 10.1 Momentum

The position of the classical particle in three-dimensional space is described by the components $x$, $y$, $z$ of the *position vector*:

$$\mathbf{r} = (r, y, z) \quad .$$

Similarly, the three components of momentum form the *momentum vector*:

$$\mathbf{p} = (p_x, p_y, p_z) \quad .$$

Following our one-dimensional description in Section 3.3, we now introduce operators for all three components of momentum:

$$\hat{p}_x = \frac{\hbar}{i}\frac{\partial}{\partial x} \quad , \qquad \hat{p}_y = \frac{\hbar}{i}\frac{\partial}{\partial y} \quad , \qquad \hat{p}_z = \frac{\hbar}{i}\frac{\partial}{\partial z} \quad .$$

The three operators form the *vector operator of momentum*,

$$\hat{\mathbf{p}} = (\hat{p}_x, \hat{p}_y, \hat{p}_z) = \frac{\hbar}{i}\left(\frac{\partial}{\partial x}, \frac{\partial}{\partial y}, \frac{\partial}{\partial z}\right) = \frac{\hbar}{i}\nabla \quad ,$$

which is the differential operator $\nabla$, called *nabla* or *del*, multiplied by $\hbar/i$.

The three-dimensional stationary plane wave

$$
\begin{aligned}
\varphi_{\mathbf{p}}(\mathbf{r}) &= \frac{1}{(2\pi\hbar)^{1/2}}\exp\left(\frac{i}{\hbar}p_x x\right)\frac{1}{(2\pi\hbar)^{1/2}}\exp\left(\frac{i}{\hbar}p_y y\right) \\
&\quad \times \frac{1}{(2\pi\hbar)^{1/2}}\exp\left(\frac{i}{\hbar}p_z z\right) \\
&= \frac{1}{(2\pi\hbar)^{3/2}}\exp\left(\frac{i}{\hbar}\mathbf{p}\cdot\mathbf{r}\right)
\end{aligned}
$$

with

$$\mathbf{p}\cdot\mathbf{r} = p_x x + p_y y + p_z z$$

S. Brandt and H.D. Dahmen, *The Picture Book of Quantum Mechanics*,
DOI 10.1007/978-1-4614-3951-6_10, © Springer Science+Business Media New York 2012

is simply the product of three one-dimensional stationary waves of the momentum components $p_x$, $p_y$, and $p_z$ corresponding to the three directions $x$, $y$, and $z$ in space. The surfaces of constant phase $\delta$ are given by

$$\frac{i}{\hbar}\mathbf{p}\cdot\mathbf{r} = \delta \quad .$$

They are planes perpendicular to the *wave vector*

$$\mathbf{k} = \frac{\mathbf{p}}{\hbar} \quad .$$

The wave vector is the three-dimensional generalization of wave number $k$ in one dimension, as introduced in Section 2.1. It determines the wavelength through the relation

$$\lambda = \frac{2\pi}{|\mathbf{k}|} \quad .$$

The three-dimensional stationary plane wave is a simultaneous solution – also called a simultaneous eigenfunction – of the three equations

$$\hat{p}_x\varphi_{\mathbf{p}}(\mathbf{r}) = p_x\varphi_{\mathbf{p}}(\mathbf{r}) \quad , \quad \hat{p}_y\varphi_{\mathbf{p}}(\mathbf{r}) = p_y\varphi_{\mathbf{p}}(\mathbf{r}) \quad , \quad \hat{p}_z\varphi_{\mathbf{p}}(\mathbf{r}) = p_z\varphi_{\mathbf{p}}(\mathbf{r}) \quad .$$

The three numbers $p_x$, $p_y$, and $p_z$ forming the vector $\mathbf{p}$ are called the momentum eigenvalues of the plane wave $\varphi_{\mathbf{p}}(\mathbf{r})$.

The three-dimensional time-dependent wave function, like the one-dimensional, is obtained by multiplying the stationary eigenfunction $\varphi_{\mathbf{p}}(\mathbf{r})$ by the energy-dependent phase factor,

$$\exp\left(-\frac{i}{\hbar}Et\right) \quad , \quad E = \frac{\mathbf{p}^2}{2M} = \frac{1}{2M}(p_x^2 + p_y^2 + p_z^2) \quad ,$$

that is,

$$\begin{aligned}
\psi_{\mathbf{p}}(\mathbf{r},t) &= \frac{1}{(2\pi\hbar)^{3/2}}\exp\left(-\frac{i}{\hbar}Et\right)\exp\left(\frac{i}{\hbar}\mathbf{p}\cdot\mathbf{r}\right) \\
&= \psi_{p_x}(x,t)\psi_{p_y}(y,t)\psi_{p_z}(z,t) \quad .
\end{aligned}$$

Here $M$ is the mass of the particle. This time-dependent expression for the three-dimensional harmonic wave also factors into exponentials corresponding to the three dimensions.

The three-dimensional free, unaccelerated motion of a particle is again described by a superposition of these plane waves with a spectral function,

$$\begin{aligned}
f(\mathbf{p}) &= f_x(p_x)f_y(p_y)f_z(p_z) \quad , \\
f_a(p_a) &= \frac{1}{(2\pi)^{1/4}\sqrt{\sigma_{p_a}}}\exp\left[-\frac{(p_a - p_{a0})^2}{4\sigma_{p_a}^2}\right] \quad , \quad a = x, y, z \quad ,
\end{aligned}$$

which is the product of three Gaussian spectral functions centered around the expectation values $(p_{x0}, p_{y0}, p_{z0}) = \mathbf{p}_0$ with the widths $\sigma_{p_x}, \sigma_{p_y}, \sigma_{p_z}$ as introduced in Section 3.2. The superposition of the functions $\psi_{\mathbf{p}}(\mathbf{r} - \mathbf{r}_0, t)$ with the spectral function $f(\mathbf{p})$ is given by

$$\psi(\mathbf{r}, t) = \int f(\mathbf{p}) \psi_{\mathbf{p}}(\mathbf{r} - \mathbf{r}_0, t) \, d^3 p \quad .$$

It represents the moving wave packet that starts at $t = 0$ around point $\mathbf{r}_0$ with the average momentum $\mathbf{p}_0$. Because of the product forms of $f(\mathbf{p})$ and $\psi_{\mathbf{p}}(\mathbf{r} - \mathbf{r}_0, t)$, the equation can also be written in product form,

$$\psi(\mathbf{r}, t) = M_x(x, t) e^{i\phi_x(x,t)} M_y(y, t) e^{i\phi_y(y,t)} M_z(z, t) e^{i\phi_z(z,t)} \quad ,$$

where the meaning of the symbols can easily be inferred from the one-dimensional wave packet of Section 3.2.

The set of the upper three plots of Figure 10.1 shows the probability distribution $|\psi(x, y, 0, t)|^2$ in the $x, y$ plane of the moving wave packet for the initial moment in time, $t_0 = 0$ and two later ones. The straight line in the $x, y$ plane marks the classical trajectory that has been chosen to lie in this plane. The dots indicate the positions of the corresponding classical particle at the three moments in time. The probability distribution shown is a two-dimensional Gaussian dispersing in time. The ellipse encircling the bell-shaped bump comprises a certain fraction of the total probability. It is the covariance ellipse, which was already discussed in Section 3.5. As the wave packet disperses, this ellipse grows in size. For a Gaussian wave packet this ellipse completely characterizes the position and the degree of localization of the particle in the $x, y$ plane. The complete three-dimensional Gaussian wave packet is then characterized by a *covariance ellipsoid*. The lowest plot of Figure 10.1 shows the ellipsoids that correspond to the three situations of Figure 10.1.

## 10.2 Quantile Motion, Probability Transport

In Section 7.1 the quantile motion was introduced for one-dimensional problems. We consider two different probabilities $P_1 < P_2$ with the two quantile trajectories $x_{P_1}(t), x_{P_2}(t)$. Then, the difference $P_2 - P_1$ is the time-independent probability contained in the interval $x_{P_2}(t) < x < x_{P_1}(t)$, i.e.,

$$\int_{x_{P_2}(t)}^{x_{P_1}(t)} \varrho(x, t) \, dx = P_2 - P_1 \quad .$$

For three-dimensional systems a corresponding statement holds true. We denote by $V_t$ the image at time $t$ of the volume $V_{t_0}$ at time $t_0$ under the transformation

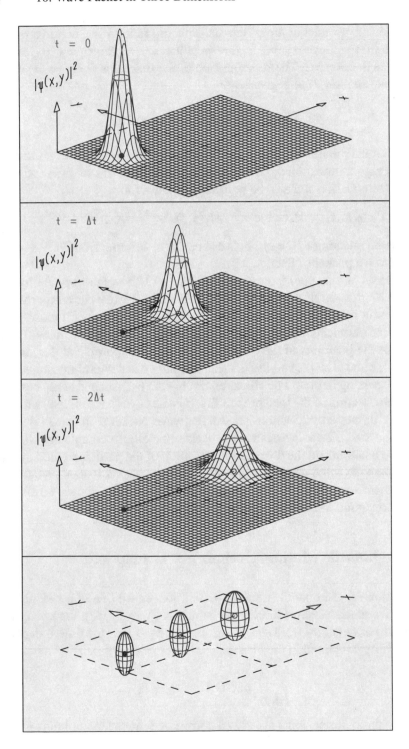

**Fig. 10.1.**

$$\mathbf{r} = \mathbf{r}(t, \mathbf{x})$$

of every point $\mathbf{x}$ in $V_{t_0}$ into $\mathbf{r} \in V_t$. The transformation $\mathbf{r}(t, \mathbf{x})$ is the solution of the differential equation

$$\frac{\partial \mathbf{r}(t, \mathbf{x})}{\partial t} = \mathbf{v}(\mathbf{r}(t, \mathbf{x}), t)$$

with the initial condition $\mathbf{r}(t_0, \mathbf{x}) = \mathbf{x}$. Here the velocity field $\mathbf{v}(\mathbf{r}, t)$ is the quotient $\mathbf{v}(\mathbf{r}, t) = \mathbf{j}(\mathbf{r}, t)/\varrho(\mathbf{r}, t)$ of the probability current density $\mathbf{j}(\mathbf{r}, t)$ and the probability density $\varrho(\mathbf{r}, t)$ of the quantum-mechanical system. With these provisions the statement reads: The probability $P$ contained in the volume $V_{t_0}$ at time $t_0$ is contained in the volume $V_t$ at time $t$,

$$\int_{V_t} \varrho(\mathbf{r}, t)\, d^3 r = P \quad .$$

We consider the force-free Gaussian wave packet of the last section. If for simplicity we choose all momentum widths equal, $\sigma_{p_x} = \sigma_{p_y} = \sigma_{p_z} = \sigma_p$, it has the probability density

$$\varrho(\mathbf{r}, t) = \frac{1}{(2\pi)^{3/2} \sigma^3(t)} \exp\left( -\frac{(\mathbf{r} - \mathbf{r}_0 - \mathbf{v}_0 t)^2}{2\sigma^2(t)} \right)$$

with the initial expectation values of position $\mathbf{r}_0 = (x_0, y_0, z_0)$ and of velocity $\mathbf{v}_0 = (v_{0x}, v_{0y}, v_{0z})$ at $t = 0$. The square of its time-dependent width is $\sigma^2(t) = \sigma_0^2 + (\sigma_p t/m)^2$, where $\sigma_0 = \hbar/(2\sigma_p)$.

In Figure 10.2 we show quantile trajectories for this wave packet. They are curved lines, even though the motion of the wave packet is force-free. This is due to the fact that the dispersion of the Gaussian wave packet follows the width $\sigma(t)$, which is a non-linear function of time.

In fact, the quantile trajectories for the force-free Gaussian wave have the form

$$\mathbf{r}(t, \mathbf{x}) = \mathbf{r}_0 + \mathbf{v}_0 t + \frac{\sigma(t)}{\sigma_0}(\mathbf{x} - \mathbf{r}_0) \quad .$$

**Fig. 10.1. A three-dimensional Gaussian wave packet moves freely in space. Its position expectation value moves on a straight line in the $x, y$ plane. The first three illustrations show for three equidistant moments in time the probability density in the $x, y$ plane as a bell-shaped surface, the expectation value as a dot on the plane, and the trajectory of the corresponding classical particle as a straight line in the plane. The covariance ellipse encircling the surface comprises a fixed fraction of the total probability. It contains the complete probability density information for the $x, y$ plane. The complete information for the three-dimensional probability distribution is given by the probability ellipsoid. It is centered around the position expectation value and shown at the bottom for the three moments in time that are depicted separately in the first three plots. The classical trajectory in space is also shown.**

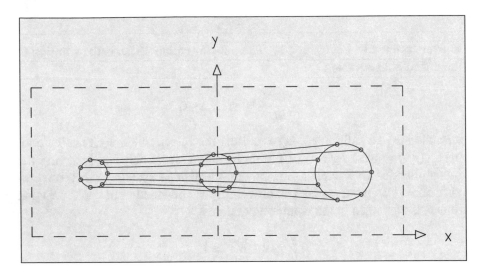

**Fig. 10.2. The expectation value of a free three-dimensional spherically symmetric Gaussian wave packet, which initially (at time $t = t_0$) lies in the $x, y$ plane, moves in the positive $x$ direction. As initial volume $V_{t_0}$ comprising the probability $P$ a sphere around the expectation value is chosen. Quantile trajectories $\mathbf{x}_P(t)$ of points which at $t = t_0$ lie on the surface of $V_{t_0}$ at later times lie on the surface of volumes which also comprise the same probability $P$. In this simple example all volumes $V_{t_i}$ are spheres. The plot shows the cuts $z = 0$ through three spheres $V_{t_0}, V_{t_1}, V_{t_2}$ which are circles and trajectories in the $x, y$ plane. All parameters are as in Figure 10.1.**

For a given probability $P$ the initial sphere $V_{t_0}$ with the radius $R_{0,P}$ consists of all points $\mathbf{x}$ which satisfy the inequality $|\mathbf{x} - \mathbf{r}_0| = x < R_{0,P}$. At times $t > 0$ the points $\mathbf{x}$ are mapped into the points $\mathbf{r}(t, \mathbf{x})$. The mapping $\mathbf{r}(t, \mathbf{x})$ satisfies the above differential equation. Also images $V_t$ of $V_{t_0}$ are spheres. They contain all the points $\mathbf{r}(t, \mathbf{x})$ with $\mathbf{x} \in V_{t_0}$.

The initial radius $R_{0,P}$ of the sphere is determined through the quantile condition, i.e.,

$$\int_0^{R_{0,P}} \int_{-1}^{1} \int_0^{2\pi} \frac{1}{(2\pi)^{3/2}\sigma_{x0}^3} \exp\left(-\frac{x^2}{2\sigma_0^2}\right) x^2 \, dx \, d\cos\vartheta \, d\varphi = P \quad ,$$

which leads to the equation

$$1 - \operatorname{erfc}\left(\frac{R_{0,P}}{\sqrt{2}\sigma_0}\right) - \sqrt{\frac{2}{\pi}} \frac{R_{0,P}}{\sigma_0} \exp\left(-\frac{R_{0,P}^2}{2\sigma_0^2}\right) = P \quad ,$$

where $\operatorname{erfc} x$ denotes the error function discussed in Section 7.5. The time-dependent radius of the sphere is determined by

$$R_P(t) = \frac{\sigma(t)}{\sigma_0} R_{0,P} \quad .$$

## 10.3 Angular Momentum, Spherical Harmonics

Three-dimensional motion is further characterized by *angular momentum*. For a classical particle it is simply the vector product of the position vector and the momentum vector,

$$\mathbf{L} = \mathbf{r} \times \mathbf{p} \quad,$$

or in components,

$$L_x = yp_z - zp_y \quad, \qquad L_y = zp_x - xp_z \quad, \qquad L_z = xp_y - yp_x \quad.$$

The quantum-mechanical analog is obtained by inserting the operator of momentum $\hat{\mathbf{p}} = (\hbar/i)\nabla$ into the classical expression for $\mathbf{L}$. This yields the *vector operator of angular momentum*,

$$\hat{\mathbf{L}} = \mathbf{r} \times \hat{\mathbf{p}} = \frac{\hbar}{i}\mathbf{r} \times \nabla \quad,$$

or in components,

$$\hat{L}_x = \frac{\hbar}{i}\left(y\frac{\partial}{\partial z} - z\frac{\partial}{\partial y}\right), \ \hat{L}_y = \frac{\hbar}{i}\left(z\frac{\partial}{\partial x} - x\frac{\partial}{\partial z}\right), \ \hat{L}_z = \frac{\hbar}{i}\left(x\frac{\partial}{\partial y} - y\frac{\partial}{\partial x}\right).$$

Whereas the components of momentum commute with each other, that is, $[\hat{p}_x, \hat{p}_y] = \hat{p}_x\hat{p}_y - \hat{p}_y\hat{p}_x = 0$, and so on, the components of angular momentum do not. In fact, the *commutation relations* are

$$[\hat{L}_x, \hat{L}_y] = i\hbar\hat{L}_z \quad, \qquad [\hat{L}_y, \hat{L}_z] = i\hbar\hat{L}_x \quad, \qquad [\hat{L}_z, \hat{L}_x] = i\hbar\hat{L}_y \quad.$$

Because the commutators do not vanish, an eigenfunction of $\hat{L}_z$ cannot in general be an eigenfunction of $\hat{L}_y$ as well. If, in addition to the eigenvalue equation

$$\hat{L}_z Y = \ell_z Y \quad,$$

the relation

$$\hat{L}_y Y = \ell_y Y$$

would also hold, we would in general have a contradiction to the commutator relation $[\hat{L}_y, \hat{L}_z] = i\hbar\hat{L}_x$ when applied to the eigenfunction $Y$:

$$(\hat{L}_y\hat{L}_z - \hat{L}_z\hat{L}_y)Y = (\ell_y\ell_z - \ell_z\ell_y)Y = 0 \neq i\hbar\hat{L}_x Y \quad.$$

This observation is tantamount to the statement that noncommuting operators do *not* have simultaneous eigenfunctions, except for trivial ones.

There is, however, another operator, the square of the vector operator of angular momentum,

$$\hat{\mathbf{L}}^2 = \hat{L}_x^2 + \hat{L}_y^2 + \hat{L}_z^2 \quad,$$

which does commute with any of the components:

$$[\hat{\mathbf{L}}^2, \hat{L}_a] = 0 \quad , \qquad a = x, y, z \quad .$$

This relation is easily verified with the help of the commutation relations, for example,

$$
\begin{aligned}
&[\hat{L}_x^2 + \hat{L}_y^2 + \hat{L}_z^2, \hat{L}_z] \\
&= [\hat{L}_x^2, \hat{L}_z] + [\hat{L}_y^2, \hat{L}_z] \\
&= \hat{L}_x[\hat{L}_x, \hat{L}_z] + [\hat{L}_x, \hat{L}_z]\hat{L}_x + \hat{L}_y[\hat{L}_y, \hat{L}_z] + [\hat{L}_y, \hat{L}_z]\hat{L}_y \\
&= \hat{L}_x(-i\hbar\hat{L}_y) - i\hbar\hat{L}_y\hat{L}_x + \hat{L}_y(i\hbar)\hat{L}_x + i\hbar\hat{L}_x\hat{L}_y = 0 \quad .
\end{aligned}
$$

Thus, simultaneous eigenfunctions for $\hat{\mathbf{L}}^2$ and any of the components, for example, $\hat{L}_z$, can be found. For the following discussion it is convenient to use *polar coordinates* $r$, $\vartheta$, and $\phi$ rather than Cartesian coordinates $x$, $y$, and $z$. In a polar coordinate system a point is given by its distance $r$ from the origin, its *polar angle* $\vartheta$, and its *azimuth* $\phi$. The relations between the coordinates of the two systems are

$$
\begin{aligned}
x &= r\sin\vartheta\cos\phi \quad , \\
y &= r\sin\vartheta\sin\phi \quad , \\
z &= r\cos\vartheta \quad .
\end{aligned}
$$

In polar coordinates the operators of angular momentum are

$$
\begin{aligned}
\hat{L}_x &= i\hbar\left(\sin\phi\frac{\partial}{\partial\vartheta} + \cotan\vartheta\cos\phi\frac{\partial}{\partial\phi}\right) \quad , \\
\hat{L}_y &= -i\hbar\left(\cos\phi\frac{\partial}{\partial\vartheta} - \cotan\vartheta\sin\phi\frac{\partial}{\partial\phi}\right) \quad , \\
\hat{L}_z &= -i\hbar\frac{\partial}{\partial\phi} \quad , \\
\hat{\mathbf{L}}^2 &= -\hbar^2\left[\frac{1}{\sin\vartheta}\frac{\partial}{\partial\vartheta}\left(\sin\vartheta\frac{\partial}{\partial\vartheta}\right) + \frac{1}{\sin^2\vartheta}\frac{\partial^2}{\partial\phi^2}\right] \quad .
\end{aligned}
$$

We can write eigenvalue equations for the two operators $\hat{\mathbf{L}}^2$ and $\hat{L}_z$:

$$
\begin{aligned}
\hat{\mathbf{L}}^2 Y_{\ell m} &= \ell(\ell+1)\hbar^2 Y_{\ell m} \quad , \\
\hat{L}_z Y_{\ell m} &= m\hbar Y_{\ell m} \quad .
\end{aligned}
$$

Both operators have as eigenfunctions the *spherical harmonics* $Y_{\ell m}(\vartheta, \phi)$, which are discussed in the next paragraphs. The eigenvalues of the square

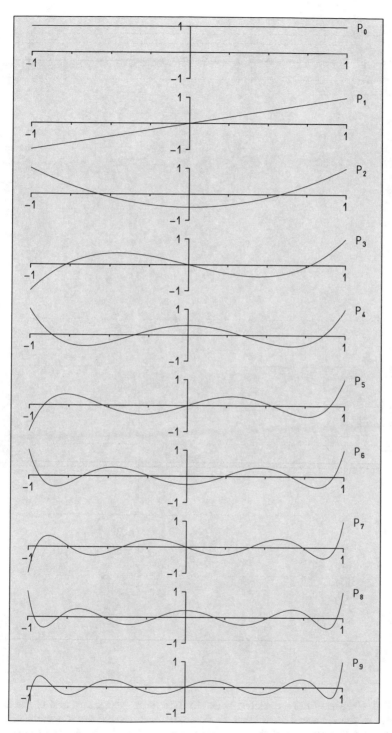

**Fig. 10.3. The first ten Legendre polynomials** $P_\ell(u) = \frac{1}{2^\ell \ell!} \frac{d^\ell}{du^\ell}\left[(u^2-1)^\ell\right]$.

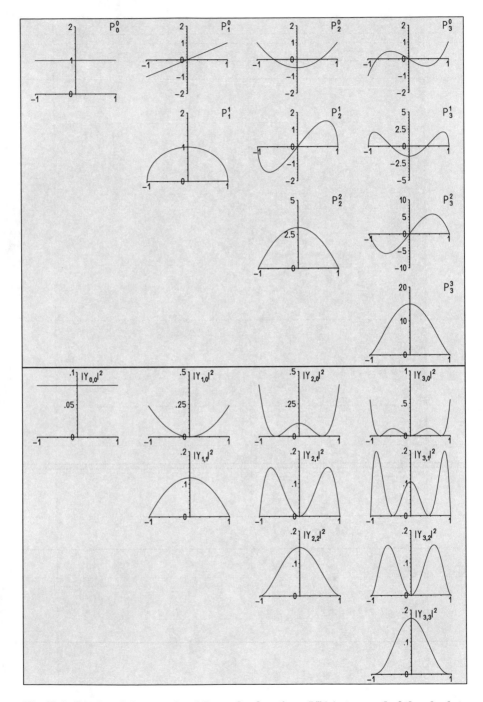

**Fig. 10.4. Graphs of the associated Legendre functions $P_\ell^m(u)$, top, and of the absolute squares of the spherical harmonics $Y_{\ell m}(\vartheta,\phi)$, bottom. Except for a normalization factor, the absolute squares of the spherical harmonics are the squares of the associated Legendre functions.**

of angular momentum are $\ell(\ell+1)\hbar^2$. This *quantum number of angular momentum* $\ell$ can take on only integer values $\ell = 0, 1, 2, \dots$. Thus, in contrast to classical mechanics, the square of angular momentum can take only discrete values that are integer multiples of $\hbar^2$. Correspondingly, the eigenvalues of the *z component $L_z$ of angular momentum* are $m\hbar$. The quantum number $m$ can vary only in the range $-\ell \le m \le \ell$. In fact, $m$ takes on only integer numbers in this range. For historical reasons quantum number $m$ is sometimes called *magnetic quantum number*.

The spherical harmonics $Y_{\ell m}(\vartheta, \phi)$ have an explicit representation which is commonly based on the *Legendre polynomials*

$$P_\ell(u) = \frac{1}{2^\ell \ell!} \frac{\mathrm{d}^\ell}{\mathrm{d}u^\ell} \left[(u^2 - 1)^\ell\right] \quad .$$

Figure 10.3 shows the plots of these polynomials for $\ell = 0, 1, 2, \dots, 9$, and the domain $-1 \le u \le 1$.

The Legendre polynomials are special cases of the *associated Legendre functions $P_\ell^m$*, which are defined by

$$P_\ell^m(u) = (1 - u^2)^{m/2} \frac{\mathrm{d}^m}{\mathrm{d}u^m} P_\ell(u) \quad , \qquad m = 0, 1, 2, \dots, \ell \quad .$$

The top part of Figure 10.4 gives their graphs for $\ell = 0, 1, 2, 3$.

Finally, for $m \ge 0$, the spherical harmonics $Y_{\ell m}$ have the representation

$$Y_{\ell m}(\vartheta, \phi) = (-1)^m \sqrt{\frac{2\ell+1}{4\pi} \cdot \frac{(\ell-m)!}{(\ell+m)!}} P_\ell^m(\cos\vartheta) e^{im\phi} \quad .$$

For negative $m = -1, -2, \dots, -\ell$ the spherical harmonics are

$$Y_{\ell,-m}(\vartheta, \phi) = (-1)^m Y_{\ell m}^*(\vartheta, \phi) \quad .$$

Whereas the Legendre polynomials $P_\ell(u)$ and the associated Legendre functions $P_\ell^m(u)$ are real functions of the argument $u$, the spherical harmonics $Y_{\ell m}$ are complex functions of their arguments. As an example, Figure 10.5 shows the real and imaginary parts as well as the absolute square of $Y_{3m}(\vartheta, \phi)$. As the definition and the plots indicate, $|Y_{\ell m}|^2$ depends only on $\vartheta$. In fact, except for the normalization factor, it is equal to $[P_\ell^m(\cos\vartheta)]^2$. For comparison, the bottom of Figure 10.4 plots $|Y_{\ell m}|^2$ below $P_\ell^m$ for $\ell = 0, 1, 2, 3$.

Since the variables of the spherical harmonics are the polar angle $\vartheta$ and the azimuth $\phi$ of a spherical coordinate system, it is advantageous to represent $|Y_{\ell m}|^2$ in such a coordinate system. This is done in Figure 10.6 where $|Y_{\ell m}(\vartheta, \phi)|^2$ is the length of the radius subtended under the angles $\vartheta$ and $\phi$ from the origin to the surface. In this way $|Y_{00}|^2 = 1/(4\pi)$ turns out to be a

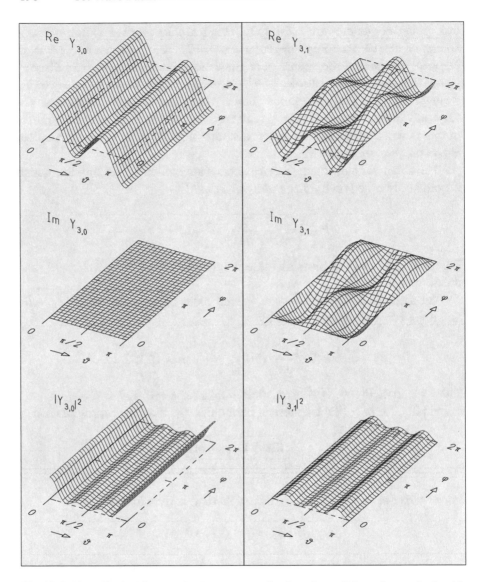

**Fig. 10.5. The spherical harmonics $Y_{\ell m}$ are complex functions of the polar angle $\vartheta$, with $0 \leq \vartheta \leq \pi$, and the azimuth $\phi$, with $0 \leq \phi < 2\pi$. They can be visualized by showing their real and imaginary parts and their absolute square over the $\vartheta, \phi$ plane. Such graphs are shown here for $\ell = 3$ and $m = 0, 1, 2, 3$.**

sphere. For all possible values $\ell$ and $m$ the functions $|Y_{\ell m}|^2$ are rotationally symmetric around the $z$ axis. They can vanish for certain values of $\vartheta$. These are called $\vartheta$ nodes if they occur for values of $\vartheta$ other than zero or $\pi$. It should be noted that $|Y_{\ell \ell}|^2$ does not have nodes, whereas $|Y_{\ell m}|^2$ possesses $\ell - |m|$ nodes.

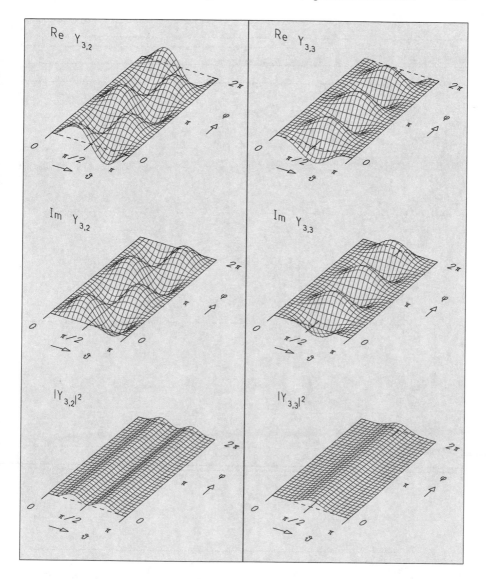

**Fig. 10.5. (continued)**

The Legendre polynomials possess the following *orthonormality proper-*
*ties*:

$$\int_{-1}^{1} P_\ell(u) P_{\ell'}(u) \, du = \frac{2}{2\ell+1} \delta_{\ell\ell'} \quad .$$

Here $\delta_{\ell\ell'}$ is the *Kronecker symbol*

$$\delta_{\ell\ell'} = \begin{cases} 1 \;, & \ell = \ell' \\ 0 \;, & \ell \neq \ell' \end{cases} \quad .$$

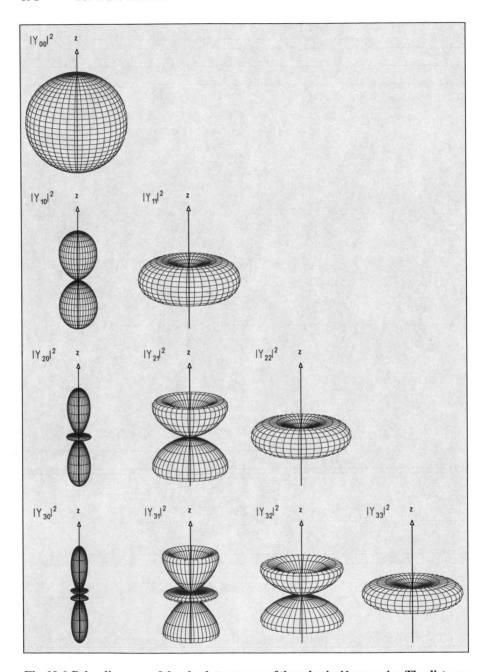

**Fig. 10.6. Polar diagrams of the absolute squares of the spherical harmonics. The distance from the origin of the coordinate system to a point on the surface seen under the angles $\vartheta$ and $\phi$ is equal to $|Y_{\ell m}(\vartheta, \phi)|^2$. Different scales are used for the individual parts of the figure.**

(The expression *orthonormality* stems from the similarity of the integral with a scalar product, cf. Appendix A, so that Legendre polynomials with different index can be considered orthogonal to each other.)

For the spherical harmonics the orthonormality relation reads

$$\int_{\cos\vartheta=1}^{-1}\int_{\phi=0}^{2\pi} Y_{\ell m}^*(\vartheta,\phi)Y_{\ell'm'}(\vartheta,\phi)\,\mathrm{d}\cos\vartheta\,\mathrm{d}\phi = \delta_{\ell\ell'}\delta_{mm'} \quad .$$

Since the integral is extended over all possible angles $\vartheta$ and $\phi$ one can say that integration is performed over the full *solid angle* $\Omega = 4\pi$ and one writes the above integral in the somewhat abbreviated form

$$\int Y_{\ell m}^*(\vartheta,\phi)Y_{\ell m}(\vartheta,\phi)\,\mathrm{d}\Omega = \delta_{\ell\ell'}\delta_{mm'} \quad .$$

## 10.4 Means and Variances of the Components of Angular Momentum

In Section 10.3 we discussed the eigenvalue equations for the spherical harmonics

$$\begin{aligned}
\hat{\mathbf{L}}^2 Y_{\ell m} &= \hbar^2 \ell(\ell+1)Y_{\ell m} \quad , \\
\hat{L}_z Y_{\ell m} &= \hbar m Y_{\ell m} \quad .
\end{aligned}$$

Application of the operators $\hat{L}_x$ and $\hat{L}_y$ yields

$$\begin{aligned}
\hat{L}_x Y_{\ell m} &= \frac{\hbar}{2}\sqrt{\ell(\ell+1)-m(m+1)}Y_{\ell m+1} \\
&\quad -\frac{\hbar}{2}\sqrt{\ell(\ell+1)-m(m-1)}Y_{\ell m-1} \quad , \\
\hat{L}_y Y_{\ell m} &= \frac{\hbar}{2i}\sqrt{\ell(\ell+1)-m(m+1)}Y_{\ell m+1} \\
&\quad -\frac{\hbar}{2i}\sqrt{\ell(\ell+1)-m(m-1)}Y_{\ell m-1} \quad ,
\end{aligned}$$

showing that the $Y_{\ell m}$ are not eigenfunctions of $\hat{L}_x$, $\hat{L}_y$.

With the help of the orthonormality relations of the spherical harmonics given at the end of Section 10.3 we calculate the expectation values of the three components and of the square of angular momentum,

$$\begin{aligned}
\langle L_x \rangle_{\ell m} &= \int Y_{\ell m}^*(\vartheta,\phi)\hat{L}_x Y_{\ell m}(\vartheta,\phi)\,\mathrm{d}\Omega = 0 \quad , \\
\langle L_y \rangle_{\ell m} &= \int Y_{\ell m}^*(\vartheta,\phi)\hat{L}_y Y_{\ell m}(\vartheta,\phi)\,\mathrm{d}\Omega = 0 \quad ,
\end{aligned}$$

$$\langle L_z \rangle_{\ell m} = \int Y_{\ell m}^*(\vartheta,\phi) \hat{L}_z Y_{\ell m}(\vartheta,\phi) \, d\Omega = m\hbar \quad ,$$

$$\langle \mathbf{L}^2 \rangle_{\ell m} = \int Y_{\ell m}^*(\vartheta,\phi) \hat{\mathbf{L}}^2 Y_{\ell m}(\vartheta,\phi) \, d\Omega = \ell(\ell+1)\hbar^2 \quad .$$

Obviously, the expectation values of the three components $(0,0,m)$ cannot be interpreted as the three components of a vector, since the modulus square of such a vector is $m^2$, which is always smaller than the expectation value $\ell(\ell+1)\hbar^2$ of $\hat{\mathbf{L}}^2$,

$$\langle L_x \rangle_{\ell m}^2 + \langle L_y \rangle_{\ell m}^2 + \langle L_z \rangle_{\ell m}^2 = m^2\hbar^2 \le \ell(\ell+1)\hbar^2 \quad .$$

The reason for this astonishing result becomes obvious if we calculate the expectation values of the squares of the angular momentum components. Since $Y_{\ell m}$ is an eigenfunction of $\hat{L}_z$, we find

$$\langle L_z^2 \rangle_{\ell m} = \int Y_{\ell m}^*(\vartheta,\phi) \hat{L}_z^2 Y_{\ell m}(\vartheta,\phi) \, d\Omega = \hbar^2 m^2 \quad .$$

For the two other components we make use of the equations for $\hat{L}_x Y_{\ell m}$ and $\hat{L}_y Y_{\ell m}$ given above and get

$$\langle L_{x,y}^2 \rangle = \int Y_{\ell m}^*(\vartheta,\phi) \hat{L}_{x,y}^2 Y_{\ell m}(\vartheta,\phi) \, d\Omega = \frac{\hbar^2}{2} \left[ \ell(\ell+1) - m^2 \right] \quad .$$

The non-vanishing of these two expectation values $\langle L_{x,y}^2 \rangle_{\ell m}$ resolves the above difference,

$$\langle L_x^2 \rangle_{\ell m} + \langle L_y^2 \rangle_{\ell m} + \langle L_z^2 \rangle_{\ell m}$$
$$= \frac{\hbar^2}{2} \left[ \ell(\ell+1) - m^2 \right] + \frac{\hbar^2}{2} \left[ \ell(\ell+1) - m^2 \right] + \hbar^2 m^2 = \ell(\ell+1) \quad .$$

With the help of the results for the expectation values of the squares of the components we calculate the variances of the angular-momentum components

$$(\mathrm{var}(L_z))_{\ell m} = \left\langle L_z^2 - \langle L_z \rangle^2 \right\rangle_{\ell m} = \langle L_z^2 \rangle_{\ell m} - \hbar^2 m^2 = 0 \quad ,$$

$$(\mathrm{var}(L_x))_{\ell m} = \left\langle L_x^2 - \langle L_x \rangle^2 \right\rangle_{\ell m} = \langle L_x^2 \rangle_{\ell m} = \frac{\hbar^2}{2} \left[ \ell(\ell+1) - m^2 \right] \quad ,$$

$$\left(\mathrm{var}(L_y)\right)_{\ell m} = \left\langle L_y^2 - \langle L_y \rangle^2 \right\rangle_{\ell m} = \langle L_y^2 \rangle_{\ell m} = \frac{\hbar^2}{2} \left[ \ell(\ell+1) - m^2 \right] \quad .$$

The uncertainties

$$(\Delta L_{x,y,z})_{\ell m} = (\mathrm{var}(L_{x,y,z}))_{\ell m}^{1/2}$$

of the three components of angular momentum for the eigenfunctions turn out to be

$$(\Delta L_x)_{\ell m} = (\Delta L_y)_{\ell m} = \hbar \frac{1}{\sqrt{2}} \left[ \ell(\ell+1) - m^2 \right]^{1/2} \quad , \qquad (\Delta L_z)_{\ell m} = 0 \quad .$$

This shows that the eigenfunction $Y_{\ell\ell}$ belonging to the eigenvalue $m = \ell$ plays a particular role among the set $-\ell \leq m \leq \ell$:

(i) For $Y_{\ell\ell}(\vartheta, \varphi)$ the value of the $z$ component $m = \ell$ is closest to the expectation value of the modulus $\sqrt{\ell(\ell+1)}$.

(ii) The uncertainties of the three angular-momentum components are smallest.

For these reasons we shall take the eigenfunction $Y_{\ell\ell}(\vartheta, \varphi)$ as the quantum-mechanical state corresponding most closely to the classical vector

$$\mathbf{L} = \langle L_z \rangle_{\ell\ell} \mathbf{e}_z = \hbar\ell\mathbf{e}_z$$

of angular momentum. Here $\mathbf{e}_z$ is a vector of unit length pointing in the $z$ direction.

## 10.5 Interpretation of the Eigenfunctions of Angular Momentum

In Section 10.3 we found the eigenfunctions of angular momentum to be completely specified by the eigenvalue $\ell(\ell+1)\hbar^2$ of the square $\hat{\mathbf{L}}^2$ of the angular-momentum vector operator $\hat{\mathbf{L}} = (\hat{L}_x, \hat{L}_y, \hat{L}_z)$ and by the eigenvalue $m\hbar$ of $\hat{L}_z$, the $z$ component of $\hat{\mathbf{L}}$. The choice of this particular coordinate frame is of no special significance. In order to distinguish this frame from others we shall indicate the $z$ direction $\mathbf{e}_z = (0, 0, 1)$ explicitly in the corresponding spherical harmonics by replacing the notation,

$$Y_{\ell m}(\vartheta, \phi) \rightarrow Y_{\ell m}(\vartheta, \phi, \mathbf{e}_z) \quad .$$

We choose another direction denoted by the unit vector $\mathbf{n} = (n_x, n_y, n_z)$ as the $z$ direction of another coordinate system. In this frame the polar angle is denoted by $\vartheta'$ and the azimuth by $\phi'$. The eigenfunctions of $\hat{\mathbf{L}}^2$ and $\hat{L}'_z = \mathbf{n} \cdot \hat{\mathbf{L}} = n_x\hat{L}_x + n_y\hat{L}_y + n_z\hat{L}_z$ are then $Y_{\ell m}(\vartheta', \phi', \mathbf{n})$. We shall denote the polar and azimuthal angles of the direction $\mathbf{n}$ in the original coordinate system by $\Theta$ and $\Phi$,

$$\mathbf{n} = (\sin\Theta \cos\Phi, \sin\Theta \sin\Phi, \cos\Theta) \quad .$$

In Section 10.4 we found that the eigenfunction $Y_{\ell\ell}(\vartheta, \phi, \mathbf{n})$ is the quantum-mechanical state which most closely resembles the classical angular-momentum vector

$$\mathbf{L} = \langle L'_z \rangle_{\ell\ell} \mathbf{n} = \hbar\ell\mathbf{n} \quad .$$

We now analyze the wave function $Y_{\ell m}(\vartheta,\phi,\mathbf{e}_z)$ of total angular momentum $\ell\hbar$ and $z$ component $m\hbar$ by the wave function $Y_{\ell\ell}(\vartheta,\phi,\mathbf{n})$. At this moment the reader is encouraged to turn to the discussion in Appendix C about the analysis of a wave function by another wave function. In the present case the analyzing amplitude is

$$
\begin{aligned}
a &= N \int_0^{2\pi} \int_{-1}^{1} Y_{\ell\ell}^*(\vartheta,\phi,\mathbf{n}) Y_{\ell m}(\vartheta,\phi,\mathbf{e}_z) \, d\cos\vartheta \, d\phi \\
&= N D_{m\ell}^{(\ell)}(\Phi,\Theta,0) \quad .
\end{aligned}
$$

These functions and the notation $D_{m\ell}^{(\ell)}$ were introduced by Eugene P. Wigner and are known in the literature as *Wigner functions*. The normalization constant $N$ will be determined in the sequel.

We consider the absolute square of the analyzing amplitude

$$
|a|^2 = f_{\ell m}(\Theta,\Phi) = |N|^2 \left| D_{m\ell}^{(\ell)}(\Phi,\Theta,0) \right|^2 = |N|^2 \left[ d_{m\ell}^{(\ell)}(\Theta) \right]^2 \quad .
$$

Here, $d_{m\ell}^{(\ell)}(\Theta)$ is also referred to as a *Wigner function* in the literature. It has the explicit representation

$$
d_{m\ell}^{(\ell)}(\Theta) = \sqrt{\frac{(2\ell)!}{(\ell+m)!(\ell-m)!}} \left( \cos\frac{\Theta}{2} \right)^{\ell+m} \left( \sin\frac{\Theta}{2} \right)^{\ell-m} \quad .
$$

These functions are shown in Figures 10.7 and 10.8. In our discussion in Appendix C we found that $|a|^2$ is a probability density describing the result of the measurement performed on a physical state described by one wave function with a detector characterized by another wave function. In the particular case at hand the physical state is described by the spherical harmonic $Y_{\ell m}(\vartheta,\phi,\mathbf{e}_z)$. The detector with which we want to measure the direction $\mathbf{n}$ of angular momentum is characterized by the spherical harmonic $Y_{\ell\ell}(\vartheta,\phi,\mathbf{n})$.

By an appropriate choice of the normalization constant

$$
|N|^2 = \frac{(2\ell+1)(\ell+1)}{4\pi\ell} \quad \text{for} \quad \ell = 1,2,3,\dots,
$$

the quantity

$$
f_{\ell\ell}(\Theta,\Phi) = \frac{(2\ell+1)(\ell+1)}{4\pi\ell} \left[ d_{\ell\ell}^{(\ell)}(\Theta) \right]^2
$$

is turned into a *directional distribution* fulfilling the normalization condition

$$
\langle \mathbf{n} \rangle_{\ell\ell} = \int_0^{2\pi} \int_{-1}^{1} \mathbf{n}(\Theta,\Phi) f_{\ell\ell}(\Theta,\Phi) \, d\cos\Theta \, d\Phi = \mathbf{e}_z \quad .
$$

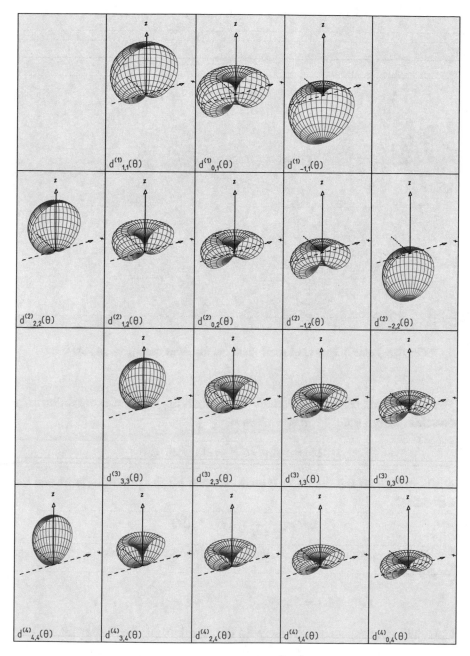

**Fig. 10.7. Polar diagrams of the Wigner functions** $d^{(\ell)}_{m\ell}(\Theta)$**. Lines 1 and 2 show the functions for** $\ell = 1, 2$ **and** $m = \ell, \ell-1, \ldots, -\ell$**. Lines 3 and 4 give them for** $\ell = 3, 4$ **and** $m = \ell, \ell-1, \ldots, 0$**. The functions are independent of** $\Phi$**. They have large values only in a restricted region of** $\Theta$**. That region is near** $\Theta = 0$ **for** $m = \ell$ **and decreases in regular steps via** $\Theta = \pi/2$ **for** $m = 0$ **to** $\Theta = \pi$ **for** $m = -\ell$**.**

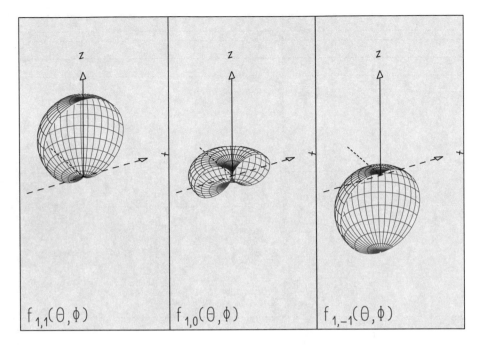

**Fig. 10.8. Polar diagrams of the directional distributions $f_{\ell m}(\Theta, \Phi)$ for $\ell = 1$.**

The distribution $f_{\ell\ell}(\Theta, \Phi)$ defines the probability for the observation of a direction within the *solid-angle element*

$$\mathrm{d}\Omega = \mathrm{d}\cos\Theta\,\mathrm{d}\Phi = \sin\Theta\,\mathrm{d}\Theta\,\mathrm{d}\Phi$$

positioned about the direction $\mathbf{n}$ characterized by the polar angle $\Theta$ and the azimuth $\Phi$,

$$\mathrm{d}P = \frac{\ell}{\ell+1} f_{\ell\ell}(\Theta, \Phi)\mathrm{d}\Omega \quad .$$

We now introduce the "classical" angular-momentum vector with integer length $\ell\hbar$:

$$\mathbf{L}_\ell(\Theta, \Phi) = \hbar\ell\mathbf{n}(\Theta, \Phi) \quad , \qquad \ell = 1,2,3,\ldots \quad .$$

Calculating its expectation value with the distribution $f_{\ell\ell}(\Theta, \Phi)$ yields

$$\langle\mathbf{L}_\ell\rangle_{\ell\ell} = \int \mathbf{L}_\ell(\Theta, \Phi) f_{\ell\ell}(\Theta, \Phi)\mathrm{d}\Omega = \ell\hbar\mathbf{e}_z \quad .$$

Normalizing the functions $f_{\ell m}(\Theta, \Phi)$ with the same constant $|N|^2$,

$$f_{\ell m}(\Theta, \Phi) = \frac{(2\ell+1)(\ell+1)}{4\pi\ell}\left[d_{m\ell}^{(\ell)}(\Theta)\right]^2 \quad ,$$

we find that

$$\langle \mathbf{L}_\ell \rangle_{\ell m} = \int \mathbf{L}_\ell(\Theta, \Phi) f_{\ell m}(\Theta, \Phi) \, d\Omega = m\hbar \mathbf{e}_z \quad .$$

Therefore, we can also interpret the functions $f_{\ell m}(\Theta, \Phi)$ as a measure for angular-momentum probability densities. The probability of detecting in the state $Y_{\ell m}$ the angular-momentum vector $\mathbf{L}_\ell$ within the solid angle $d\Omega$ positioned about the direction characterized by $\Theta$ and $\Phi$ is

$$dP = \frac{\ell}{\ell + 1} f_{\ell m}(\Theta, \Phi) \, d\Omega \quad .$$

The "classical" average values $\langle \mathbf{L}_\ell \rangle_{\ell m} = m\hbar \mathbf{e}_z$ are the same as the quantum-mechanically calculated expectation values,

$$\langle \mathbf{L} \rangle_{\ell m} = \int Y_{\ell m}^*(\vartheta, \phi, \mathbf{e}_z) \hat{\mathbf{L}} Y_{\ell m}(\vartheta, \phi, \mathbf{e}_z) \, d\Omega = m\hbar \mathbf{e}_z \quad ,$$

as obtained in components in Section 10.4. Also the expectation value

$$\langle \mathbf{L}^2 \rangle_{\ell m} = \ell(\ell + 1)\hbar^2$$

of the square of the angular-momentum operator $\hat{\mathbf{L}}$ is reproduced by the distribution $f_{\ell m}(\Theta, \Phi)$ of angular momentum,

$$\int \mathbf{L}_\ell^2(\Theta, \Phi) f_{\ell m}(\Theta, \Phi) \, d\Omega = \ell(\ell + 1)\hbar^2 \quad .$$

We may now ask what angle $\Theta$ the angular-momentum vector $\mathbf{L}_\ell$ forms with the $z$ axis in the state $Y_{\ell m}$. As a first step we form the marginal distribution with respect to $\cos \Theta$ of the distribution $f_{\ell m}(\Theta, \Phi)$,

$$f_{\ell m \cos \Theta}(\cos \Theta) = \int_0^{2\pi} f_{\ell m}(\Theta, \Phi) \, d\Phi = 2\pi f_{\ell m}(\Theta, 0) \quad .$$

We turn this into an *angular distribution* in $\Theta$ by using the transformation

$$f_{\ell m \Theta}(\Theta) = f_{\ell m \cos \Theta}(\cos \Theta) \left| \frac{d \cos \Theta}{d\Theta} \right| = 2\pi f_{\ell m}(\Theta, 0) \sin \Theta \quad .$$

Polar diagrams of these densities are shown in Figure 10.9.

The figures show clearly that the distribution $f_{\ell m \Theta}(\Theta)$ is concentrated about a maximum value at $\Theta_{\ell m}$. With the explicit form of the $d_{m\ell}^{(\ell)}(\Theta)$ given before, the angle $\Theta_{\ell m}$ can be calculated from the above formula in the form of

$$\cos \Theta_{\ell m} = \frac{m}{\ell + \frac{1}{2}} \quad .$$

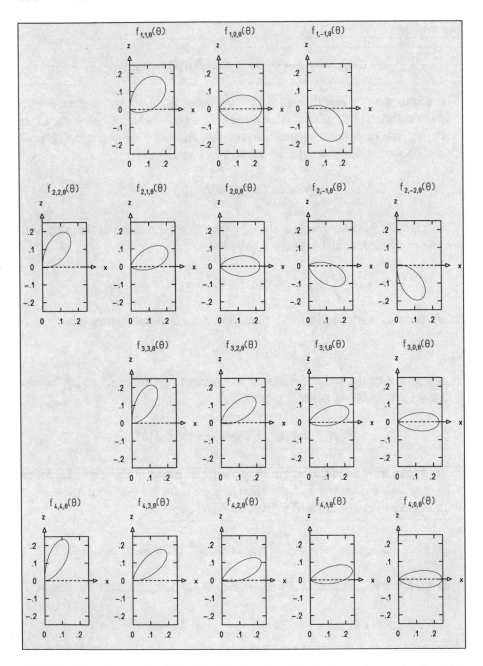

**Fig. 10.9. Polar diagrams of the distribution** $f_{\ell m \Theta}(\Theta)$ **for the polar angle** $\Theta$ **of the direction of angular momentum. Lines 1 and 2 show the functions for** $\ell = 1, 2$ **and** $m = \ell, \ell - 1, \dots, -\ell$. **Lines 3 and 4 give them for** $\ell = 3, 4$ **and** $m = \ell, \ell - 1, \dots, 0$.

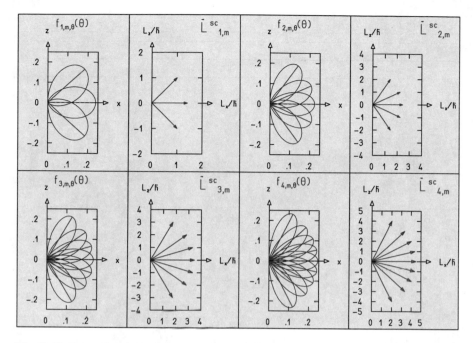

**Fig. 10.10. Polar diagrams of the angular distribution** $f_{\ell m \Theta}(\Theta)$. **The top left plot contains all polar diagrams for** $\ell = 1$. **In addition each polar diagram contains a line from the origin to the point** $f_{\ell m \Theta}(\Theta_{\ell m})$, **where** $\Theta_{\ell m}$ **is the angle for which** $f_{\ell m \Theta}$ **has its maximum. The second plot from the left in the top row shows the semiclassical angular-momentum vectors** $\mathbf{L}_{1,m}^{sc}$ **which have polar angles similar to** $\Theta_{\ell m}$. **Pairs of plots of** $f_{\ell m \Theta}(\Theta)$ **and** $\mathbf{L}_{\ell m}^{sc}$ **are also shown for** $\ell = 2, 3,$ **and** 4.

We compare this with the angles of the *semiclassical vector model* as introduced by Arnold Sommerfeld before the advent of quantum mechanics to account for the quantization of angular momentum. He postulated the angular-momentum vector in atomic physics to be of length $\sqrt{\ell(\ell+1)}\hbar$, and $z$ component $m\hbar$ as shown in Figure 10.10. The angles $\Theta_{\ell m}^{sc}$ of the various semiclassical vectors of $z$ component $m\hbar$ are determined by

$$\cos\Theta_{\ell m}^{sc} = \frac{m}{\sqrt{\ell(\ell+1)}} \quad ,$$

which for $\ell \gg 1$ approaches the above formula for $\Theta_{\ell m}$ if one neglects terms of order $(1/\ell)^2$ and higher in the denominator. Also for small values of $\ell \approx 1$ the angles $\Theta_{\ell m}$ and $\Theta_{\ell m}^{sc}$ do not differ very much; for $\ell = 1$, $m = 1$, we find $\Theta_{\ell m}^{sc} = 45°$ as compared to $\Theta_{\ell m} \approx 48°$.

To conclude this section we turn back to the directional distribution $f_{\ell m}(\Theta, \Phi)$. In Figure 10.11 we show polar diagrams of $f_{\ell \ell}(\Theta, \Phi)$ for increasing values of $\ell$. The distributions become more and more concentrated around

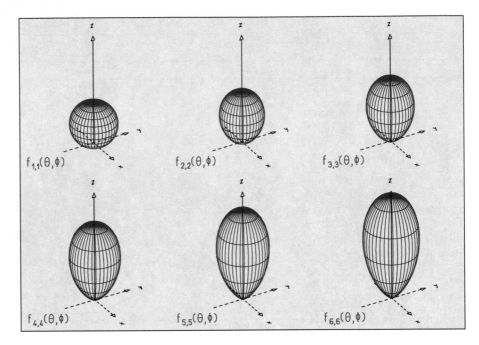

**Fig. 10.11. Polar diagrams of the directional distribution** $f_{\ell\ell}(\Theta,\Phi)$ **for** $\ell = 1,2,\ldots,6$.

the $z$ direction. In the classical limit $\ell \to \infty$ the distribution is different from zero only in the $z$ direction.

## 10.6 Schrödinger Equation

As we did for the one-dimensional harmonic wave in Section 3.2, let us compare time and spatial derivatives of the three-dimensional harmonic wave $\varphi_\mathbf{p}(\mathbf{r},t)$, which was introduced in Section 10.1. They are

$$i\hbar\frac{\partial}{\partial t}\psi_\mathbf{p}(\mathbf{r},t) = E\psi_\mathbf{p}(\mathbf{r},t) \quad,$$

$$-\frac{\hbar^2}{2M}\nabla^2\psi_\mathbf{p}(\mathbf{r},t) = \frac{\mathbf{p}^2}{2M}\psi_\mathbf{p}(\mathbf{r},t) \quad.$$

Here $M$ is the mass of the particle.[1] The *Laplace operator* $\nabla^2$ is simply the sum of the three second-order derivatives with respect to the coordinates:

$$\nabla^2 = \frac{\partial^2}{\partial x^2} + \frac{\partial^2}{\partial y^2} + \frac{\partial^2}{\partial z^2} \quad.$$

---

[1]From Chapter 10 onward we denote the mass of a particle by the capital letter $M$. This is done to avoid confusion with magnetic quantum number $m$.

Making use of the relation $E = \mathbf{p}^2/(2M)$ between energy $E$, momentum $\mathbf{p}$, and mass $M$ of a free particle, we obtain the Schrödinger equation for three-dimensional unaccelerated motion,

$$i\hbar \frac{\partial}{\partial t} \psi_\mathbf{p}(\mathbf{r},t) = -\frac{\hbar^2}{2M} \nabla^2 \psi_\mathbf{p}(\mathbf{r},t) \quad .$$

We may consider the operator on the right-hand side of this equation as the operator of kinetic energy,

$$
\begin{aligned}
T &= \frac{\hat{\mathbf{p}}^2}{2M} = \frac{1}{2M}(\hat{p}_x^2 + \hat{p}_y^2 + \hat{p}_z^2) \\
&= \frac{1}{2M}\left(-\hbar^2 \frac{\partial^2}{\partial x^2} - \hbar^2 \frac{\partial^2}{\partial y^2} - \hbar^2 \frac{\partial^2}{\partial z^2}\right) \\
&= -\frac{\hbar^2}{2M} \nabla^2 \quad .
\end{aligned}
$$

Thus the Schrödinger equation for three-dimensional free motion has the simple form

$$i\hbar \frac{\partial}{\partial t} \psi_\mathbf{p}(\mathbf{r},t) = T \psi_\mathbf{p}(\mathbf{r},t) \quad .$$

The equation can be extended to motion in a force field represented by a potential energy $V(\mathbf{r})$ by substituting for the operator of kinetic energy $T$ the Hamiltonian operator of total energy,

$$H = T + V \quad .$$

The Schrödinger equation for motion under the influence of a force therefore reads

$$i\hbar \frac{\partial}{\partial t} \psi_\mathbf{p}(\mathbf{r},t) = H \psi_\mathbf{p}(\mathbf{r},t) = \left[-\frac{\hbar^2}{2M} \nabla^2 + V(\mathbf{r})\right] \psi_\mathbf{p}(\mathbf{r},t) \quad .$$

With the *ansatz*

$$\psi_\mathbf{p}(\mathbf{r},t) = \exp\left[-\frac{i}{\hbar} E t\right] \varphi_E(\mathbf{r}) \quad ,$$

which factors the wave function $\psi_\mathbf{p}(\mathbf{r},t)$ into a time-dependent exponential and the time-independent, stationary wave function $\varphi_E(\mathbf{r})$, we obtain the stationary Schrödinger equation

$$\left[-\frac{\hbar^2}{2M} \nabla^2 + V(\mathbf{r})\right] \varphi_E(\mathbf{r}) = E \varphi_E(\mathbf{r}) \quad .$$

## 10.7 Solution of the Schrödinger Equation of Free Motion

Besides the solutions $\psi_{\mathbf{p}}(\mathbf{r},t)$ of the free Schrödinger equation, which represent harmonic plane waves with momentum $\mathbf{p}$, there are equivalent solutions which are determined by the quantum numbers $\ell$ and $m$ of angular momentum and energy $E$. To find these solutions, we express the Laplace operator in polar coordinates $r$, $\vartheta$, and $\phi$:

$$\nabla^2\varphi(r) = \frac{1}{r}\frac{\partial^2}{\partial r^2}r\varphi(r) - \frac{1}{r^2}\frac{1}{\hbar^2}\hat{\mathbf{L}}^2\varphi(r) \quad .$$

Since the operator $\hat{\mathbf{L}}^2$ of the square of angular momentum, as discussed in Section 10.3, depends only on $\vartheta$ and $\phi$, we now solve the Schrödinger equation using an *ansatz*,

$$\varphi_{E\ell m}(\mathbf{r}) = R(r)Y_{\ell m}(\vartheta,\phi) \quad ,$$

which is a product of two functions. The first function $R(r)$ depends only on the radial coordinate. The second function is the spherical harmonic $Y_{\ell m}(\vartheta,\phi)$, which was recognized in Section 10.3 as the eigenfunction for $\hat{\mathbf{L}}^2$. We obtain

$$
\begin{aligned}
-\frac{\hbar^2}{2M}\nabla^2\varphi_{E\ell m}(\mathbf{r}) &= -\frac{\hbar^2}{2M}\left[\frac{1}{r}\frac{\partial^2}{\partial r^2}rR(r) - \frac{\ell(\ell+1)}{r^2}R(r)\right]Y_{\ell m}(\vartheta,\phi) \\
&= ER(r)Y_{\ell m}(\vartheta,\phi)
\end{aligned}
$$

and conclude that

$$-\frac{\hbar^2}{2M}\left[\frac{1}{r}\frac{\partial^2}{\partial r^2}r - \frac{\ell(\ell+1)}{r^2}\right]R_{E\ell}(r) = ER_{E\ell}(r)$$

is the eigenvalue equation for the *radial wave function* $R_{E\ell}(r)$ for positive values of $r$. Here we explicitly indicate the dependence of the radial wave function on energy $E$ and total angular momentum $\ell$. We call $\varphi_{E\ell m}(\mathbf{r}) = R_{E\ell}(r)Y_{\ell m}(\vartheta,\phi)$ a *partial wave* of angular momentum $\ell$ and $z$ component $m$. The solutions of this "free radial Schrödinger equation" are discussed in some detail in the next section.

## 10.8 Spherical Bessel Functions

Let us consider the solutions of the linear differential equation that depends on the integer parameter $\ell$,

$$\left[\frac{1}{\rho}\frac{d^2}{d\rho^2}\rho - \frac{\ell(\ell+1)}{\rho^2} + 1\right]f_\ell(\rho) = 0 \quad .$$

For $\rho = kr$, $k = (1/\hbar)\sqrt{2ME}$, it is equivalent to the free radial Schrödinger equation.

The complex solutions of this linear differential equation are the *spherical Hankel functions* of the first $(+)$ and second $(-)$ kind,

$$h_\ell^{(\pm)}(\rho) = C_\ell^\pm \frac{e^{\pm i\rho}}{\rho} \quad ,$$

where the complex coefficients $C_\ell$ are polynomials of $\rho^{-1}$ of the form

$$C_\ell^\pm = (\mp i)^\ell \sum_{s=0}^{\ell} \frac{1}{2^s s!}\frac{(\ell+s)!}{(\ell-s)!}(\mp i\rho)^{-s} \quad .$$

The first few of the Hankel functions are

$$h_0^{(\pm)} = \frac{e^{\pm i\rho}}{\rho} \quad , \qquad h_1^{(\pm)} = \left(\mp i + \frac{1}{\rho}\right)\frac{e^{\pm i\rho}}{\rho} \quad .$$

An equivalent set of solutions are the *spherical Bessel functions*, which are simply the linear combinations

$$j_\ell(\rho) = \frac{1}{2i}\left[h_\ell^{(+)}(\rho) - h_\ell^{(-)}(\rho)\right] \quad .$$

The *spherical Neumann functions*

$$n_\ell(\rho) = \frac{1}{2}\left[h_\ell^{(+)}(\rho) + h_\ell^{(-)}(\rho)\right]$$

are also solutions of the linear differential equation. In terms of the spherical Bessel and Neumann functions, the spherical Hankel functions can be expressed as

$$h_\ell^{(\pm)}(\rho) = n_\ell(\rho) \pm i j_\ell(\rho) \quad .$$

The first few spherical Bessel and Neumann functions are

$$j_0(\rho) = \frac{\sin\rho}{\rho} \quad , \qquad j_1(\rho) = \frac{\sin\rho}{\rho^2} - \frac{\cos\rho}{\rho} \quad ,$$

$$n_0(\rho) = \frac{\cos\rho}{\rho} \quad , \qquad n_1(\rho) = \frac{\cos\rho}{\rho^2} + \frac{\sin\rho}{\rho} \quad .$$

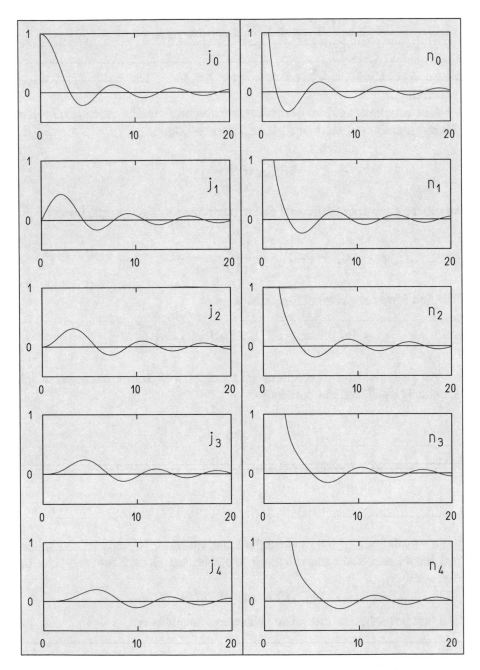

**Fig. 10.12. Spherical Bessel functions** $j_\ell(\rho)$ **and spherical Neumann functions** $n_\ell(\rho)$ **for** $\ell = 0, 1, \ldots, 4.$

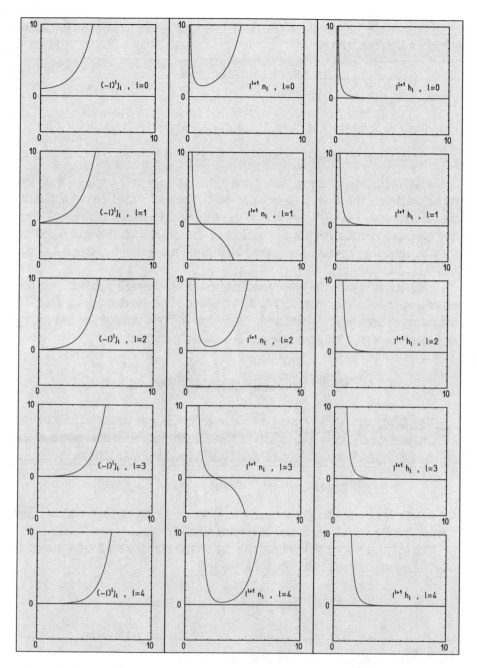

**Fig. 10.13. For purely imaginary arguments** $i\eta$, $\eta$ **real, the spherical Bessel functions** $j_\ell$, **the spherical Neumann functions** $n_\ell$, **and the spherical Hankel functions** $h_\ell^{(+)}$ **are either purely real or purely imaginary. The functions shown, that is,** $(-i)^\ell j_\ell(\eta)$, $i^{\ell+1} n_\ell(\eta)$, **and** $i^{\ell+1} h_\ell^{(+)}(\eta)$, **are purely real.**

The behavior of the spherical Bessel and Neumann functions for small values of the argument is

$$j_\ell(\rho) \sim \rho^\ell \quad , \qquad n_\ell(\rho) \sim \rho^{-(\ell+1)} \quad ,$$

and for large $\rho$,

$$j_\ell(\rho) \quad \xrightarrow[\rho\to\infty]{} \quad \tfrac{1}{\rho}\sin(\rho - \tfrac{1}{2}\ell\pi) \quad ,$$
$$n_\ell(\rho) \quad \xrightarrow[\rho\to\infty]{} \quad \tfrac{1}{\rho}\cos(\rho - \tfrac{1}{2}\ell\pi) \quad .$$

Since the spherical Neumann functions $n_\ell(\rho)$ diverge at the origin, only the spherical Bessel functions $j_\ell(\rho)$ are physical solutions of the free radial Schrödinger equation. The $n_\ell(\rho)$ as well as the spherical Hankel functions $h_\ell^{(\pm)}(\rho)$, however, are needed for the discussion of the radial Schrödinger equation for a square-well potential. Figure 10.12 plots the $j_\ell(\rho)$ and the $n_\ell(\rho)$ for $\ell = 0, \ldots, 4$.

In connection with the wave functions for a square-well potential, we encounter negative energies $E_i$ that make values of wave number $k_i = \sqrt{2mE_i}/\hbar$ imaginary. Therefore the functions $j_\ell$, $n_\ell$, and $h_\ell^{(+)}$ are needed for imaginary arguments $\rho = i\eta$. Using the original definition, we can write

$$h_\ell^{(\pm)}(i\eta) = (\mp)^{\ell\pm1} \sum_{s=0}^{\ell} \frac{1}{2^s s!} \frac{(\ell+s)!}{(\ell-s)!} (\pm\eta)^{-s} \frac{e^{\mp\eta}}{\eta} \quad .$$

The $j_\ell(i\eta)$ and $n_\ell(i\eta)$ are again given by the linear combinations of the $h_\ell^{(+)}(i\eta)$ and the $h_\ell^{(-)}(i\eta)$. The values of these functions for such arguments are either real or purely imaginary. Figure 10.13 presents the functions

$$(-i)^\ell j_\ell(i\eta) \quad , \qquad i^{\ell+1} n_\ell(i\eta) \quad , \qquad i^{\ell+1} h_\ell^{(+)}(i\eta)$$

for $\ell = 0, 1, \ldots, 4$. The powers of i in front of $j_\ell(i\eta)$, $n_\ell(i\eta)$, and $h_\ell^{(+)}(i\eta)$ ensure that the functions plotted in Figure 10.13 are real.

The $h_\ell^{(+)}(i\eta)$ play a role in describing bound states outside the potential well. Their asymptotic behavior for large $\eta$ is

$$i^{\ell+1} h_\ell^{(+)}(i\eta) \sim \frac{e^{-\eta}}{\eta} \quad , \qquad \eta \to \infty \quad .$$

## 10.9 Harmonic Plane Wave in Angular-Momentum Representation

The spherical waves $j_\ell(kr)Y_{\ell m}(\vartheta, \phi)$, like the harmonic plane waves, form a complete set of functions which can also be used for constructing wave packets by superposition or for decomposing stationary solutions into spherical

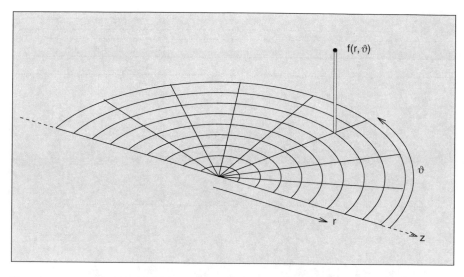

**Fig. 10.14. The polar coordinate system used throughout the book for functions of the type $f = f(r, \vartheta)$. The admissible range of variables, $0 \leq r < \infty, 0 \leq \vartheta \leq \pi$, corresponds to a half-plane. Here a half-circle around the origin, $r = 0$, is viewed perspectively from a point outside the half-plane. The polar angle $\vartheta$ is measured against the $z$ axis, which points to the lower right. Lines of constant $\vartheta$ are straight lines beginning at the origin. Lines of constant $r$ are half-circles. Using the direction perpendicular to the half-plane to define an $f$ coordinate, we can represent a function $f(r, \vartheta)$ as a surface in $r, \vartheta, f$ space. Figures 10.15 and 10.16 show lines of constant $r$ and constant $\vartheta$ on this surface.**

waves. In particular, we decompose the stationary harmonic plane wave into *partial waves*,

$$e^{i\mathbf{k}\cdot\mathbf{r}} = e^{ikz} = e^{ikr\cos\vartheta} = \sum_{\ell=0}^{\infty} (2\ell + 1)i^{\ell} j_{\ell}(kr) P_{\ell}(\cos\vartheta) \quad ,$$

where the $z$ axis was chosen to be parallel to $\mathbf{k}$ and $\vartheta$ is therefore the angle between $\mathbf{k}$ and $\mathbf{r}$. Since the left-hand side of this relation does not depend on the azimuth $\phi$, only the spherical harmonic function $Y_{\ell 0} = \sqrt{(2\ell + 1)/4\pi}\, P_{\ell}$ occurs in the sum.

Figures 10.15 and 10.16 illustrate this decomposition. The polar coordinates $r$ and $\vartheta$ are used to plot functions over the $r, \vartheta$ half-plane. The polar coordinate system used throughout the book for functions of the type $f = f(r, \vartheta)$ is explained in Figure 10.14.

In the top right corner of Figure 10.15, the function $\cos(kz) = \text{Re}\{e^{ikz}\}$ is presented. The left column contains the functions

$$(2\ell + 1)i^{\ell} j_{\ell}(kr) P_{\ell}(\cos\vartheta) \quad , \qquad \ell = 0, 2, \dots, 8 \quad ,$$

which are the first few real terms in this decomposition. The right column shows the sums of the first two terms, three terms, and so on. In the neigh-

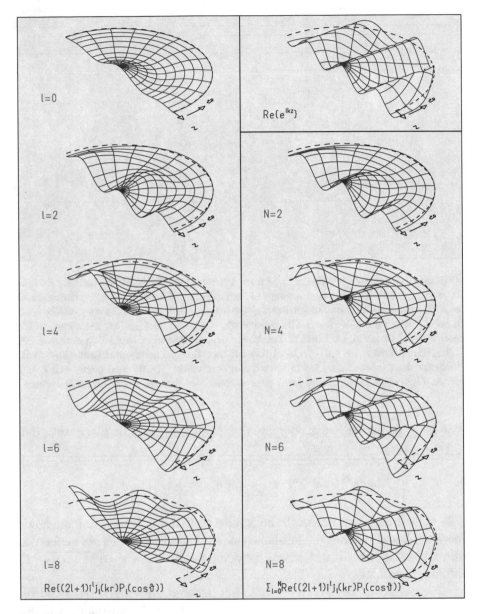

**Fig. 10.15. Decomposition of a plane wave into spherical waves. The real part** $\mathrm{Re}\{e^{ikz}\} = \cos(kz)$ **of a plane wave is shown in the top right corner. The left column contains the terms of the decomposition that are purely real. The right column contains the sums of the first two terms** ($N = 2$), **three terms** ($N = 4$), **and so on, of the left column.**

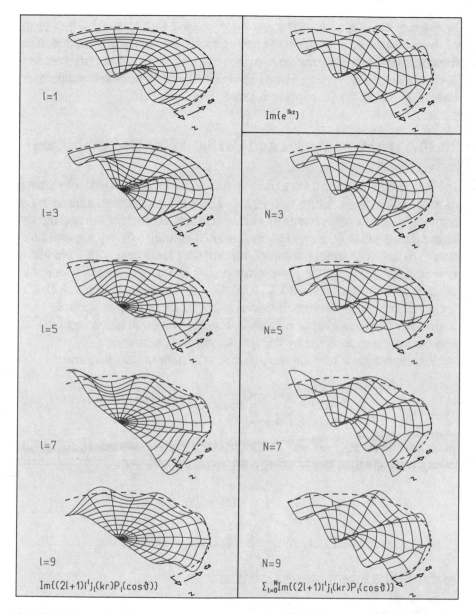

**Fig. 10.16. Decomposition of a plane wave into spherical waves. The imaginary part** $\mathrm{Im}\{e^{ikz}\} = \sin(kz)$ **of a plane wave is shown in the top right corner. The left column contains the terms of the decomposition that are purely imaginary. The right column contains the sums of the first two terms** ($N = 3$), **three terms** ($N = 5$), **and so on, of the left column.**

borhood of the origin, the plane wave is described well by the first few terms of the sum. Farther away from the origin, more terms have to be added. Near the origin the first few terms are adequate because the functions $j_\ell(\rho)$ are suppressed there for increasing $\ell$ (see Figure 10.12). A similar illustration for the imaginary part of the plane wave is given in Figure 10.16.

## 10.10 Free Wave Packet and Partial-Wave Decomposition

In Section 10.1 we discussed a three-dimensional unaccelerated wave packet moving with group velocity $\mathbf{v}_0 = \mathbf{p}_0/M$. The wave packet was represented as a superposition of plane waves that are eigenfunctions for the momentum operator. The details of the superposition were determined by the spectral function $f(\mathbf{p})$ which specifies the contribution of the plane wave with wave vector $\mathbf{k} = \mathbf{p}/\hbar$. Analogously, the same wave packet can be understood as a superposition of the eigenfunctions $Y_{\ell m}(\vartheta, \phi)$ for angular momentum multiplied by appropriately chosen weight functions $a_{\ell m}(r, 0)$ for the radius variable $r$ and at time $t = 0$. In this kind of representation, the weight function regulates the relative weight contributed by the various angular momenta.

The representation of the wave packet at initial time has the form

$$\psi(\mathbf{r}, 0) = \sum_{\ell=0}^{\infty} \sum_{m=-\ell}^{\ell} a_{\ell m}(r, 0) Y_{\ell m}(\vartheta, \phi) \quad .$$

In an additional step we may decompose the radial functions $a_{\ell m}(r, 0)$ into purely wave number, that is, energy-dependent, coefficients,

$$b_{\ell m}(k) = \int_0^{\infty} j_\ell(kr) a_{\ell m}(r, 0) r^2 \, dr \quad ,$$

$$a_{\ell m}(r, 0) = \frac{2}{\pi} \int_0^{\infty} b_{\ell m}(k) j_\ell(kr) k^2 \, dk \quad ,$$

so that the free wave packet at $t = 0$ is now

$$\psi(\mathbf{r}, 0) = \frac{2}{\pi} \sum_{\ell=0}^{\infty} \sum_{m=-\ell}^{\ell} \int_0^{\infty} b_{\ell m}(k) j_\ell(kr) Y_{\ell m}(\vartheta, \phi) k^2 \, dk \quad .$$

In this decomposition of the free wave packet in terms of the eigenfunctions of the free Schrödinger equation for the eigenvalues $E$, $\ell$, and $m$, the functions $b_{\ell m}(k)$ play the role of spectral coefficients for angular momentum and spectral functions for energy $E = \hbar^2 k^2/2M$. In Section 10.1 the spectral function $f(\mathbf{p})$ played a similar role in decomposition of the wave packet in terms of eigenfunctions of the three momentum components.

The moving wave packet is described by the time-dependent wave function $\psi(\mathbf{r},t)$ which is obtained from the initial wave function by taking into account the time-dependent phase factor $\exp(-iEt/\hbar)$, that is,

$$\psi(\mathbf{r},t) = \frac{2}{\pi} \sum_{\ell=0}^{\infty} \sum_{m=-\ell}^{\ell} \int_0^{\infty} b_{\ell m}(k) \exp\left[-\frac{i}{\hbar}Et\right] j_\ell(kr) Y_{\ell m}(\vartheta,\phi) k^2 \, dk \quad .$$

The angular-momentum content of the free wave packet is given by the spectral coefficients $b_{\ell m}(k)$. They are time independent because angular momentum is conserved.

If we ask for the contribution having angular-momentum quantum number $\ell$ and magnetic quantum number $m$ irrespective of wave number $k$, we have to integrate the probabilities $b_{\ell m}^*(k)b_{\ell m}(k)k^2 \, dk$ over all wave numbers:

$$W_{\ell m} = \frac{2}{\pi} \int_0^{\infty} b_{\ell m}^*(k)b_{\ell m}(k)k^2 \, dk \quad .$$

The probabilities $W_{\ell m}$ fulfill the normalization condition

$$\sum_{\ell=0}^{\infty} \sum_{m=-\ell}^{\ell} W_{\ell m} = 1 \quad .$$

As an example, we consider the wave packet shown in Figure 10.17a. Its center moves with constant velocity in the negative $x$ direction keeping a constant distance $b$ from the $x$ axis. That is, it behaves like a classical particle with time-dependent position vector

$$\mathbf{r}(t) = (x(t),b,0) \quad ,$$

and constant momentum vector

$$\mathbf{p}(t) = (-p,0,0) \quad .$$

The angular-momentum vector of the classical particle,

$$\mathbf{L} = \mathbf{r} \times \mathbf{p} = (0,0,bp) \quad ,$$

is independent of time and is oriented along the $z$ direction. The absolute value of the angular momentum is

$$L = |\mathbf{L}| = bp \quad .$$

We now consider a particle of constant momentum $\mathbf{p}$ which travels along an arbitrary straight line. The shortest distance of this line from the origin is

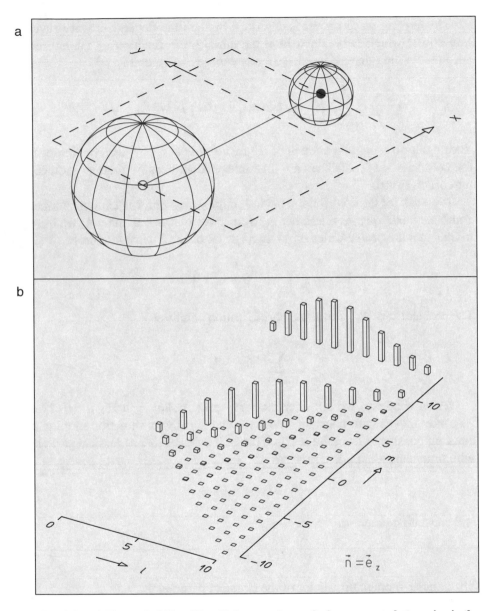

**Fig. 10.17. (a)** The probability ellipsoid, here a sphere, of a free wave packet moving in the $x, y$ plane antiparallel to the $x$ axis, shown at two moments of time. The dispersion of the wave packet is apparent through the growth of the sphere with time. **(b)** Decomposition of the wave packet shown in part a into angular-momentum states. The height of the column drawn at point $(\ell, m)$ is proportional to the probability $W_{\ell m}$ that the particle, which is described by the wave packet, has angular-momentum quantum number $\ell$ and quantum number $m$ for the component of angular momentum along the quantization axis n. In this figure n was chosen to be the $z$ axis. Also shown, on the upper margin, are the probabilities $W_{\ell}$, that the particle possesses quantum number $\ell$ irrespective of the value of $m$.

called the *impact parameter b*. Obviously, for reasons of symmetry, the absolute value of angular momentum for this particle is again $L = bp$.

Let us now study the probabilities $W_{\ell m}$ for the wave packet of Figure 10.17a. We quantize the angular momentum in the $z$ direction, that is, we use the eigenfunctions of $\hat{\mathbf{L}}^2$ and $\hat{L}_z$ in decomposing the wave packet. In Figure 10.17b the probabilities $W_{\ell m}$ are plotted for various values of $\ell$ and $m$. In the graphical representation each probability is proportional to the height of the column sitting on top of point $(\ell, m)$ in the coordinate grid. Obviously, the probabilities can be different from zero only in points lying within a sector between two straight lines, for which $m = \ell$ and $m = -\ell$. We note that, in contrast to the classical point particle, various angular momenta contribute to the wave packet. In fact, for the quantization axis chosen, the probabilities at points $\ell = m$ are by far the largest for every $\ell$. This is not surprising since the angular momentum of the corresponding particle has only a $z$ component. Nevertheless, values $m < \ell$ also contribute. The contributions $W_{\ell m}$ for $m = \ell - 1, \ell - 3, \dots$ vanish. Because of the mirror symmetry of the wave packet with respect to the $x, y$ plane, functions $Y_{\ell m}(\vartheta, \phi)$ with $m = \ell - 1, \ell - 3, \dots$ do not contribute. They are antisymmetric in $\vartheta$ with respect to point $\vartheta = \pi/2$.

The probabilities that a certain quantum number $\ell$ will contribute irrespective of $m$ are

$$W_\ell = \sum_{m=-\ell}^{\ell} W_{\ell m} \quad .$$

They are plotted on the upper margin of Figure 10.17b. As a function of $\ell$, the probabilities $W_\ell$ have a bell-shaped envelope reminiscent of a Gaussian. The maximum of the marginal distribution corresponds to the angular momentum of the classical particle.

We now study the dependence of the $W_{\ell m}$ distribution on the quantization axis. Instead of the $z$ axis, we first choose an axis **n** that forms an angle of $\pi/4$ with the $y$ axis in the $z, y$ plane. Figure 10.18a shows that many more $m$ values now participate in the superposition of the wave packet. The marginal distribution, however, remains unchanged. There are changes in the $m$ distribution because the new quantization axis does not point in the direction of the classical angular-momentum vector. The distribution $W_\ell$ of the modulus of angular momentum is independent of the quantization axis.

Finally, Figure 10.18b shows the probabilities $W_{\ell m}$ for the $y$ axis as the quantization direction of angular momentum, and Figure 10.18c shows them for the $x$ axis as the quantization direction. Since in both figures the quantization direction is perpendicular to the classical angular-momentum vector, we foresee that the expectation value of $m$ will vanish. Indeed, the two distributions are symmetric around $m = 0$.

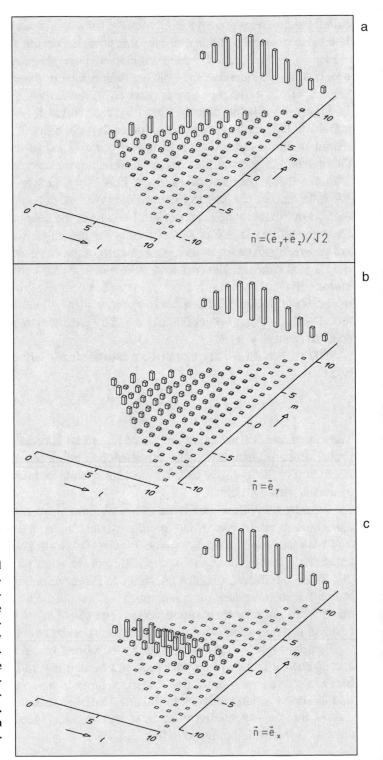

**Fig. 10.18.** All three figures apply to the situation of Figure 10.17a. Like Figure 10.17b, they show the decomposition of the wave packet into angular-momentum states. The quantization axes are different, however.

# Problems

10.1. Assuming that the components of the momentum operator in the three spatial dimensions are given by

$$\hat{p}_i = \frac{\hbar}{i} \frac{\partial}{\partial x_i} \quad , \qquad \hat{\mathbf{p}} = (\hat{p}_1, \hat{p}_2, \hat{p}_3) \quad ,$$

show that the simultaneous stationary eigenfunction for the three momentum operators $\hat{p}_i$ is the product of three one-dimensional momentum eigenfunctions.

10.2. Calculate the probability density $\rho(\mathbf{r},t) = \psi^*(\mathbf{r},t)\psi(\mathbf{r},t)$ of the three-dimensional Gaussian wave packet of Section 10.1, using the explicit form of $M(x,t)$ as given in Section 3.2. In which direction does the wave packet move? What is the square of its velocity? What determines in which direction the wave packet disperses fastest?

10.3. Verify the commutation relations of the components of angular momentum as given at the beginning of Section 10.3.

10.4. The spatial reflection is the transformation $\mathbf{r} \to -\mathbf{r}$. How is this transformation expressed in spherical coordinates? How do the spherical harmonics $Y_{\ell m}(\vartheta,\phi)$ behave under reflections?

10.5. Calculate the commutators of the angular-momentum-component operators $\hat{L}_x$, $\hat{L}_y$, $\hat{L}_z$, and $\hat{\mathbf{L}}^2$ with the coordinate operators $x$, $y$, and $z$ and with the momentum-component operators $\hat{p}_x$, $\hat{p}_y$, and $\hat{p}_z$.

10.6. Calculate the commutators of $L_x$, $L_y$, $L_z$, and $\hat{\mathbf{L}}^2$ with the radial coordinate $r = \sqrt{x^2 + y^2 + z^2}$, and with $\hat{\mathbf{p}}^2$. Use the results to compute the commutators of angular momentum with a Hamiltonian for a spherically symmetric potential,

$$H = \frac{\hat{\mathbf{p}}^2}{2M} + V(r) \quad .$$

10.7. Show that the three-dimensional free wave packet, as given by its spectral representation in Section 10.1,

$$\psi(\mathbf{r},t) = \int f(\mathbf{p})\psi_{\mathbf{p}}(\mathbf{r} - \mathbf{r}_0, t)\mathrm{d}^3\mathbf{p} \quad ,$$

is a solution of the Schrödinger equation for three-dimensional free motion.

10.8. What is the difference between the classical and the quantum-mechanical centrifugal term $L^2/2Mr^2$ in the Hamiltonian for a given angular momentum?

10.9. Verify that the explicit expressions of the spherical Bessel functions $j_0(\rho)$ and $j_1(\rho)$ and the spherical Neumann functions $n_0(\rho)$ and $n_1(\rho)$ satisfy the free radial Schrödinger equation given at the beginning of Section 10.8.

10.10. The explicit form of the spherical Hankel functions $h_\ell^{(\pm)}(\rho)$ is given at the beginning of Section 10.8. Show that the asymptotic forms of the spherical Bessel and Neumann functions for $\rho \to 0$ and $\rho \to \infty$ are as given in this section.

10.11. Calculate the expression $\hat{\mathbf{L}}\varphi_{\mathbf{p}}(\mathbf{r})$. Explain why the result does not imply that $\varphi_{\mathbf{p}}(\mathbf{r})$ is a simultaneous eigenfunction of the angular-momentum operators $\hat{L}_x, \hat{L}_y, \hat{L}_z$.

10.12. Calculate the expressions $\hat{\mathbf{L}}\{j_\ell(kr)Y_{\ell m}(\vartheta,\phi)\}$ for $\ell = 0, 1$. What distinguishes the two cases $\ell = 0$ and $\ell = 1$?

10.13. What is the expectation value of angular momentum of a Gaussian wave packet as given in Section 10.1? Explain why the result is time independent.

10.14. Why is the $m$ distribution in Figure 10.18b wider than that in Figure 10.18c? To find the answer, consider the $y$ and $z$ components of angular momentum for the classical assembly of particles imitating the wave packet.

# 11. Solution of the Schrödinger Equation in Three Dimensions

In Section 10.6 the time-dependent Schrödinger equation for three-dimensional motion under the influence of a potential was separated with respect to time and space coordinates with the help of the *ansatz*

$$\psi(\mathbf{r},t) = \exp\left[-\frac{i}{\hbar}Et\right]\varphi_E(\mathbf{r}) \quad .$$

The three-dimensional stationary Schrödinger equation for the function $\varphi_E(\mathbf{r})$ obtained at the end of that section is

$$\left[-\frac{\hbar^2}{2M}\nabla^2 + V(\mathbf{r})\right]\varphi_E(\mathbf{r}) = E\varphi_E(\mathbf{r}) \quad .$$

We now restrict ourselves to spherically symmetric systems, those in which the potential $V(\mathbf{r})$ depends only on the radial coordinate $r$. Following the same line of thought used in Section 10.7, we separate radial and angular coordinates,

$$\varphi_{E\ell m}(\mathbf{r}) = R(r)Y_{\ell m}(\vartheta,\phi) \quad ,$$

and arrive at the radial Schrödinger equation for the radial wave functions $R_\ell(k,r)$:

$$-\frac{\hbar^2}{2M}\left[\frac{1}{r}\frac{d^2}{dr^2}r - \frac{\ell(\ell+1)}{r^2} - \frac{2M}{\hbar^2}V(r)\right]R_\ell(k,r) = ER_\ell(k,r) \quad .$$

Because the potential has spherical symmetry, this equation does not depend on quantum number $m$ of the $z$ component of angular momentum. Therefore the $R_\ell(k,r)$ do not depend on $m$. Besides the kinetic and potential energies, the terms on the left-hand side of this equation represent the *centrifugal potential*

$$\frac{\hbar^2}{2M}\frac{\ell(\ell+1)}{r^2} \quad ,$$

which is attributable to the angular momentum. This and the potential term $V(r)$ are often combined to give the *effective potential* for a given angular momentum $\ell$,

S. Brandt and H.D. Dahmen, *The Picture Book of Quantum Mechanics*, DOI 10.1007/978-1-4614-3951-6_11, © Springer Science+Business Media New York 2012

$$V_\ell^{\text{eff}}(r) = \frac{\hbar^2}{2M} \frac{\ell(\ell+1)}{r^2} + V(r) \quad .$$

The radial Schrödinger equation then reads

$$\left[ -\frac{\hbar^2}{2M} \frac{1}{r} \frac{\mathrm{d}^2}{\mathrm{d}r^2} r + V_\ell^{\text{eff}}(r) \right] R_\ell(k,r) = E R_\ell(k,r) \quad .$$

This equation is a differential equation with one variable. Its solution for potential functions that are simple in structure proceeds along the same lines used to solve the one-dimensional Schrödinger equation in Chapter 4. Since the radius variable $r$ assumes positive values only, here we are looking for solutions $R_\ell(k,r)$ only on the positive half-axis. At the origin the solution $R_\ell(k,r)$ must be finite. Again, we have to distinguish the two types of solutions, those for scattering processes and those for bound states.

In contrast to the three-dimensional Schrödinger equation, which does not refer to a particular angular momentum, the radial Schrödinger equation describes a particle of a given angular-momentum quantum number $\ell$. The centrifugal potential acts as a repulsive potential, also called a *centrifugal barrier*, and keeps the particle of momentum $p$ sufficiently distant from the origin of the polar coordinate system. This way the impact parameter $b$ – see Figure 10.17 – remains sufficiently large to guarantee that angular momentum $L = bp$ is conserved.

## 11.1 Stationary Scattering Solutions

As in Section 4.2, we have to formulate the boundary conditions for the solutions that describe elastic scattering of a particle on a potential. In Sections 10.7 and 10.8 we have seen that the solutions of the radial Schrödinger equation of free motion are the spherical Bessel and Neumann functions $j_\ell(k,r)$ and $n_\ell(k,r)$. From Section 4.2 we learned that for forces of finite range the particles move force free at distances far from the range of the force. For elastic scattering on a potential of finite range $d$, the radial wave functions $R_\ell(k,r)$ must therefore approach a linear combination of spherical Bessel and Neumann functions for values of $r$ large compared to range $d$:

$$R_\ell(k,r) \to A_\ell j_\ell(kr) + B_\ell n_\ell(kr) \quad , \qquad r \gg d \quad .$$

For some potentials the solution of the radial Schrödinger equation can be given explicitly. As a particularly instructive example, we consider a square-well potential:

$$V(r) = \begin{cases} V_{\text{I}} & , & 0 \leq r < d_1 & , & \text{region I} \\ V_{\text{II}} & , & d_1 \leq r < d_2 & , & \text{region II} \\ V_{\text{III}} = 0 & , & d_2 \leq r < \infty & , & \text{region III} \end{cases} \quad .$$

Since the potential vanishes in region III, we say that it has the finite range $d = d_2$.

Scattering solutions of the radial Schrödinger equation have energy $E > 0$. The solution in inner region I consists of $j_\ell(k_I r)$ only, since $n_\ell(k_I r)$ is singular for $r = 0$. In regions II and III the solution can be written as a superposition of $j_\ell$ and $n_\ell$:

$$R_\ell(k,r) = \begin{cases} R_{\ell I} & = & A_{\ell I} j_\ell(k_I r) \\ R_{\ell II} & = & A_{\ell II} j_\ell(k_{II} r) + B_{\ell II} n_\ell(k_{II} r) \\ R_{\ell III} & = & A_{\ell III} j_\ell(kr) + B_{\ell III} n_\ell(kr) \end{cases} \quad .$$

Here the wave numbers $k_i$ in regions $i = $ I, II are

$$k_i = \frac{1}{\hbar}\sqrt{2M(E - V_i)} \quad .$$

In region III

$$k = \frac{1}{\hbar}\sqrt{2ME}$$

is the wave number of the incident particles.

For every value of $\ell$, four of the coefficients $A_{\ell N}$ and $B_{\ell N}$ are determined in terms of the fifth by the continuity conditions for the wave function and its derivative at $r = d_1$ and $r = d_2$:

$$R_{\ell I}(k,d_1) = R_{\ell II}(k,d_1) \quad , \quad \frac{dR_{\ell I}}{dr}(k,d_1) = \frac{dR_{\ell II}}{dr}(k,d_1)$$

and

$$R_{\ell II}(k,d_2) = R_{\ell III}(k,d_2) \quad , \quad \frac{dR_{\ell II}}{dr}(k,d_2) = \frac{dR_{\ell III}}{dr}(k,d_2) \quad .$$

The coefficient $A_{\ell III}$ can be chosen equal to unity, thus fixing a normalization for the incident wave. For this choice the four coefficients $A_{\ell I}$, $A_{\ell II}$, $B_{\ell II}$, and $B_{\ell III}$ calculated from the continuity conditions as functions of the incident wave number $k$ are real coefficients. Therefore the radial wave function $R_\ell(k,r)$ is real. Figure 11.1a presents the solutions $R_\ell(k,r)$ as functions of $r$ for a fixed-energy value, that is, a fixed value of $E_0$, and for a number of angular-momentum values $\ell$.

Figure 11.1b shows for comparison the functions $j_\ell(kr)$, which are in fact the functions $R_\ell(k,r)$ for a vanishing potential, that is, for the undisturbed plane waves. Because $j_\ell(kr) \sim (kr)^\ell$ is strongly suppressed near the origin for high $\ell$, the wave function $R_{\ell III}$ for high enough $\ell$ is approximated well by the term $A_{\ell III} j_\ell(kr)$ so that $B_{\ell III}$ is numerically small. Therefore the cases with and without a potential do not differ substantially for high enough $\ell$. We obtain a rough idea of the size of the value of $\ell$ above which the radial function $R_\ell$ is only slightly changed by the potential.

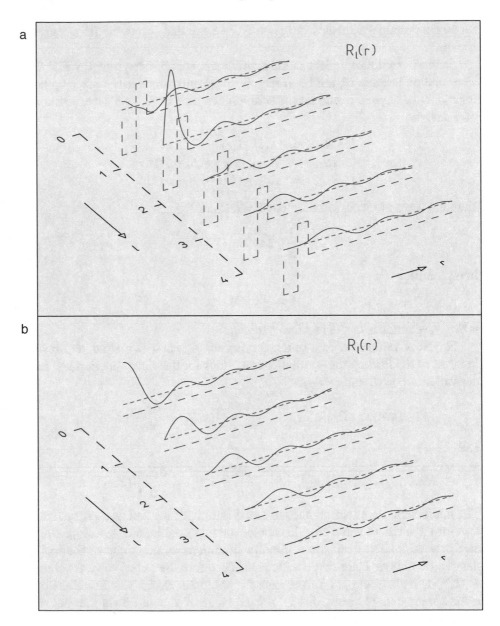

**Fig. 11.1. (a)** Solutions $R_\ell(k,r)$ of the radial Schrödinger equation for a potential that is negative in region I, $V_{\mathrm{I}} < 0$; is positive and larger than the particle energy in region II, $V_{\mathrm{II}} > E$; and vanishes in region III. The shape of the potential $V(r)$ is indicated by the long-dash line, the particle energy $E$ by the short-dash line. The short-dash lines also serve as zero lines for the functions $R_\ell(k,r)$. The energy is kept constant. The various curves correspond to different angular-momentum quantum numbers $\ell$. **(b)** The situation is the same as that in part a except that the potential is zero everywhere, $V(r) \equiv 0$. Here the solutions $R_\ell(k,r)$ are identical to the spherical Bessel functions $j_\ell(kr)$, $k = \sqrt{2ME}/\hbar$.

The argument rests on the discussion in Section 10.10. In Figure 10.17b we showed the distribution of the angular-momentum components of a Gaussian wave packet representing a classical particle moving with momentum $p = \hbar k$ and impact parameter $b$. The classical angular momentum has the value $L = pb$. We found that the spectral distribution of the angular momenta of the wave packet peaks at this classical value $L$. If the impact parameter $b$ is larger than the range $d$ of the potential, $b > d$, that is, if the classical angular momentum $L$ is high enough,

$$L > L_0 \quad , \qquad L_0 = \hbar k d \quad ,$$

the trajectory of the classical particle will not be changed by the potential. By implication, the radial wave functions $R_\ell(k, r)$ with angular momenta $\hbar\ell > L_0$, that is, $\ell > kd$, are essentially unaffected by the potential. Comparing the wave functions of Figures 11.1a and 11.1b shows that they are very similar for high values of $\ell$.

## 11.2 Stationary Bound States

The bound-state solutions occur for discrete values of negative energies $E$. Let us study the "spherical square-well" potential, the simplest situation:

$$V(r) = \begin{cases} V_{\mathrm{I}} < 0 \quad , & 0 \leq r < d \quad , & \text{region I} \\ V_{\mathrm{II}} = 0 \quad , & d \leq r < \infty \quad , & \text{region II} \end{cases}$$

The wave number

$$k_i = \sqrt{2M(E - V_i)}/\hbar$$

is real in region I for $E > V_{\mathrm{I}}$ and imaginary in region II for $E < 0$:

$$k_{\mathrm{II}} = i\kappa_{\mathrm{II}} \quad , \qquad \kappa_{\mathrm{II}} = \sqrt{-2ME}/\hbar \quad .$$

The wave function has to be proportional to $j_\ell(k_{\mathrm{I}}r)$ in region I, again because $n_\ell(k_{\mathrm{I}}r)$ is singular at $r = 0$. In region II the solution has to be proportional to $h_\ell^{(+)}(i\kappa_{\mathrm{II}}r)$, for only this function converges toward zero for large distances $r$:

$$R_\ell(k, r) = \begin{cases} R_{\ell\mathrm{I}}(k, r) & = A_{\ell\mathrm{I}} j_\ell(k_{\mathrm{I}}r) \\ R_{\ell\mathrm{II}}(k, r) & = A_{\ell\mathrm{II}} h_\ell^{(+)}(i\kappa_{\mathrm{II}}r) \end{cases} \quad .$$

The coefficient $A_{\ell\mathrm{II}}$ is determined in terms of $A_{\ell\mathrm{I}}$ as a function of energy by the two continuity conditions for the wave function and its derivative at $r = d$. The continuity can be achieved only for certain discrete bound-state energies. The constant $A_{\ell\mathrm{I}}$ is fixed by the normalization of the wave function,

$$\int_0^\infty |R_\ell(k, r)|^2 r^2 \, \mathrm{d}r = 1 \quad .$$

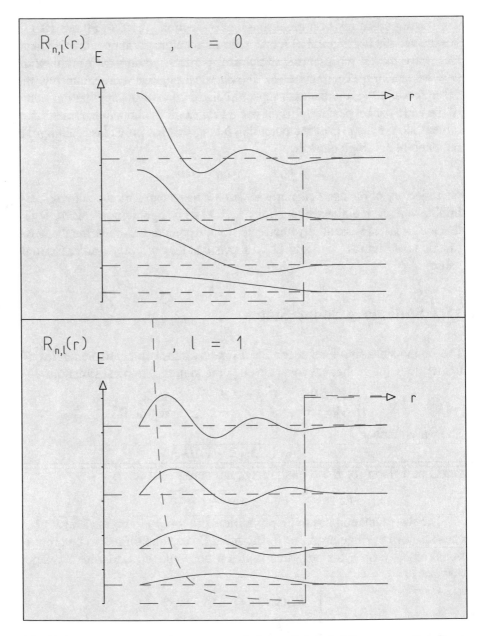

**Fig. 11.2.** Bound-state solutions $R_{n\ell}(r)$ of the radial Schrödinger equation for a square-well potential for two angular-momentum values, $\ell = 0$, $\ell = 1$. The form of the potential $V(r)$ is indicated by the long-dash line. On the left side an energy scale is drawn, and to the right of it the energies $E_n$ of the bound states are indicated by horizontal lines. These lines are repeated as short-dash lines on the right. They serve as zero lines for the solutions $R_{n\ell}(r)$. For $\ell \neq 0$ the radial dependence of the "effective potential" $V_\ell^{\mathrm{eff}}(r)$ shown as a short-dash curve indicates the influence of angular momentum (see Section 13.1).

Because the wave function falls off exponentially in region II, the particle is essentially confined to region I, the region of the potential. This confinement is the typical signature of a bound state. Figure 11.2 shows for low angular momenta the wave functions for the bound states in the square-well potential described.

## Problems

11.1. Explain by a wave-mechanical argument resting on the potential of the centrifugal barrier why the radial wave functions $R_\ell(k,r)$ for higher values of $\ell$ do not penetrate into the potential region in Figure 11.1.

11.2. Show by direct calculation that the spherical wave $\varphi(r) = \sin(kr)/r$ is a solution of the three-dimensional Schrödinger equation

$$-\frac{\hbar^2}{2M}\nabla^2\varphi(r) = E\varphi(r) \quad , \qquad E = \frac{\hbar^2 k^2}{2M} \quad .$$

11.3. A bound-state solution of vanishing angular momentum in a square-well potential of finite depth $V_0$ is given by

$$\varphi(\mathbf{r}) = \frac{A}{2\mathrm{i}}\left(\frac{\mathrm{e}^{\mathrm{i}kr}}{r} - \frac{\mathrm{e}^{-\mathrm{i}kr}}{r}\right) = A\frac{\sin(kr)}{r} \quad ,$$

$$k = \frac{1}{\hbar}\sqrt{2M(V_0 - E)} \quad .$$

Outside the well the $r$ dependence of the wave function is given by $\exp(-\kappa r)/r$, $\kappa = (1/\hbar)\sqrt{2ME}$. Therefore the function $\sin(kr)$ must have negative or zero slope at the edge $r = d$ of the well. Use this information about the slope to find a minimum value for the potential $V_0$ within the well so that there is at least one bound state. Explain why there is always at least one bound state in a one-dimensional square well.

# 12. Three-Dimensional Quantum Mechanics: Scattering by a Potential

## 12.1 Diffraction of a Harmonic Plane Wave. Partial Waves

In Section 11.1 we found the solutions $R_\ell(k,r)$ of the radial stationary Schrödinger equation for spherical square-well potentials. Since the radial Schrödinger equation is linear, its solutions are determined up to an arbitrary complex normalization constant, which has to be inferred from the boundary conditions of the three-dimensional problem we want to solve. As we have found in Section 5.5, a harmonic plane wave is an appropriately chosen idealization of an incoming wave packet representing a particle with sharp momentum. We want to apply this finding to the three-dimensional case, that is, the scattering or *diffraction* of a three-dimensional harmonic plane wave which represents a particle of sharp momentum. Then the normalization of the radial wave function has to be chosen in such a way that, for great distances from the region of the potential, the three-dimensional wave function consists of an incoming plane wave $\exp(\mathrm{i}\mathbf{k}\cdot\mathbf{r})$ and an outgoing wave.

In Section 11.1 the solutions $R_\ell(k,r)$ of the radial Schrödinger equation for spherical well potentials were chosen to be real so that in particular the coefficients $A_{\ell\,\mathrm{III}}$ and $B_{\ell\,\mathrm{III}}$ are real. To help us find the correct normalization, we turn to the physical interpretation of the solution $R_\ell(k,r)$ in region III,

$$R_{\ell\,\mathrm{III}}(k,r) = A_{\ell\,\mathrm{III}}\,j_\ell(kr) + B_{\ell\,\mathrm{III}}\,n_\ell(kr) \quad ,$$

using the decomposition of the spherical Bessel functions $j_\ell$ and $n_\ell$ into the spherical Hankel functions $h_\ell^{(\pm)}$ of Section 10.8,

$$j_\ell(kr) = \frac{1}{2\mathrm{i}}\left[h_\ell^{(+)}(kr) - h_\ell^{(-)}(kr)\right] \quad ,$$

$$n_\ell(kr) = \frac{1}{2}\left[h_\ell^{(+)}(kr) + h_\ell^{(-)}(kr)\right] \quad .$$

The spherical Hankel functions have the asymptotic behavior of complex spherical waves,

S. Brandt and H.D. Dahmen, *The Picture Book of Quantum Mechanics*, 232
DOI 10.1007/978-1-4614-3951-6_12, © Springer Science+Business Media New York 2012

$$h_\ell^{(\pm)}(kr)\xrightarrow[kr\to\infty]{}\frac{1}{kr}\exp\left[\pm i(kr-\ell\frac{\pi}{2})\right]=(\mp i)^\ell\frac{1}{kr}\exp(\pm ikr)\quad.$$

In Section 4.2 we learned that wave packets formed with a stationary wave $\exp(ikx)$ move in the direction of increasing $x$, whereas those with $\exp(-ikx)$ move in the direction of decreasing $x$ values. For spherical waves this implies that a stationary wave $\exp(-ikr)$ describes a particle moving from large values of $r$ toward the origin $r=0$, that is, an incoming particle. By the same token $\exp(ikr)$ describes an outgoing particle. Thus, except for an $r$-independent factor, the decomposition of $R_{\ell\mathrm{III}}$ into spherical Hankel functions

$$R_{\ell\mathrm{III}}=\frac{i}{2}\left[(A_{\ell\mathrm{III}}-iB_{\ell\mathrm{III}})h_\ell^{(-)}-(A_{\ell\mathrm{III}}+iB_{\ell\mathrm{III}})h_\ell^{(+)}\right]\cdot$$

describes an incoming, $h_\ell^{(-)}$, and an outgoing, $h_\ell^{(+)}$, part of the wave function. We now divide the radial functions $R_\ell$ by $A_{\ell\mathrm{III}}-iB_{\ell\mathrm{III}}$ and obtain

$$R_\ell^{(+)}(k,r)=\frac{1}{A_{\ell\mathrm{III}}-iB_{\ell\mathrm{III}}}R_\ell(k,r)\quad,$$

which takes in region III the explicit form

$$R_{\ell\mathrm{III}}^{(+)}(k,r)=-\frac{1}{2i}h_\ell^{(-)}(kr)+\frac{1}{2i}S_\ell(k)h_\ell^{(+)}(kr)\quad.$$

Here $S_\ell(k)$ is the *scattering-matrix element* of the $\ell$th partial wave

$$S_\ell(k)=\frac{A_{\ell\mathrm{III}}+iB_{\ell\mathrm{III}}}{A_{\ell\mathrm{III}}-iB_{\ell\mathrm{III}}}\quad.$$

Now, finally, we have achieved the decomposition of $R_{\ell\mathrm{III}}^{(+)}$ into the $\ell$th component $j_\ell(kr)$ of the plane wave and the outgoing spherical wave $h_\ell^{(+)}(kr)$. This structure becomes obvious if we add to the first term and subtract from the second $(1/2i)h_\ell^{(+)}(kr)$:

$$\begin{aligned}R_{\ell\mathrm{III}}^{(+)}&=\frac{1}{2i}\left[h_\ell^{(+)}(kr)-h_\ell^{(-)}(kr)\right]+\frac{1}{2i}(S_\ell-1)h_\ell^{(+)}(kr)\\&=j_\ell(kr)+f_\ell(k)h_\ell^{(+)}(kr)\quad.\end{aligned}$$

Here $f_\ell$ is the *partial scattering amplitude*

$$f_\ell(k)=\frac{1}{2i}(S_\ell(k)-1)\quad.$$

It determines the amplitude of the outgoing spherical wave in relation to the $\ell$th component $j_\ell(kr)$ of the incoming plane wave.

The recipe for constructing the three-dimensional stationary wave function is indicated by the formula for decomposing the plane wave $\exp(i\mathbf{k}\cdot\mathbf{r})$ into partial waves:

$$e^{i\mathbf{k}\cdot\mathbf{r}} = \sum_{\ell=0}^{\infty}(2\ell+1)i^{\ell}j_{\ell}(kr)P_{\ell}(\cos\vartheta) \quad , \qquad \cos\vartheta = \mathbf{k}\cdot\mathbf{r}/(kr) \quad .$$

Replacing the free radial wave function $j_{\ell}(kr)$ by the solution $R_{\ell}^{(+)}(k,r)$ of the radial Schrödinger equation for a potential $V(r)$, we obtain

$$\varphi_{\mathbf{k}}^{(+)}(\mathbf{r}) = \sum_{\ell=0}^{\infty}(2\ell+1)i^{\ell}R_{\ell}^{(+)}(k,r)P_{\ell}(\cos\vartheta) \quad .$$

Figure 12.1 gives the real and imaginary parts and the absolute square of $\varphi_{\mathbf{k}}^{(+)}$ for the scattering of a plane wave from a repulsive potential that is constant within a sphere around the origin:

$$V(r) = \begin{cases} V_0 > 0 \quad , & 0 \leq r < d \\ 0 \quad , & r \geq d \end{cases} \quad .$$

The energy $E$ of the wave is two-thirds of the height of the potential, that is, $E = 2V_0/3$. The two upper plots of Figure 12.1 for the real and the imaginary parts show that the plane wave coming in from the left is strongly suppressed within the sphere of the repulsive potential and that its pattern is modified, particularly in the forward direction, by interference with the outgoing scattered spherical wave. The patterns in these figures bear a certain resemblance to those of water waves diffracted on a cylindrical obstacle in a *ripple tank*. The real and imaginary parts of $\varphi_{\mathbf{k}}^{(+)}$ are dominated by the incident plane wave $\exp(i\mathbf{k}\cdot\mathbf{r})$. The pattern of the absolute square $|\varphi_{\mathbf{k}}^{(+)}|^2$, however, stems entirely from the superposition of the incident and scattered waves, since the absolute square of the unscattered incident wave $|\exp(i\mathbf{k}\cdot\mathbf{r})|^2 = 1$ would produce a flat sheet. In particular, the ripples to the left of center in the bottom plot of Figure 12.1 are caused by the interference of the incident wave and the scattered wave in the backward direction. This interference pattern accordingly exhibits a wavelength half that of the incident wave. It tapers off with $1/r$ because the outgoing spherical wave itself falls off with $1/r$. There are no such ripples in the forward direction because the exponentials in the scattered spherical wave and the incident wave are identical.

## 12.2 Scattered Wave and Scattering Cross Section

If we insert into the right-hand side of the formula for $\varphi_{\mathbf{k}}^{(+)}(\mathbf{r})$ the function $R_{\ell\,\mathrm{III}}^{(+)}$ in terms of $j_{\ell}(kr)$ and the outgoing spherical wave $h_{\ell}^{(+)}$, we obtain the superposition

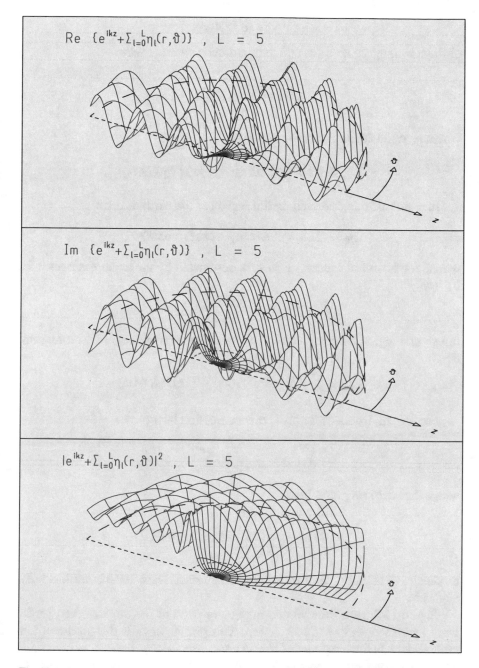

**Fig. 12.1. Scattering of a plane wave incident from the left, that is, along the $z$ direction, by a repulsive potential. The potential is confined to the region $r < d$, indicated by the small half-circle marked off by a short-dash line. The energy $E$ of the plane wave is two-thirds the height of the potential in this region. Shown are the real part, the imaginary part, and the absolute square of the wave function $\varphi_{\mathbf{k}}^{(+)}$.**

$$\varphi_{\mathbf{k}}^{(+)}(\mathbf{r}) = e^{i\mathbf{k}\cdot\mathbf{r}} + \eta_{\mathbf{k}}(\mathbf{r})$$

of the incoming plane wave and the scattered spherical wave

$$\eta_{\mathbf{k}}(\mathbf{r}) = \sum_{\ell=0}^{\infty} \eta_\ell(r,\vartheta) \quad ,$$

where $\eta_\ell$ is the $\ell$th *scattered partial wave*:

$$\eta_\ell = (2\ell+1)i^\ell \left[ R_\ell^{(+)}(kr) - j_\ell(kr) \right] P_\ell(\cos\vartheta) \quad .$$

In region III this scattered partial wave has the explicit form

$$\eta_\ell = (2\ell+1)i^\ell f_\ell(k) h_\ell^{(+)}(kr) P_\ell(\cos\vartheta) \quad ,$$

which, for far-out distances, $kr \gg 1$, is dominated by the asymptotic term for $h_\ell^{(+)}(kr)$,

$$\eta_\ell = (2\ell+1) f_\ell(k) \frac{e^{ikr}}{r} P_\ell(\cos\vartheta) \quad .$$

In external region III the scattered spherical wave has the explicit representation

$$\eta_{\mathbf{k}III}(\mathbf{r}) = \sum_{\ell=0}^{\infty} (2\ell+1)i^\ell f_\ell(k) h_\ell^{(+)}(kr) P_\ell(\cos\vartheta) \quad .$$

For far-out distances, $kr \gg 1$, this expression becomes

$$\eta_{\mathbf{k}III}(\mathbf{r}) \xrightarrow[kr \gg 1]{} f(\vartheta) \frac{e^{ikr}}{r} \quad ,$$

where the *scattering amplitude*

$$f(\vartheta) = \frac{1}{k} \sum_{\ell=0}^{\infty} (2\ell+1) f_\ell(k) P_\ell(\cos\vartheta)$$

modulates the amplitude of the scattered spherical wave for the various polar angles $\vartheta$.

Figure 12.2 plots the real and imaginary parts of the $\ell$th scattered partial wave $\eta_\ell$ for values $\ell = 0, 1, 2, 3, 4, 5$. This partial wave is the product of an $r$-dependent factor responsible for the variation along lines $\vartheta = $ const and a $\vartheta$-dependent factor responsible for the variation along lines $r = $ const. As expected from the Legendre polynomials $P_\ell(\cos\vartheta)$, there is no $\vartheta$ variation for $\ell = 0$, whereas the increasing complexity of the higher $P_\ell$ is signaled by their $\ell$ nodes in $\vartheta$. The pictures indicate a $1/r$ falloff for large values of $r$, as expected from the asymptotic form of the $\eta_\ell$. As already mentioned in Section 11.1, the deviations of the radial wave functions from the free radial

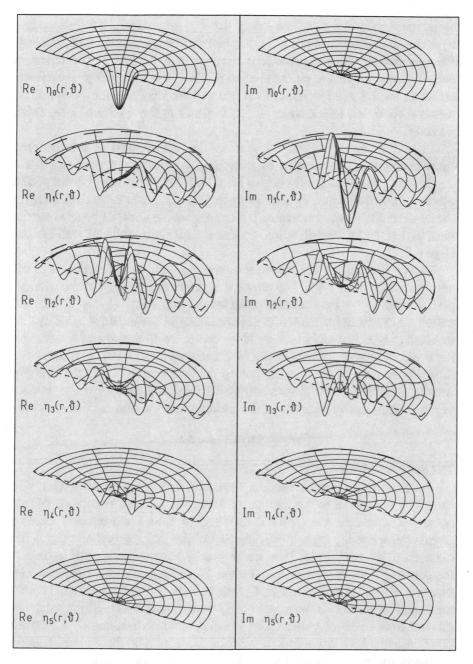

**Fig. 12.2.** Real and imaginary parts of the scattered partial waves $\eta_\ell$, resulting from the scattering of a plane wave by a repulsive potential, as shown in Figure 12.1.

wave function $j_\ell$ are substantial only for $\ell \leq kd$. Indeed, we observe that $\eta_5$ is essentially zero, as are $\eta_6$, $\eta_7$, and so on. In our example $kd$ equals 4. We may wonder why the scattered partial waves $\eta_\ell$ for low $\ell$ have important contributions within the potential sphere. They are expected to contribute, however, for the superposition of the $\eta_\ell$ has to compensate for the harmonic plane wave in this region since $\varphi_k^{(+)}(\mathbf{r})$ is small in the sphere of a repulsive potential.

Figures 12.3a and b give the real and imaginary parts of the scattered spherical wave $\eta_k(\mathbf{r})$ obtained by summing the scattered partial waves for $0 \leq \ell \leq 5$. The $\eta_\ell$ essentially vanish for $\ell > 5$. Whereas the scattered partial waves $\eta_\ell$ have the symmetry of the corresponding $P_\ell$, their superposition $\eta_k(\mathbf{r})$ shows a definite forward structure, indicating that the scattering occurs for the most part in the forward direction. Obviously, $\eta_k(\mathbf{r})$ also falls off with $1/r$ for large $r$.

Figure 12.3c gives the absolute square $|\eta_k(\mathbf{r})|^2$. This function falls off asymptotically with $1/r^2$. The physical significance of $|\eta_k|^2$ is the average particle density for the scattered particles moving with velocity $v = \hbar k/M$ radially away from the center. In experiments the scattered particles can be detected only at distances that are large compared to the size of the scattering center. The average number $\Delta n$ of scattered particles passing through the sensitive area $\Delta a$ of the detector during the time interval $\Delta t$ is the quantity usually measured. For a given sensitive area $\Delta a$, this number is the product of the current density $|\eta_k(\mathbf{r})|^2 v$ of the particles and the area $\Delta a$ times $\Delta t$:

$$\Delta n = v|\eta_k(\mathbf{r})|^2 \, \Delta a \, \Delta t \quad .$$

The detector is located at $\mathbf{r}$.

For fixed experimental conditions $\Delta a$, $\Delta t$, and $v$, the quantity $|\eta_k(\mathbf{r})|^2$ is directly proportional to the number of scattered particles observed. We assume that many detectors are distributed evenly along a half-circle of radius $r$ around the scattering center. The direction of the incident particles forms the diameter of the half-circle. Then we have only to compute $|\eta_k(\mathbf{r})|^2$ to predict the counting rates in all detectors. Figure 12.4a illustrates this situation. The function $|\eta_k(\mathbf{r})|^2$ is plotted in a half-circle band in the region where the detectors could be placed. To overcome the $1/r^2$ suppression in $|\eta_k|^2$ the values of this function have been blown up by a scaling factor. It becomes obvious from this figure that $|\eta_k|^2$ depends to a considerable extent on the scattering angle $\vartheta$, that is, the angle between the incident and the scattered particle.

Actually, the asymptotic form of $\eta_k$, that is, of $\eta_{k\mathrm{III}}$, which we found to be

$$\eta_{k\mathrm{III}} \xrightarrow[kr \gg 1]{} f(\vartheta)\frac{e^{ikr}}{r} \quad ,$$

shows that the quantity

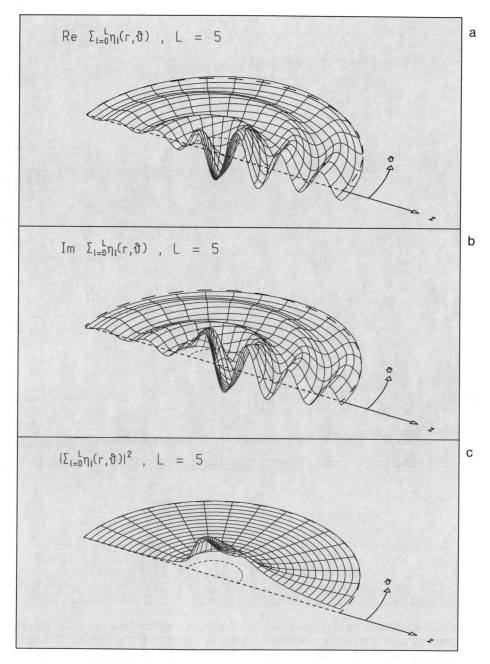

Fig. 12.3. Real part, imaginary part, and absolute square of the scattered spherical wave $\eta_k$ resulting from the scattering of a plane wave by a repulsive potential, as shown in Figure 12.1.

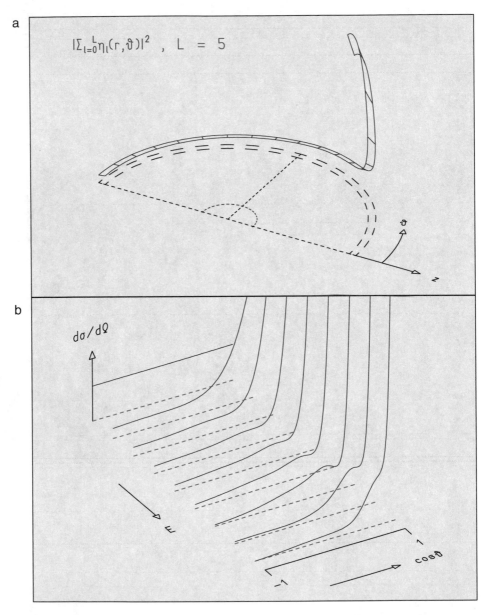

**Fig. 12.4. (a) Intensity of the scattered spherical wave resulting from the scattering of a plane wave by a repulsive potential, as shown in Figure 12.1. The intensity at a fixed radius far outside the scattering region and for a given scattering angle $\vartheta$ is indicated by the height of the band. The band corresponds to the outer rim of Figure 12.3c, blown up by a scale factor. (b) Energy dependence of the differential scattering cross section $d\sigma(\vartheta)/d\Omega$ for the scattering of a plane wave by a repulsive potential. The differential cross section is proportional to the intensity of the scattered wave, as we can see by comparing the curve in the middle of part b with the band in part a. Both correspond to the same energy.**

$$|r\,\eta_{kIII}|^2 \xrightarrow[kr \gg 1]{} |f(\vartheta)|^2$$

is dependent only on the scattering angle $\vartheta$. Its physical interpretation in terms of the counting rate $\Delta n$ becomes clear if we observe that $\Delta a / r^2 = \Delta\Omega$ is the sensitive solid angle of the detector. Furthermore, the incident current density is equal to the incident average particle density times the average velocity,

$$j = |e^{i\mathbf{k}\cdot\mathbf{r}}|^2 v = v \quad .$$

Thus the number of scattered particles $\Delta n$ can be re-expressed in the form

$$
\begin{aligned}
\Delta n &= j\frac{\Delta a}{r^2}|r\,\eta_{kIII}|^2\,\Delta t \\
&= j|f(\vartheta)|^2\,\Delta\Omega\,\Delta t \quad ,
\end{aligned}
$$

which shows that $|f(\vartheta)|^2$ has the following physical meaning. It is the average number of particles from an incident particle current of density 1 scattered per second at angle $\vartheta$ per unit solid angle,

$$|f(\vartheta)|^2 = \frac{1}{j}\frac{\Delta n}{\Delta\Omega\,\Delta t} \quad .$$

In a classical experiment in which a particle beam is incident on a hard sphere, the quantity on the right-hand side is the *differential scattering cross section*. This notion is derived from the elastic scattering of a beam of point particles with current density $j$ incident on a rigid sphere of radius $d$. As Figure 12.5 indicates, the impact parameter $b$ is related to the scattering angle $\vartheta$ by

$$b = d\cos\frac{\vartheta}{2} \quad .$$

The number $\Delta n$ of particles incident during $\Delta t$ in the azimuthal sector $\Delta\phi$ with an impact parameter between $b$ and $b + \Delta b$ is

$$\Delta n = j\,\Delta t\,b\,\Delta b\,\Delta\phi \quad .$$

This number of particles is scattered into the solid angle $\Delta\Omega = \Delta\cos\vartheta\,\Delta\phi$ where $\Delta\cos\vartheta$ corresponds to $\Delta b$ through the relation

$$\frac{db}{d\cos\vartheta} = \frac{d\vartheta}{d\cos\vartheta}\frac{db}{d\vartheta} = \frac{d}{4}\frac{1}{\cos\vartheta/2} \quad .$$

The number of particles scattered at angle $\vartheta$ per unit solid angle and unit time is then

$$\frac{\Delta n}{\Delta\Omega\,\Delta t} = \frac{1}{4}d^2 j \quad .$$

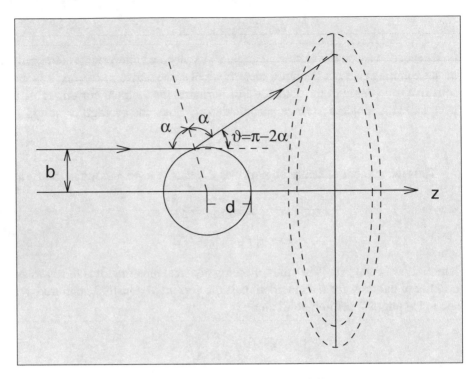

**Fig. 12.5. The classical elastic scattering of a point particle by a rigid sphere.**

This rate of particles for a current density one is completely determined by the properties of the scattering center, here a rigid sphere. The rate per unit current density is the *differential scattering cross section* $d\sigma/d\Omega$. For a rigid sphere it is

$$\frac{d\sigma}{d\Omega} = \frac{1}{4}d^2 \quad .$$

In general, the differential scattering cross section is not constant but depends on the direction of the scattered particle. When the scattering center is spherically symmetric, the differential scattering cross section is a function only of the scattering angle $\vartheta$. Integration over the full solid angle $4\pi$ yields the *total scattering cross section*. When classical particles are scattered off a hard sphere, it is obtained by multiplying $\frac{1}{4}d^2$ by the full solid angle $4\pi$,

$$\sigma_{\text{tot}} = \pi d^2 \quad .$$

As expected, it is the geometrical cross section of the rigid sphere.

Coming back to our quantum-mechanical discussion, we can identify the differential scattering cross section as

$$\frac{d\sigma}{d\Omega} = |f(\vartheta)|^2 \quad .$$

The function $f(\vartheta)$ as calculated earlier has the form

$$f(\vartheta) = \frac{1}{k} \sum_{\ell=0}^{\infty} (2\ell+1) f_\ell(k) P_\ell(\cos\vartheta) \quad ,$$

which shows that it depends not only on the scattering angle but also on the energy $E = (\hbar k)^2/2M$ of the incident particles. It is customary to plot $d\sigma/d\Omega = |f(\vartheta)|^2$ as a function of $\cos\vartheta$ rather than $\vartheta$. In Figure 12.4b this is done for a range of energies and for the potential used in the earlier figures of this chapter. For very low energy the differential cross section is constant in $\cos\vartheta$. With increasing energy it acquires a more complicated angular dependence. This dependence is easily explained by observing that for very low energy only the lowest partial wave, $\ell = 0$, contributes to the scattering amplitude $f(\vartheta)$ through the Legendre polynomial $P_0(\cos\vartheta)$, which is a constant. With increasing energy more and more partial waves contribute, allowing a richer structure in $\cos\vartheta$.

The *total cross section* is obtained by an integration over the full solid angle

$$\sigma_{\text{tot}} = \int \frac{d\sigma}{d\Omega} d\Omega = 2\pi \int_{-1}^{+1} |f(\vartheta)|^2 d\cos\vartheta \quad .$$

For the following we need the orthogonality of different Legendre polynomials,

$$\int_{-1}^{+1} P_\ell(\cos\vartheta) P_{\ell'}(\cos\vartheta) d\cos\vartheta = \frac{2}{2\ell+1} \delta_{\ell\ell'} \quad ,$$

which can be inferred from the orthonormality of the spherical harmonics $Y_{\ell 0}$ and their relation to the $P_\ell$, as discussed in Section 10.3. If we insert the series for $f(\vartheta)$ into the integral for $\sigma_{\text{tot}}$, we obtain

$$\sigma_{\text{tot}} = \frac{4\pi}{k^2} \sum_{\ell=0}^{\infty} (2\ell+1) |f_\ell(k)|^2 = \sum_{\ell=0}^{\infty} \sigma_\ell \quad .$$

The terms in this sum are called *partial cross sections*,

$$\sigma_\ell = \frac{4\pi}{k^2} (2\ell+1) |f_\ell(\vartheta)|^2 \quad .$$

Figure 12.6 shows the various partial cross sections as functions of energy. We notice that the partial cross section for $\ell > 0$ starts at zero for $k = 0$. Furthermore, the contribution of the cross sections for increasing $\ell$ sets in with increasing energy so that for a given energy the sum over the partial cross sections can be truncated at $l_{\max} \gtrsim kd$, the maximum value of the classical angular momentum at which scattering takes place. The total cross section obtained by the summation is plotted as the topmost diagram of Figure 12.6.

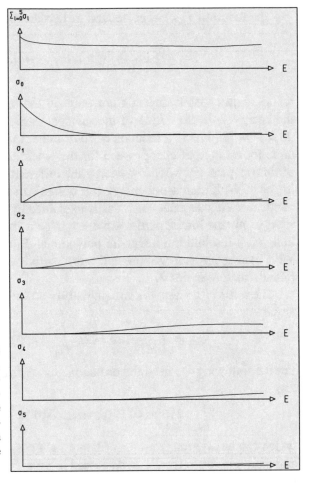

**Fig. 12.6. The partial cross sections** $\sigma_\ell(E)$ **for** $\ell = 0, 1,$ $\ldots, 5,$ **and the total cross section** $\sigma_{\text{tot}}(E)$, **which is approximated by the sum over the first five partial cross sections for the scattering of a plane wave by a repulsive potential.**

## 12.3 Scattering Phase and Amplitude, Unitarity, Argand Diagrams

In Section 12.1 we obtained as a representation for the radial wave function $R^{(+)}_{\ell\,\mathrm{III}}(k,r)$ the form

$$R^{(+)}_{\ell\,\mathrm{III}}(k,r) = \frac{\mathrm{i}}{2}\left[h^{(-)}_\ell(kr) - S_\ell(k)h^{(+)}_\ell(kr)\right] \quad .$$

We interpreted this solution as the superposition of the incoming spherical wave $h^{(-)}_\ell$ and the outgoing spherical wave $h^{(+)}_\ell$, which is multiplied by the $S$-matrix element $S_\ell$. Potential scattering conserves particle number, angular momentum, and energy $E = (\hbar k)^2/2M$ so that the magnitude of velocity $\hbar k/M$ remains unaltered. Therefore the current density of the incoming spherical wave has the same size as the current density of the outgoing spherical wave. As a consequence, the particle densities in the incoming and outgoing spherical waves in region III have to be the same. Since particle densities are determined by the absolute squares of amplitudes, the scattering-matrix element $S_\ell$, representing the relative factor between the incoming and outgoing spherical waves must have absolute value one.

In fact, the representation for $S_\ell$, found in Section 12.1,

$$S_\ell = \frac{A_{\ell\,\mathrm{III}} + iB_{\ell\,\mathrm{III}}}{A_{\ell\,\mathrm{III}} - iB_{\ell\,\mathrm{III}}} \quad ,$$

satisfies this requirement,

$$S_\ell^* S_\ell = \frac{A_{\ell\,\mathrm{III}} - iB_{\ell\,\mathrm{III}}}{A_{\ell\,\mathrm{III}} + iB_{\ell\,\mathrm{III}}} \cdot \frac{A_{\ell\,\mathrm{III}} + iB_{\ell\,\mathrm{III}}}{A_{\ell\,\mathrm{III}} - iB_{\ell\,\mathrm{III}}} = 1 \quad ,$$

which is called the *unitarity relation for the S-matrix elements*. Thus $S_\ell$ can be represented as a complex phase factor

$$S_\ell(k) = \frac{A_{\ell\,\mathrm{III}} + iB_{\ell\,\mathrm{III}}}{A_{\ell\,\mathrm{III}} - iB_{\ell\,\mathrm{III}}} = e^{2i\delta_\ell(k)} \quad .$$

The *scattering phase* $\delta_\ell$ determining $S_\ell$ can be calculated directly from $A_{\ell\,\mathrm{III}}$ and $B_{\ell\,\mathrm{III}}$ if we observe that

$$e^{\pm i\delta_\ell} = \frac{A_{\ell\,\mathrm{III}} \pm iB_{\ell\,\mathrm{III}}}{\sqrt{A_{\ell\,\mathrm{III}}^2 + B_{\ell\,\mathrm{III}}^2}}$$

allows the identification

$$\cos\delta_\ell = \frac{A_{\ell\,\mathrm{III}}}{\sqrt{A_{\ell\,\mathrm{III}}^2 + B_{\ell\,\mathrm{III}}^2}} \quad , \qquad \sin\delta_\ell = \frac{B_{\ell\,\mathrm{III}}}{\sqrt{A_{\ell\,\mathrm{III}}^2 + B_{\ell\,\mathrm{III}}^2}} \quad .$$

These relations can be used to show that $\delta_\ell$ is a phase shift produced by the potential. To this end we use the asymptotic representations, $kr \gg 1$, for $j_\ell(kr)$ and $n_\ell(kr)$, as given in Section 10.8. In fact, the solution $R_{\ell\,\mathrm{III}}$ of the stationary Schrödinger equation presented at the beginning of Section 12.1 has the form

$$R_{\ell\,\mathrm{III}} = \sqrt{A_{\ell\,\mathrm{III}}^2 + B_{\ell\,\mathrm{III}}^2} \left[\cos\delta_\ell\, j_\ell(kr) + \sin\delta_\ell\, n_\ell(kr)\right] \quad ,$$

which asymptotically becomes

$$R_{\ell\,\mathrm{III}} \xrightarrow[kr \gg 1]{} \sqrt{A_{\ell\,\mathrm{III}}^2 + B_{\ell\,\mathrm{III}}^2}\, \frac{1}{kr} \left[\cos\delta_\ell \sin\left(kr - \ell\frac{\pi}{2}\right)\right.$$
$$\left. + \sin\delta_\ell \cos\left(kr - \ell\frac{\pi}{2}\right)\right] \quad ,$$

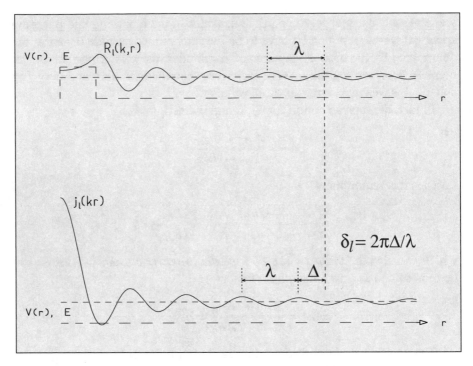

**Fig. 12.7. Definition of the scattering phase shift $\delta_\ell$. The solution $R_\ell$ of the radial Schrödinger equation for a given $\ell$, here $\ell = 0$, is shown for the scattering of a wave of energy $E$ by a repulsive potential (top) and for vanishing potential (bottom). Asymptotically, that is, far outside the potential region, both solutions differ only by a phase shift $\delta_\ell$.**

i.e.,

$$R_{\ell\,\mathrm{III}} \xrightarrow[kr\gg1]{} \sqrt{A_{\ell\,\mathrm{III}}^2 + B_{\ell\,\mathrm{III}}^2}\,\frac{1}{kr}\sin\left(kr - \ell\frac{\pi}{2} + \delta_\ell\right) \quad .$$

Figure 12.7 plots $R_\ell$, together with $j_\ell$, the $\ell$th partial wave of the harmonic plane wave. The scattering phase shift $\delta_\ell$ is easily recognized as the phase difference between the two in the asymptotic region. In Figure 12.8 the energy dependence of the various phase shifts $\delta_\ell$ is shown for the repulsive square-well potential used as our example. We have chosen the phases $\delta_\ell$ so that they are equal to 0 for $E = 0$. For the potential of our example, they fall off smoothly with increasing energy.

In terms of the scattering phase $\delta_\ell(k)$ the partial scattering amplitude can be represented as

$$\begin{aligned}
f_\ell(k) &= \frac{1}{2i}[S_\ell(k) - 1] = \frac{1}{2i}(e^{2i\delta_\ell} - 1) = e^{i\delta_\ell}\left[\frac{1}{2i}(e^{i\delta_\ell} - e^{-i\delta_\ell})\right] \\
&= e^{i\delta_\ell}\sin\delta_\ell \quad .
\end{aligned}$$

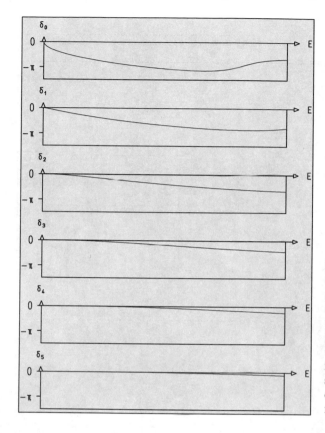

**Fig. 12.8.** Energy dependence of the phase shifts $\delta_0(E)$, $\delta_1(E) \ldots$, $\delta_5(E)$ for scattering by a repulsive potential. There is an ambiguity in the definition of $\delta_\ell$, which is resolved by choosing $\delta_\ell(0) = 0$. All phase shifts vary slowly with energy for scattering by a repulsive potential.

The relation expressing that $S_\ell$ has absolute value one, reflects itself in the equivalent *unitarity relation for the partial scattering amplitude* $f_\ell$,

$$\operatorname{Im} f_\ell(k) = |f_\ell(k)|^2 \quad .$$

As a complex number, the partial scattering amplitude can be plotted in an Argand diagram similar to the one in Section 5.5. Here, however, $f_\ell$ stays on the circumference of the circle with radius $1/2$ centered at $i/2$ in the complex plane because the unitarity relation can be written as

$$(\operatorname{Re} f_\ell)^2 + \left(\operatorname{Im} f_\ell - \frac{1}{2}\right)^2 = \frac{1}{4} \quad .$$

This relation is the equation for a circle of radius $1/2$ centered at $i/2$ in the complex plane. It is shown in Figure 12.9a.

As the wave number $k = (1/\hbar)\sqrt{2ME}$ of the incident wave changes, $f_\ell$ moves on the circle. The scattering phase $\delta_\ell$ is the angle between the arrow representing the complex number $f_\ell$, and the real axis. The energy dependence of the complex scattering amplitude $f_\ell$ is shown in detail in Figure 12.9b. The real and imaginary parts of $f_\ell$ as a function of the energy are

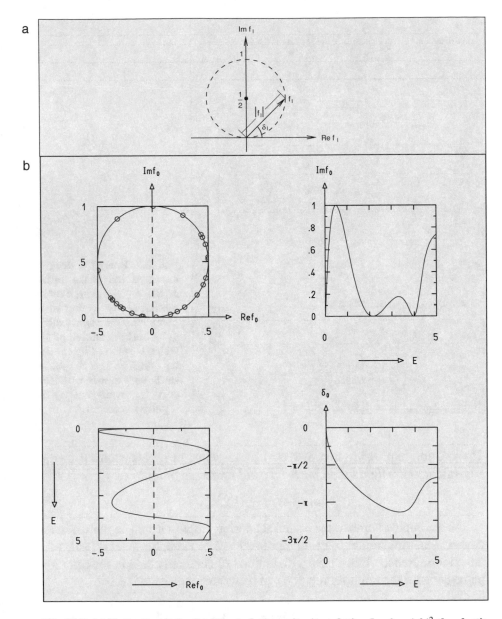

**Fig. 12.9. (a) Unitarity circle. (b) Through the unitarity relation** $\operatorname{Im} f_\ell = |f_\ell|^2$ **the elastic partial-wave amplitude is confined to a circle in an Argand diagram. The angle between the vector** $f_\ell$ **in the complex plane and the real axis is the phase shift** $\delta_\ell$**. As the energy** $E$ **increases the point** $f_\ell(E)$ **moves on the circle starting at** $f_\ell(0) = 0$**. Points equidistant in energy are marked off by small circles (top left). Projections onto a vertical and horizontal axis yield graphs of** $\operatorname{Im} f_\ell(E)$ **(top right) and** $\operatorname{Re} f_\ell(E)$ **(bottom left), respectively. The function** $\delta_\ell(E)$ **is also shown (bottom right).**

then obtained by projecting the Argand diagram onto the real and imaginary axes. Finally, for completeness, we also show the energy dependence of the scattering phase shift $\delta_\ell$.

There is an interesting relationship between the function $f_\ell(\vartheta)$ in the forward direction and the total cross section. We have

$$\sigma_{\text{tot}} = \frac{4\pi}{k^2} \sum_{\ell=0}^{\infty} (2\ell + 1)|f_\ell(k)|^2 \quad .$$

Using the unitarity relation for the partial scattering amplitude,

$$|f_\ell(k)|^2 = \text{Im} \, f_\ell(k) \quad ,$$

and the particular value of $P_\ell(\cos\vartheta)$ in the forward direction ($\vartheta = 0$),

$$P_\ell(1) = 1 \quad ,$$

we obtain

$$\sigma_{\text{tot}} = \frac{4\pi}{k} \frac{1}{k} \sum_{\ell=0}^{\infty} (2\ell + 1)\text{Im} \, f_\ell(k) P_\ell(1) \quad ,$$

$$\sigma_{\text{tot}} = \frac{4\pi}{k} \text{Im} \, f(0) \quad ,$$

if we use the partial-wave representation of $f(\vartheta)$ for $\vartheta = 0$.

This equation is called the *optical theorem*. It states that the total cross section is directly given by the imaginary part of the forward scattering amplitude. The optical theorem reflects the conservation of the particle current in the scattering process. In fact, the total current contained in the scattered wave has to be supplied by the incident current. That is done through the interference between the incident and the scattered waves in the forward direction.

## Problems

12.1. Why is the wave function $\varphi_{\mathbf{k}}^{(+)}(\mathbf{r})$ in Figure 12.1 suppressed beyond the potential region indicated by the dashed circle close to the center? Which effect makes it recover along the positive $z$ axis? Use Huygens' principle to draw an analogy to the scattering of light by a black disk.

12.2. Why must the scattered spherical wave $\eta_{\mathbf{k}}(\mathbf{r})$ as shown in Figure 12.3 be unequal to zero in the region of the repulsive potential and have a wave pattern there? What can be said about its wavelength within the potential region?

12.3. In Section 12.2 the classical elastic scattering of point particles by a rigid sphere of radius $d$ was discussed. Replace the point particles by spheres of radius $a$. Show that the results for the differential and total cross sections stay valid if $d$ is replaced by $d + a$.

12.4. Verify the unitarity relation for the partial scattering amplitude $f_\ell$,

$$\text{Im} \, f_\ell = f_\ell f_\ell^* \quad ,$$

using the unitarity relation for the scattering-matrix element $S_\ell$,

$$S_\ell S_\ell^* = 1 \quad ,$$

as derived in Section 12.3. Put the unitarity relation for $f_\ell$ into the form of an equation for the unitarity circle as given in Section 12.3.

# 13. Three-Dimensional Quantum Mechanics: Bound States

## 13.1 Bound States in a Spherical Square-Well Potential

Figure 11.2 has already shown the radial wave function of bound states in a three-dimensional square-well potential. Now in Figure 13.1 we plot the radial wave function $R_{n\ell}$ together with its square $R_{n\ell}^2$ and the function $r^2 R_{n\ell}^2$ for the low angular-momentum quantum numbers $\ell = 0, 1, 2$. The reason for showing $r^2 R_{n\ell}^2$ is that $r^2 R_{n\ell}^2(r)\,dr$ represents the probability that a particle is within a spherical shell of radius $r$ and thickness $dr$. Also shown in Figure 13.1 is the energy spectrum of the eigenvalues. We observe that the number of bound states is finite. The spacing between the different eigenvalues increases with increasing energy. For a given $\ell$ value the lowest-lying state has no node in $r$, the next one has one node, and so on. We can enumerate the eigenvalues $E_{n\ell}$, $n = 1, 2, \ldots$, for a given $\ell$ by the number $n - 1$ of nodes they possess. In Figure 13.1 the square-well potential $V(r)$ is drawn as a long-dash line, the effective potential as a short-dash line. The effective potential, as we have learned, is made up of the centrifugal potential and the square-well potential,

$$V_\ell^{\text{eff}}(r) = \frac{\hbar^2}{2M} \frac{\ell(\ell+1)}{r^2} + V(r) \quad .$$

The repulsive nature of the centrifugal potential suppresses the radial wave function,

$$R_{n\ell}(r) = A_\mathrm{I} j_\ell(kr) \to \frac{A_\mathrm{I}}{(2\ell+1)!!}(kr)^\ell , \quad kr \ll 1 , \ (2\ell+1)!! = 1 \cdot 3 \ldots (2\ell+1) ,$$

for small values of $r$ and $\ell > 0$. It is also responsible for the increase in energy $E_{n\ell}$ for given $n$ and increasing $\ell$. This suppression of the wave function for $\ell \geq 1$ near the origin is easily verified in Figure 13.1. For $\ell = 0$ the wave functions start with values larger than zero at the origin. For $\ell = 1$ the wave function is zero at the origin; however, it increases linearly close to $r = 0$. For $\ell = 2$ the growth of the wave function from zero at the origin is only that of a parabola. The slopes close to the origin become steeper with higher

S. Brandt and H.D. Dahmen, *The Picture Book of Quantum Mechanics*,
DOI 10.1007/978-1-4614-3951-6_13, © Springer Science+Business Media New York 2012

quantum numbers $n$ at fixed $\ell$. Thus for fixed $\ell$ the particle comes closer to the origin for higher values of $n$. Higher values of $n$ correspond to higher values of the energy and of the momentum $p$. For a given angular momentum $L$ the classical relation $L = bp$ shows that larger momenta $p$ correspond to smaller impact parameters $b$. In this respect the quantum-mechanical behavior of the particle corresponds to classical mechanics.

The plot of $r^2 R_{n\ell}^2(k,r)$ allows a particularly simple discussion. Let us start with the value $\ell = 0$. The radial wave functions within the potential region $r < d$ are

$$
\begin{aligned}
R_{n0}(r) &= A_{\mathrm{I}} j_0(k_{\mathrm{In}} r) = A_{\mathrm{I}} \frac{\sin k_{\mathrm{In}} r}{k_{\mathrm{In}} r} \quad , \\
k_{\mathrm{In}} &= \sqrt{2m(E_{n0} - V_0)} \quad ,
\end{aligned}
$$

so that the function

$$
r^2 R_{n0}^2(r) = \frac{A_{\mathrm{I}}^2}{k_{\mathrm{In}}^2} \sin^2 k_{\mathrm{In}} r
$$

behaves in a simple sine-squared manner. For the higher angular momenta we recall the asymptotic relation

$$
j_\ell(kr) \to \frac{1}{kr} \sin\left(kr - \ell\frac{\pi}{2}\right) \quad , \qquad kr \gg 1 \quad ,
$$

so that for $r \gg 1/k_{\mathrm{In}}$ the behavior is again sine-squared,

$$
r^2 R_{n\ell}^2(r) \to \frac{A_{\mathrm{I}}^2}{k_{\mathrm{In}}^2} \sin^2\left(k_{\mathrm{In}} r - \ell\frac{\pi}{2}\right) \quad , \qquad k_{\mathrm{In}} r \gg 1 \quad .
$$

Again, looking at Figure 13.1, we recognize the approach of the quantity $r^2 R_{n\ell}^2$ toward this behavior. In region I close to the edge of the potential at $r = d$, the centrifugal barrier is low for low values of $\ell$; it can therefore be neglected in a coarse approximation. Thus, close to the outer rim of the potential, the wave functions for different $\ell$, but equal $n$ should look almost alike and behave in a sine-squared manner. This is easily verified in Figure 13.1.

**Fig. 13.1. The radial eigenfunctions $R_{n\ell}(r)$ of bound states in a square-well potential for three angular-momentum values, $\ell = 0, 1, 2$, are shown as continuous lines in the left column. The form $V(r)$ of the potential is indicated by the long-dash line. Also shown for $\ell \neq 0$ is the effective potential $V_\ell^{\mathrm{eff}}(r)$ which contains the influence of angular momentum. On the left is an energy scale and to the right of it the energy eigenvalues $E_n$ are indicated by horizontal lines. These lines are repeated as short-dash lines on the right. They serve as zero lines for the plotted functions. In the middle column the squares $R_{n\ell}^2(r)$ of the radial eigenfunctions are shown. Along a fixed direction $\vartheta, \phi$ away from the origin, this quantity is proportional to the probability that the particle will be observed within a unit volume element around point $r$, $\vartheta$, $\phi$. In the right column are the functions $r^2 R_{n\ell}^2(r)$. Their values are a measure for the probability of observing the particle anywhere within a spherical shell of radius $r$ and unit thickness.**

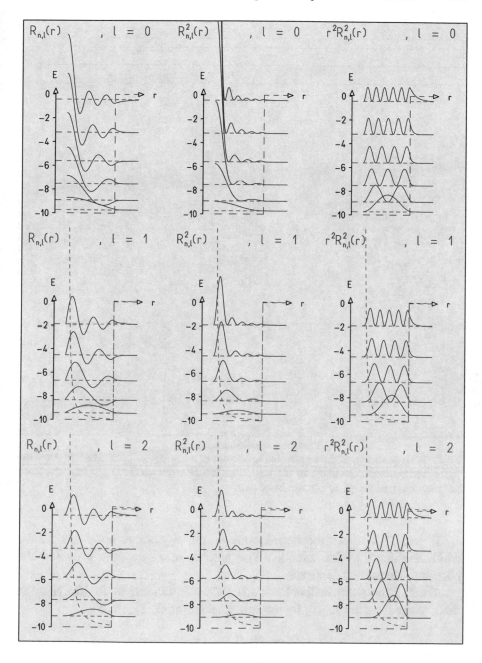

**Fig. 13.1.**

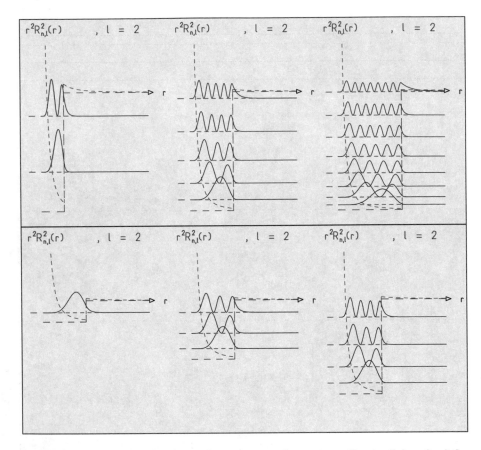

**Fig. 13.2. Dependence of the eigenvalue spectrum of a square-well potential on (top) the width and (bottom) the depth of the well. The function shown is $r^2 R_{n\ell}^2(r)$ for the fixed angular-momentum quantum number $\ell = 2$.**

Figure 13.2 shows the dependence of the eigenvalue spectrum on the width and depth of the potential. The number of eigenvalues grows as the potential widens and deepens.

The full three-dimensional wave function is obtained by multiplying the radial wave function $R_{n\ell}(r)$ by the spherical harmonic $Y_{\ell m}(\vartheta, \phi)$,

$$\varphi_{n\ell m}(\mathbf{r}) = R_{n\ell}(r) Y_{\ell m}(\vartheta, \phi) \quad .$$

Since the absolute square $\rho_{n\ell m}(r, \vartheta)$ of this wave function is independent of $\phi$,

$$\rho_{n\ell m}(r, \vartheta) \equiv |\varphi_{n\ell m}(\mathbf{r})|^2$$
$$= R_{n\ell}^2(r) \frac{2\ell+1}{4\pi} \frac{(\ell - |m|)!}{(\ell + |m|)!} \left[ P_\ell^{|m|}(\cos \vartheta) \right]^2 \quad ,$$

a                    b

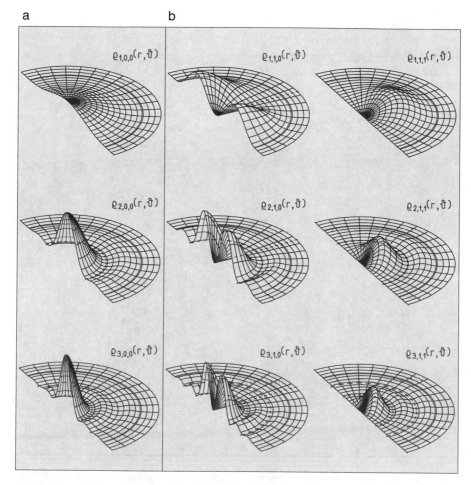

**Fig. 13.3.** Absolute squares $\rho_{n\ell m}(r,\vartheta) = |\varphi_{n\ell m}(r,\vartheta,\phi)|^2$ **of the full three-dimensional eigenfunction of a square-well potential. Here** $\rho_{n\ell m}(r,\vartheta)\,\mathrm{d}V$ **represents the probability of observing the particle in the volume element** $\mathrm{d}V$ **at location** $(r,\vartheta,\phi)$**. It is a function only of the distance** $r$ **from the origin and of the polar angle** $\vartheta$**. (a) In this figure, which applies to zero angular-momentum quantum number** $\ell$**, the function** $\varphi(r)$ **depends only on** $r$**. For values** $n = 1, 2, 3$ **of the principal quantum number it has** $n - 1 = 0, 1, 2$ **nodes in** $r$ **indicated by the dashed half-circles. Each plot gives the probability density for observing the particle at any point in a half-plane containing the** $z$ **axis. Here and in Figure 13.4 all plots have the same scale in** $r$ **and** $\vartheta$**. They do, however, have different scale factors in** $\rho$**. (b) The functions** $\rho_{n\ell m}(r,\vartheta)$ **as given in (a) but for** $\ell = 1$ **and** $m = 0, 1$**. The** $\vartheta$ **dependence is given by the Legendre functions** $P_\ell^{|m|}(\cos\vartheta)$ **which have** $\ell - |m|$ **nodes in** $\vartheta$**, indicated by the dashed lines** $\vartheta = \mathrm{const}$**.**

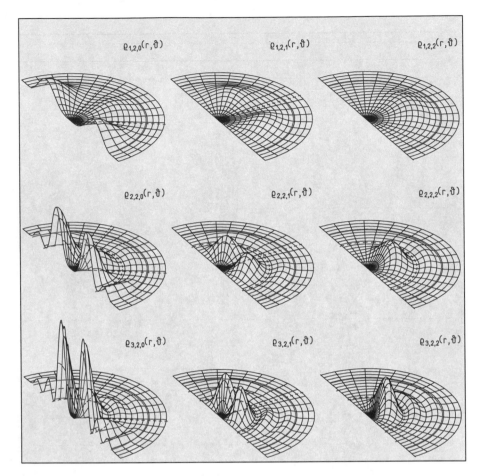

**Fig. 13.4. The functions** $\rho_{n\ell m}(r, \vartheta)$ **as given in Figure 13.3 but for** $\ell = 2$ **and** $m = 0, 1, 2$.

it can be shown in an $r, \vartheta$ plot. In Figures 13.3 and 13.4 this function is plotted as a surface over a half-circle ($0 \leq r \leq R$; $0 \leq \vartheta \leq \pi$) in the $x, z$ plane. It is the probability density for observing a particle at location $(r, \vartheta, \phi)$; that is to say

$$
\begin{aligned}
\mathrm{d}w &= |\varphi_{n\ell m}(\mathbf{r})|^2 \, \mathrm{d}V \\
&= R_{n\ell}^2 \frac{2\ell + 1}{4\pi} \frac{(\ell - |m|)!}{(\ell + |m|)!} \left[ P_\ell^{|m|}(\cos\vartheta) \right]^2 r^2 \, \mathrm{d}r \, \mathrm{d}\cos\vartheta \, \mathrm{d}\phi
\end{aligned}
$$

is the probability of finding the particle in the volume element $\mathrm{d}V = r^2 \, \mathrm{d}r$ $\mathrm{d}\cos\vartheta \, \mathrm{d}\phi$ at $(r, \vartheta, \phi)$. In Figures 13.3 and 13.4 we recognize the nodes in $r$ as half-circles in the plane at which the probability density vanishes. They are attributable to the nodes in the radial wave function $R_{n\ell}(r)$. In addition, there are $\ell - |m|$ nodes in $\vartheta$ along rays $\vartheta = $ const in the plane originating from zeros of $P_\ell^m(\cos\vartheta)$.

## 13.2 Bound States of the Spherically Symmetric Harmonic Oscillator

For many model calculations in nuclear physics, a harmonic-oscillator potential has proved to be useful. The potential energy of a spherically symmetric harmonic oscillator is

$$V(\mathbf{r}) = \frac{k}{2}r^2 = \frac{k}{2}(x_1^2 + x_2^2 + x_3^2) \quad .$$

The stationary Schrödinger equation for a particle of mass $M$ moving in this potential has the form

$$\left(-\frac{\hbar^2}{2M}\nabla^2 + \frac{k}{2}r^2\right)\varphi(\mathbf{r}) = E\varphi(\mathbf{r}) \quad .$$

Instead of the separation of variables in polar coordinates, as discussed in Section 10.7, we may just as well carry out the separation in Cartesian coordinates, for the potential is a sum of terms, each of which depends on only one of these coordinates. We start with the factorized *ansatz*

$$\varphi(\mathbf{r}) = \varphi_1(x_1)\varphi_2(x_2)\varphi_3(x_3)$$

and arrive at three Schrödinger equations for one-dimensional harmonic oscillators in the coordinates $x_1$, $x_2$, and $x_3$, which are identical to the equation discussed in Section 6.3 for the coordinate $x$,

$$\left(-\frac{\hbar^2}{2M}\frac{d^2}{dx_i^2} + \frac{M}{2}\omega^2 x_i^2\right)\varphi_i(x_i) = E_i\varphi_i(x_i) \quad , \qquad i = 1, 2, 3 \quad ,$$

$$\omega = \sqrt{k/M} \quad .$$

From Section 6.3 we know that the energy eigenvalues are

$$E_i = E(n_i) = (n_i + \frac{1}{2})\hbar\omega \quad , \qquad n_i = 0, 1, 2, \ldots \quad ,$$

with independent integer quantum numbers $n_i$ for the three oscillators. The total energy $E$ depends on the three quantum numbers $n_1$, $n_2$, and $n_3$,

$$\begin{aligned} E(n_1, n_2, n_3) &= E(n_1) + E(n_2) + E(n_3) \\ &= (n_1 + n_2 + n_3 + \frac{3}{2})\hbar\omega \quad . \end{aligned}$$

The eigenfunctions $\varphi_{n_i}(x_i)$ are normalized products of Hermite polynomials and Gaussians. They were shown in Figures 6.4 and 6.5.

The eigenfunctions of the three-dimensional harmonic oscillator are

$$\varphi'_{n_1,n_2,n_3}(x_1,x_2,x_3) = \varphi_{n_1}(x_1)\varphi_{n_2}(x_2)\varphi_{n_3}(x_3)$$

with the eigenvalue $E(n_1,n_2,n_3)$. Figure 13.5 shows, as an example, the eigenfunction

$$\varphi'_{210}(x_1,x_2,x_3) = \varphi_2(x_1)\varphi_1(x_2)\varphi_0(x_3) \quad .$$

Since it is a function of the independent coordinates $x_1$, $x_2$, and $x_3$, we represent it by plotting it for various planes $x_3 = \text{const}$ in $x_1,x_2,x_3$ space. Since the $x_3$ dependence is given by the simple Gaussian factor

$$\varphi_0(x_3) = \text{const} \cdot \exp\left(-\frac{x_3^2}{2\sigma_0^2}\right) \quad , \qquad \sigma_0^2 = \frac{\hbar}{M\omega} \quad ,$$

the function is symmetric in $x_3$ and is damped away as $x_3$ increases in magnitude (see Section 6.3). It is also symmetric in $x_1$, and antisymmetric in $x_2$.

Obviously, all the different quantum-number triplets $n_1$, $n_2$, $n_3$ having the same sum correspond to different eigenfunctions $\varphi'_{n_1n_2n_3}$, that is, to different physical states of the system. All these physical states, however, have the same energy eigenvalue. They are therefore called *degenerate states*.

The usual separation of the three-dimensional Schrödinger equation in polar coordinates yields the radial Schrödinger equation

$$\left[-\frac{\hbar^2}{2M}\frac{1}{r}\frac{d^2}{dr^2}r + V_\ell^{\text{eff}}(r)\right]R_{n\ell}(r) = E_n R_{n\ell}(r)$$

with the effective potential

$$V_\ell^{\text{eff}}(r) = \frac{\hbar^2}{2M}\frac{\ell(\ell+1)}{r^2} + \frac{k}{2}r^2 \quad .$$

The solutions of this equation are

$$R_{n\ell}(r) = N_{n\ell}\left(\frac{r^2}{\sigma_0^2}\right)^{\ell/2}\exp\left(-\frac{r^2}{2\sigma_0^2}\right)L_{n_r}^{\ell+1/2}\left(\frac{r^2}{2\sigma_0^2}\right) \quad ,$$

where the functions $L_{n_r}^{\ell+1/2}$ are the *Laguerre polynomials*,

$$L_{n_r}^{\ell+1/2}(x) = \sum_{j=0}^{n_r}(-1)^j\binom{n_r+\ell+\frac{1}{2}}{n_r-j}\frac{x^j}{j!} \quad , \qquad n_r = 0, 1, 2,\ldots \quad .$$

The normalization constants are

$$N_{n\ell} = \sqrt{\frac{n_r!2^{n+2}}{[2(\ell+n_r)+1]!!\sqrt{\pi}\sigma_0^3}} \quad .$$

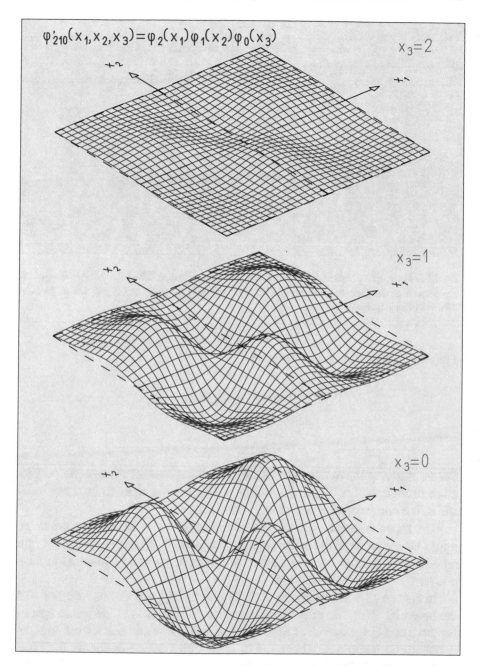

**Fig. 13.5. Eigenfunction** $\varphi'_{210}(x_1,x_2,x_3) = \varphi_2(x_1)\varphi_1(x_2)\varphi_0(x_3)$ **of the three-dimensional harmonic oscillator expressed in Cartesian coordinates** $x_1, x_2, x_3$ **and written as a product of three one-dimensional harmonic-oscillator eigenfunctions. For this figure the width parameter** $\sigma_0 = 1$ **was chosen. The function is plotted for three planes** $x_3 = 0, 1, 2$**. Because** $\varphi_0(x_3)$ **is symmetric, the plots remain unchanged if the substitution** $x_3 \to -x_3$ **is performed. This figure should be compared with Figure 6.4.**

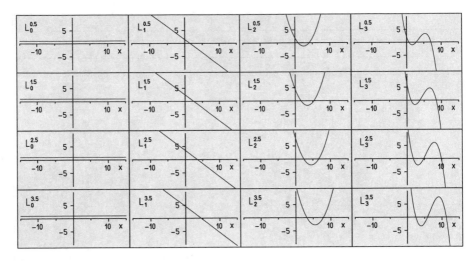

**Fig. 13.6. Laguerre polynomials of half-integer upper index. The lower index is equal to the degree of the polynomial and to the number of its zeros. All zeros are at positive values of the argument** $x$**.**

The energy eigenvalues are given by

$$E_n = (n + \frac{3}{2})\hbar\omega$$

with

$$n = 2n_r + \ell \quad .$$

Before studying the radial solutions $R_{n\ell}$, we first present the Laguerre polynomials in Figure 13.6. The degree of the polynomial is equal to its lower index and to the number of zeros on the positive $x$ axis.

The radial solutions $R_{n\ell}$ are shown in Figure 13.7. Their zeros are determined by the zeros of the corresponding Laguerre polynomial. Because of the relation between the integer quantum numbers $n$, $n_r$, and $\ell$, quantum number $n$ takes the values $\ell$, $\ell+2$, $\ell+4$, ....

In Figure 13.8 the functions $R_{n\ell}$, $R_{n\ell}^2$, and $r^2 R_{n\ell}^2$ are shown together with the potential $V(r)$, the effective potential $V_\ell^{\text{eff}}(r)$, and the eigenvalue spectra for the lowest eigenstates of the harmonic oscillator and the lowest angular-momentum quantum numbers $\ell = 0, 1, 2$. With increasing energy the functions reach out farther in $r$ since the potential increases with $r^2$. The functions are again suppressed near $r = 0$ for $\ell \neq 0$ by the centrifugal barrier. The suppression is strongest for low energy $E$ but high angular-momentum quantum numbers $\ell$.

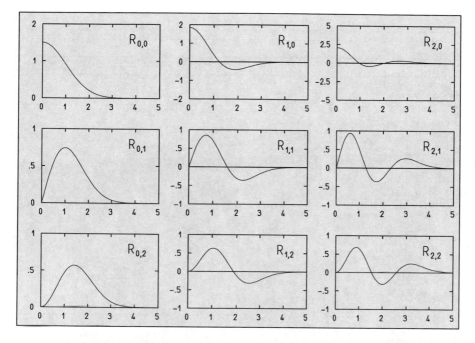

**Fig. 13.7. Radial eigenfunctions** $R_{n_r\ell}(\rho)$, $n = 2n_r + \ell$ **for the three-dimensional harmonic oscillator. Their zeros are the** $(n-1)/2$ **zeros of the Laguerre polynomial** $L_{(n-\ell)/2}^{\ell+1/2}(\rho^2)$. **The argument** $\rho$ **is the distance** $r$ **from the origin divided by the width** $\sigma_0$ **of the ground state of the oscillator. Graphs in the same column belong to the same value of** $n_r$. **Graphs in the same row belong to the same value of** $\ell$.

The three-dimensional stationary wave functions are

$$\varphi_{n\ell m}(\mathbf{r}) = R_{n\ell}(r)Y_{\ell m}(\vartheta,\phi) \quad .$$

Their absolute squares $|\varphi_{n\ell m}|^2$, which are independent of the azimuth $\phi$, are plotted in Figures 13.9 and 13.10 for low values of $n$ and $\ell = 0, 1, 2$. Since the energy eigenvalues $E_n$ depend on one quantum number only, there are again degenerate eigenfunctions. From the properties of the spherical harmonics, we know that for every $\ell$ there are $2\ell + 1$ states of different quantum number $m$. Moreover, for a given energy eigenvalue $E_n$ there are eigenstates with different angular momenta $\ell$. Because of the relation $n = 2n_r + \ell$, the number $n$ of quanta of energy $\hbar\omega$ above the energy $\frac{3}{2}\hbar\omega$ of the ground state is even or odd, depending on whether $\ell$ is even or odd. Actually there are $(n+1)(n+2)/2$ degenerate eigenfunctions to a given energy eigenvalue $E_n$.

How are the two different sets of solutions $\varphi'_{n_1,n_2,n_3}$ and $\varphi_{n\ell m}$ related? Obviously, we have to be able to describe the same physical states by either set. In fact, we are able to do so because most of the states are degenerate; that is, a large number of states have the same eigenvalue. Obviously too, a lin-

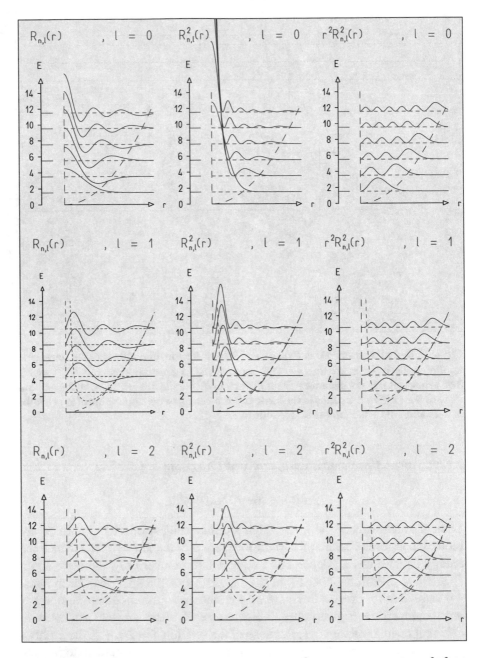

**Fig. 13.8. Radial eigenfunctions** $R_{n\ell}(r)$, **their squares** $R_{n\ell}^2(r)$, **and the functions** $r^2 R_{n\ell}^2(r)$ **for the lowest eigenstates of the harmonic oscillator and the lowest angular-momentum quantum numbers** $\ell = 0, 1, 2$. **On the left side are the eigenvalue spectra. The form of the harmonic-oscillator potential** $V(r)$ **is indicated by a long-dash line, and, for** $\ell \neq 0$, **that of the effective potential** $V_\ell^{\text{eff}}(r)$ **by a short-dash line. The eigenvalues have equidistant spacing. The eigenvalue spectra are degenerate for all even** $\ell$ **values and all odd** $\ell$ **values, except that the minimum value of the principal quantum number is** $n = \ell$.

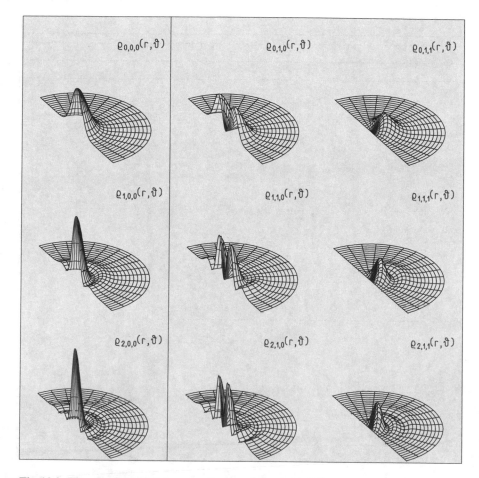

**Fig. 13.9.** The absolute squares $\rho_{n_r \ell m}(r, \vartheta) = |\varphi_{n_r \ell m}(r, \vartheta, \phi)|^2$, $n_r = (n - \ell)/2$ **of the full three-dimensional eigenfunctions for the harmonic oscillator. The absolute squares are functions only of $r$ and $\vartheta$. There are $n_r$ radial nodes, and $\ell - |m|$ polar nodes, indicated by dashed half-circles and rays, respectively. Each figure gives the probability density for observing the particle at any point in a half-plane containing the $z$ axis. All pictures have the same scale in $r$ and $\vartheta$. They do, however, have different scale factors in $\rho$. In this figure the $\rho_{n_r \ell m}$ are shown for $\ell = 0$ (left) and $\ell = 1$ (right).**

ear superposition of degenerate eigenstates is again an eigenstate of the same energy. Thus it is possible to express the eigenstates of a given energy in one set by a linear superposition of the eigenstates of the same energy in the other set. The only nondegenerate eigenstate is the ground state

$$\varphi'_{000}(\mathbf{r}) = \pi^{-3/4} \sigma_0^{-3/2} \exp\left(-\frac{r^2}{2\sigma_0^2}\right) = \varphi_{000}(\mathbf{r}) \quad ,$$

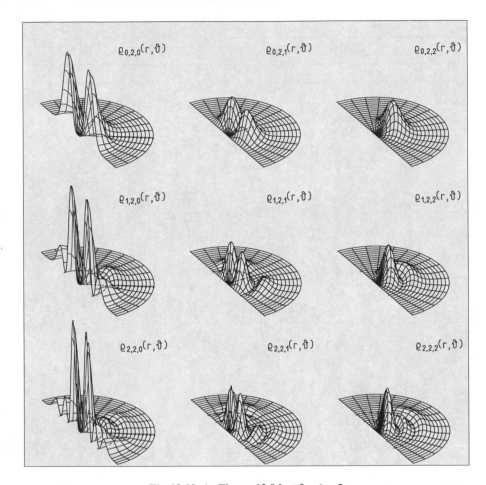

**Fig. 13.10. As Figure 13.9 but for $\ell = 2$.**

which has the ground-state energy $E_0 = \frac{3}{2}\hbar\omega$. This eigenstate is the same in both sets. All other states are degenerate. As an example of a superposition of Cartesian eigenstates $\varphi'_{n_1 n_2 n_3}$ which forms an angular-momentum eigenstate $\varphi_{n\ell m}$, we look at $n = 2$, $\ell = 2$, and $m = 0$. We have

$$\varphi_{220}(\mathbf{r}) = -\frac{1}{\sqrt{6}}\varphi'_{200}(\mathbf{r}) - \frac{1}{\sqrt{6}}\varphi'_{020}(\mathbf{r}) + \sqrt{\frac{2}{3}}\varphi'_{002}(\mathbf{r}) \quad .$$

Figure 13.11 demonstrates this particular superposition. Figures 13.11 a, b, and c give the three terms of this superposition in the $x_1, x_3$ plane. In Figure 13.11d the sum $\varphi_{220}(\mathbf{r})$ is shown. In Figure 13.11e its absolute square is plotted in the $x_1, x_3$ half-plane to facilitate comparison with the $r, \vartheta$ plot of this same function $|\varphi_{220}(\mathbf{r})|^2$, which is given in Figure 13.11f.

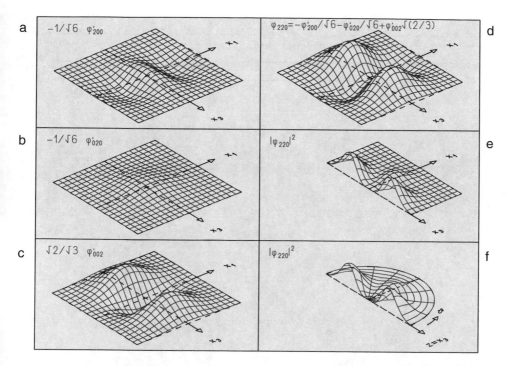

**Fig. 13.11.** An eigenstate $\varphi_{n\ell m}(r,\vartheta,\phi)$ of the harmonic oscillator can be written as a linear superposition of the degenerate eigenfunctions $\varphi'_{n_1 n_2 n_3}(x_1,x_2,x_3)$ having the same energy eigenvalue (a, b, c). The three eigenfunctions for $n=2$ in the $x_1,x_2$ plane each multiplied by the appropriate factor; (d) the sum; (e) its square; (f) the function $|\varphi_{220}|^2$ in $r,\vartheta$ representation as known from Figure 13.10. Parts e and f are identical except that part e has Cartesian coordinates, part f polar coordinates.

## 13.3 Harmonic Particle Motion in Three Dimensions

In Section 6.4 we described the motion of a Gaussian wave packet in a one-dimensional harmonic-oscillator potential. We obtained for the absolute square of the time-dependent wave functions a Gaussian distribution with an expectation value oscillating like the classical point particle. Its width oscillates with twice the oscillator frequency. We therefore anticipate that in the three-dimensional oscillator the expectation value of a three-dimensional Gaussian wave packet moves on an elliptical trajectory as a classical point particle does. The shape of the three-dimensional wave packet is completely described by its covariance ellipsoid, which we introduced in Section 10.1. The shape of the covariance ellipsoid itself oscillates, that is, it changes periodically with time, its frequency being twice the oscillator frequency.

Figure 13.12 shows two examples for such a motion. The classical trajectory is chosen to be identical for both. For simplicity the covariance el-

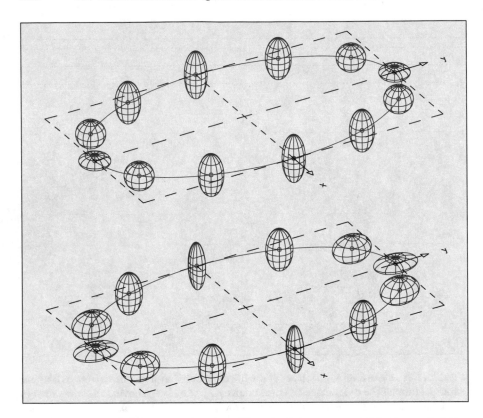

**Fig. 13.12. A three-dimensional Gaussian wave packet, represented by its probability ellipsoid, moves under the influence of an attractive force described by a harmonic-oscillator potential. Its expectation value, that is, the center of the ellipsoid, describes an elliptical trajectory. The initial conditions were chosen so that the ellipsoid does not tumble, that is, its principal axes keep constant orientations. The magnitudes of the principal axes oscillate with twice the oscillator frequency. Two examples are shown. Top: The ellipsoid stays rotationally symmetric with respect to the $z$ axis. Bottom: All three principal axes of the ellipsoid are different.**

lipsoid has two of its principal axes in the plane of motion. Moreover, the initial conditions were chosen so that the directions of its principal axes do not change while the ellipsoid is moving. Because the harmonic oscillator is spherically symmetric, the oscillation in magnitude of all three principal axes has the same frequency but may have different phases. In Figure 13.12 (top) the covariance ellipsoid stays rotationally symmetric with respect to the axis perpendicular to the plane of motion. The size of the ellipsoid changes dramatically with time. So does its shape: it oscillates between prolate and oblate. In Figure 13.12 (bottom) all three principal axes of the ellipsoid are in general different: the ellipsoid does not have rotational symmetry.

## 13.4 The Hydrogen Atom

The most fundamental application of quantum mechanics is atomic physics. The simplest atom is that of *hydrogen*; it consists of a simple nucleus, the proton, and one electron bound by the electric force acting between them. Since the mass of the proton is nearly 2000 times that of the electron, M, the center of mass of the atom, for our purposes, coincides with the position of the proton. We choose it to be the origin of our polar coordinate system. The potential energy of the electron, which carries the charge $-e$ in the electric field of the proton of charge $+e$, is given by the *Coulomb potential* of the proton,

$$U(\mathbf{r}) = \frac{e}{4\pi\varepsilon_0}\frac{1}{r} \quad ,$$

multiplied by the charge of the electron:

$$E_{\text{pot}} = V(r) = -\frac{e^2}{4\pi\varepsilon_0}\frac{1}{r} \quad .$$

Here the $\varepsilon_0 = 8.854188 \times 10^{-12}\frac{\text{C}}{\text{Vm}}$ is the electric field constant also called the permittivity of free space. The constant $e^2/(4\pi\varepsilon_0)$ has the dimension of action times velocity. It can therefore be expressed by a multiple of two fundamental constants of nature, namely Planck's constant $\hbar$ and the speed of light $c$. Inserting numbers, we obtain

$$\frac{e^2}{4\pi\varepsilon_0} = \alpha\hbar c \quad , \qquad \alpha = \frac{1}{137} \quad .$$

The dimensionless proportionality constant $\alpha$ is called the *fine-structure constant*. It was introduced by Arnold Sommerfeld in 1916.

The stationary Schrödinger equation for the hydrogen atom then has the form

$$\left(-\frac{\hbar^2}{2M}\nabla^2 - \hbar c\frac{\alpha}{r}\right)\varphi(\mathbf{r}) = E\varphi(\mathbf{r})$$

with $M$ the electron mass. We solve this equation with the separation *ansatz* in polar coordinates,

$$\varphi(\mathbf{r}) = R(r)Y_{\ell m}(\vartheta,\phi) \quad ,$$

which yields the radial Schrödinger equation for the hydrogen atom,

$$\left[-\frac{\hbar^2}{2M}\frac{1}{r}\frac{d^2}{dr^2}r + V_\ell^{\text{eff}}(r)\right]R_{n\ell}(r) = E_n R_{n\ell}(r) \quad .$$

It is an eigenvalue equation for the radial eigenfunctions $R_{n\ell}$ with the energy eigenvalues $E_n$. The effective potential is the sum of the centrifugal potential and the Coulomb potential:

$$V_\ell^{\text{eff}}(r) = \frac{\hbar^2}{2M} \frac{\ell(\ell+1)}{r^2} - \hbar c \frac{\alpha}{r} \quad .$$

The energy eigenvalues $E_n$ depend on the *principal quantum number n* only. We have

$$E_n = -\frac{1}{2}Mc^2 \frac{\alpha^2}{n^2} \quad , \qquad n = 1, 2, \ldots \quad .$$

They form an infinite set of discrete energies. The coefficient in this equation has the value $Mc^2\alpha^2/2 = 13.61\,\text{eV}$.

The normalized radial wave functions $R_{n\ell}$ have the form

$$R_{n\ell} = N_{n\ell} \left(\frac{2r}{na}\right)^\ell e^{-r/na} L_{n-\ell-1}^{2\ell+1}\left(\frac{2r}{na}\right) \quad ,$$
$$n = 1, 2, 3, \ldots \quad , \qquad \ell = 0, 1, 2, \ldots, n-1 \quad ,$$

with the normalization factor

$$N_{n\ell} = \frac{1}{a^{3/2}} \frac{2}{n^2} \sqrt{\frac{(n-\ell-1)!}{(n+\ell)!}} \quad .$$

Here the parameter

$$a = \frac{\hbar}{\alpha Mc} = 0.5292 \times 10^{-10}\,\text{m}$$

is the *Bohr radius* of the innermost orbit. In the model of the hydrogen atom that was put forward by Niels Bohr in 1913, the electron can turn around the nucleus in circular orbits. These orbits can have only certain discrete radii $r_n = n^2\hbar/(\alpha Mc)$. The innermost orbit for $n = 1$ is $r_1 = a$.

The function $L_{n-\ell-1}^{2\ell+1}(x)$ is a particular *Laguerre polynomial*,

$$L_p^k(x) = \sum_{s=0}^p (-1)^s \binom{p+k}{p-s} \frac{x^s}{s!} \quad ,$$

with the integer upper index $k = 2\ell + 1$. Some of these polynomials with low values of $p$ and $k$ are shown in Figure 13.13. We note that the number of zeros equals $p$, and that all zeros occur for positive values of the argument $x$. In Section 13.2 we found that the radial wave functions of the spherically symmetric harmonic oscillator contain the Laguerre polynomials $L_{n_r}^{\ell+1/2}(x)$ of half-integer upper index. They were plotted in Figure 13.6 for a few values of the indices. A comparison of Figures 13.13 and 13.6 reveals a strong similarity between the two sets of polynomials.

The radial wave functions $R_{n\ell}$ for the electron in the hydrogen atom are shown in Figure 13.14. Their behavior for large values of $r$ is dominated by the exponential $\exp[-r/(na)]$. Near $r = 0$ it is determined by the power

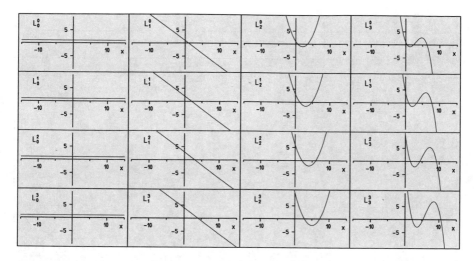

**Fig. 13.13. Laguerre polynomials of integer upper index. The lower index is equal to the degree of the polynomial and to the number of its zeros. All zeros are at positive values of the argument $x$. The graphs look rather similar to those of Figure 13.6, which shows the Laguerre polynomials of half-integer upper index. That they are in fact different can be seen, for example, from the positions of the zeros.**

$[2r/(na)]^\ell$. Their zeros are those of the corresponding Laguerre polynomial, that is, the radial wave functions $R_{n\ell}(r)$ possess $n - \ell - 1$ zeros.

Let us compare the radial wave functions of the hydrogen atom with those of the harmonic oscillator. We realize that with increasing energy eigenvalue the wave functions of the hydrogen atom spread much faster to larger radii than do those of the harmonic oscillator. Obviously, the reason is that the Coulomb potential becomes wide with total energy much more quickly than the harmonic-oscillator potential does. This difference manifests itself in the analytic form of the wave functions of the two systems. The radial wave functions of the harmonic oscillator, as presented in Section 13.2, contain the factor $\exp[-r^2/(2\sigma_0^2)]$, which varies little for $r^2/(2\sigma_0^2) \ll 1$ and approaches zero very quickly for values $r^2/(2\sigma_0^2) > 1$. The radial wave functions of the hydrogen atom contain the factor $\exp[-r/(na)]$, which varies more strongly for $r/(na) \ll 1$ and falls off to zero much more slowly in the region $r/(na) > 1$.

The spectrum of energy eigenvalues is highly degenerate because, for a given quantum number $n$, the angular-momentum quantum number $\ell$ can take any one of the values $0 \leq \ell \leq n - 1$, and, for a given $\ell$, quantum number $m$ of the $z$ component $L_z$ of angular momentum runs between $-\ell \leq m \leq \ell$. Thus for a given $n$ there are $\sum(2\ell + 1) = n^2$ different states, all having the same eigenvalue $E_n$.

In Figure 13.15 the radial wave functions $R_{n\ell}(r)$ are shown together with the Coulomb potential $V(r)$ and the effective potential $V_\ell^{\text{eff}}(r)$ for the lowest

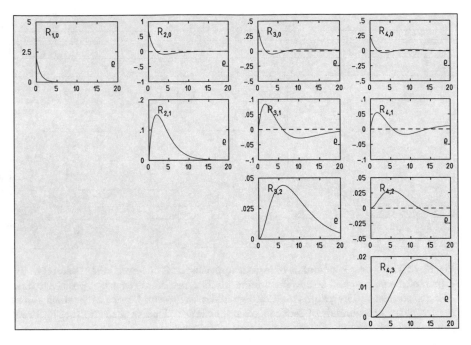

**Fig. 13.14. Radial eigenfunctions** $R_{n\ell}(\rho)$ **for the electron in the hydrogen atom. Their zeros are the** $n - \ell - 1$ **zeros of the Laguerre polynomials** $L_{n-\ell-1}^{2\ell+1}(2\rho/n)$. **Here the argument of the Laguerre polynomial is** $2\rho/n$ **with** $n$ **being the principal quantum number and** $\rho = r/a$ **the distance between electron and nucleus divided by the Bohr radius** $a$.

values of the principal quantum number, $n = 1, \ldots, 5$, and for the lowest values of the angular-momentum quantum number, $\ell = 0, 1, 2$. Figure 13.15 also contains the plots for $R_{n\ell}^2(r)$ and $r^2 R_{n\ell}^2(r)$.

It is interesting to compare the radial wave functions and the energy spectra for the three types of potentials discussed in this chapter, namely the square-well potential (Figure 13.1), the harmonic-oscillator potential (Figure 13.8), and the Coulomb potential (Figure 13.15). For the square-well potential, which vanishes in the external region, the wave functions fall off like an exponential function in this region. The Coulomb potential approaches zero with increasing $r$. The wave functions are products of polynomials and an exponential function, so that they taper off for large $r$ like a power of $r$ times the exponential function. Finally, for the quadratically increasing potential of the harmonic oscillator, the falloff of the wave functions is more pronounced. They behave like $r^n \exp(-r^2/2\sigma_0^2)$ for large $r$. This behavior reflects the intuitive expectation that an ever-increasing potential confines the particle best and that attractive potentials that approach zero better confine the particle the faster they reach zero.

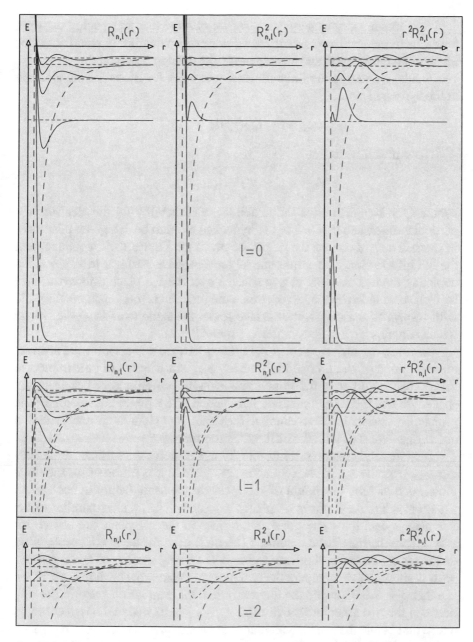

**Fig. 13.15. Radial eigenfunctions** $R_{n\ell}(r)$, **their squares** $R_{n\ell}^2(r)$, **and the functions** $r^2 R_{n\ell}^2(r)$ **for the lowest eigenstates of the electron in the hydrogen atom and the lowest angular-momentum quantum numbers** $\ell = 0, 1, 2$. **Also shown are the energy eigenvalues as horizontal dashed lines, the form of the Coulomb potential** $V(r)$, **and, for** $\ell \neq 0$, **the forms of the effective potential** $V_\ell^{\text{eff}}(r)$. **The eigenvalue spectra are degenerate for all** $\ell$ **values, except that the minimum value of the principal quantum number is** $n = \ell + 1$.

Now, looking at the energy spectra, we observe that the spacing between levels increases with energy for the square-well potential, is equidistant for the harmonic oscillator, and decreases for the Coulomb potential.

Finally, we turn to the three-dimensional wave functions for the electron in the hydrogen atom,

$$\varphi_{n\ell m}(\mathbf{r}) = R_{n\ell}(r)Y_{\ell m}(\vartheta,\phi) \quad .$$

The probability density

$$\rho(\mathbf{r}) = \rho_{n\ell m}(r,\vartheta) = |\varphi_{n\ell m}(r,\vartheta,\phi)|^2$$

contains the complete information about the probability for the electron in a particular eigenstate $(n,\ell,m)$ of the hydrogen atom to be at a given position $\mathbf{r}$ in space. Then, of course, the graphs of $\rho(r,\vartheta)$ in Figure 13.16 again contain the full information. The graphs can be understood as surfaces in $x,z,\rho$ space depicting the function $\rho(x,z)$ as a surface over the $x,z$ plane (more exactly a half-plane bounded by the $z$ axis). Since the function is rotationally symmetric with respect to the $z$ axis, the surface looks the same over any other plane which contains the $z$ axis, e.g., the $y,z$ plane.

However, the surfaces shown in Figure 13.16 are still a somewhat abstract representation of the probability density $\rho_{n\ell m}$ since they are constructed in $x,z,\rho$ space and not in the three-dimensional position space, i.e., in $x,y,z$ space. We will therefore construct plots which give a direct impression of the probability density in space although they cannot contain the full information about $\rho_{n\ell m}$. We start with the different states for $n=2$.

The left-hand column of Figure 13.17 again shows the surfaces representing $\rho_{2\ell m}$ over the $x,z$ plane. (Note that the scale in $\rho$ is different for the three plots, such that the maximum of $\rho$ appears at the same height in each plot.) Cuts through these surfaces by planes $\rho=$ const (i.e., planes parallel to the $x,z$ plane) yield lines $\rho=$ const. Such lines in the $x,z$ plane are shown for $\rho=0.02$ in the right-hand column. (The units used for the spatial probability density is $a^{-3}$, $a$ being the Bohr radius.) The interpretation of these *contour plots* is simple. If we compare the two plots in the top row of Figure 13.17 we see that $\rho_{200}>0.02$ inside the inner circle in the contour plot and in the ring between the two outer circles. Similarly, $\rho_{210}>0.02$ inside the two contours above and below the $x$ axis which are symmetrically traversed by the $z$ axis, and $\rho_{211}>0.02$ inside the two contours situated symmetrically with respect to each other to the left and to the right of the $z$ axis.

The generalization of contour lines in the $x,z$ plane to surfaces of constant probability density in three-dimensional $x,y,z$ space is now simple. The surfaces are created by rotation of the contour lines around the $z$ axis. The surfaces of constant probability are shown in Figure 13.18. In the right-hand column they are shown in all of space, in the left-hand column only in the

$$\rho_{n\ell m}(r,\vartheta) = |\varphi_{n\ell m}(r,\vartheta,\varphi)|^2$$

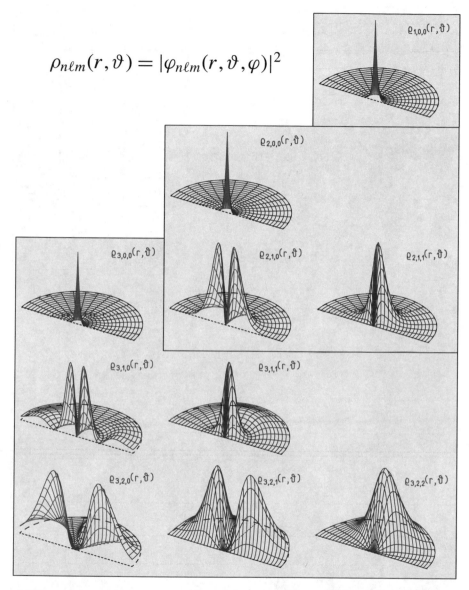

**Fig. 13.16.** The absolute squares $\rho_{n\ell m}(r,\vartheta) = |\varphi_{n\ell m}(r,\vartheta,\phi)|^2$ of the full three-dimensional wave functions for the electron in the hydrogen atom. They are functions only of $r$ and $\vartheta$. All eigenstates having the same principal quantum number have the same energy eigenvalue $E_n$. The possible angular-momentum quantum numbers are $\ell = 0, 1, \ldots, n-1$. The wave functions have $n - \ell - 1$ nodes in $r$ and $\ell - |m|$ nodes in $\vartheta$, indicated by dashed half-circles and rays, respectively. Each figure gives the probability density for observing the electron at any point in a half-plane containing the $z$ axis. All pictures have the same scale in $r$ and $\vartheta$. They do, however, have different scale factors in $\rho$.

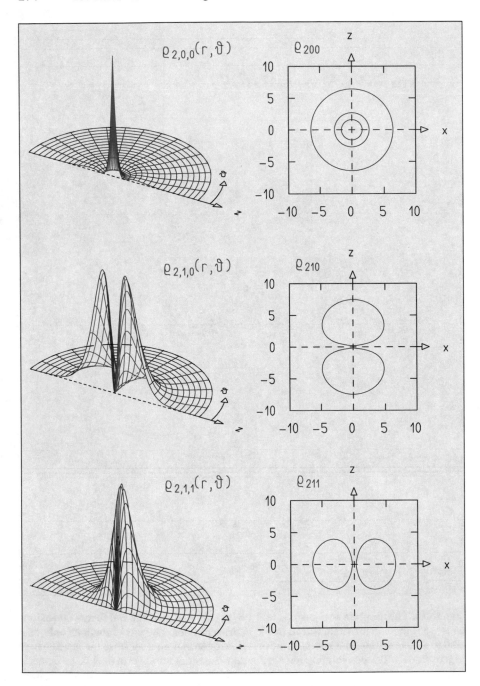

**Fig. 13.17. Left: Spatial probability density** $\rho_{2\ell m}$ **for an electron in the hydrogen atom shown over a half-plane bounded by the** $z$ **axis. Different scales in** $\rho_{2\ell m}$ **are used in the three plots. Right: Contour lines** $\rho_{2\ell m} = 0.002$ **in the** $x, z$ **plane. Numbers are in units of the Bohr radius.**

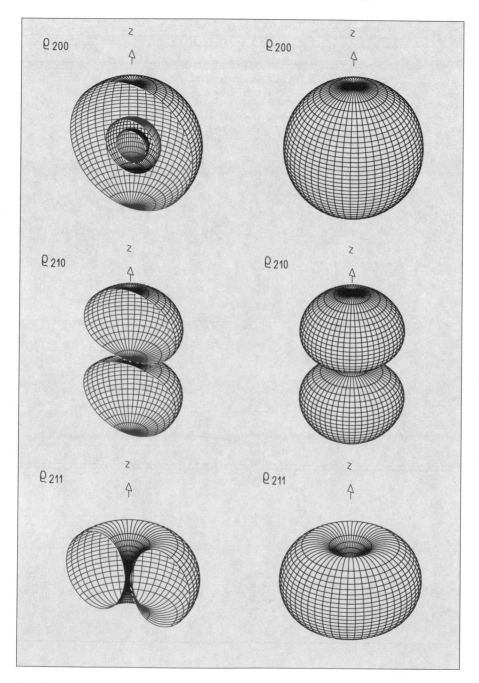

**Fig. 13.18. Surfaces of constant probability density $\rho_{2\ell m} = 0.002$ in full $x, y, z$ space (right) and in the half-space $x > 0$ (left).**

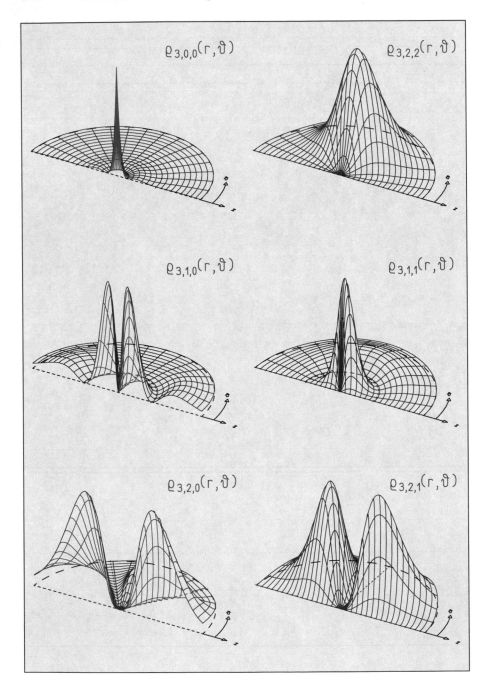

**Fig. 13.19. Spatial probability density $\rho_{3\ell m}$ for an electron in the hydrogen atom shown over a half-plane bounded by the $z$ axis. Different scales in $\rho_{3\ell m}$ are used.**

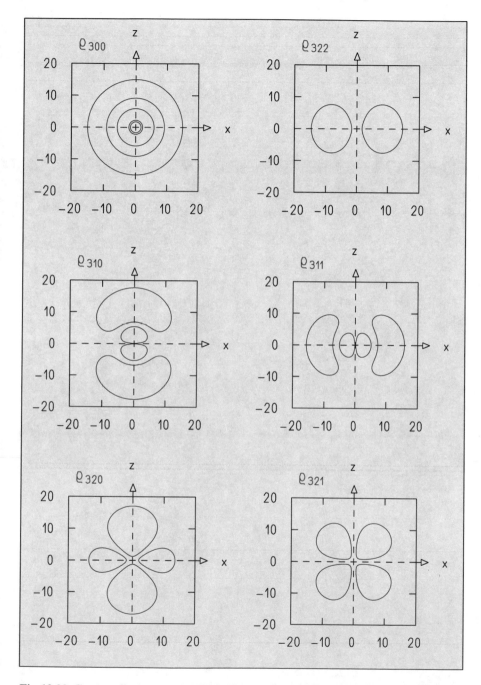

**Fig. 13.20. Contour lines** $\rho_{3\ell m} = 0.0002$ **in the** $x, z$ **plane. Numbers are in units of the Bohr radius.**

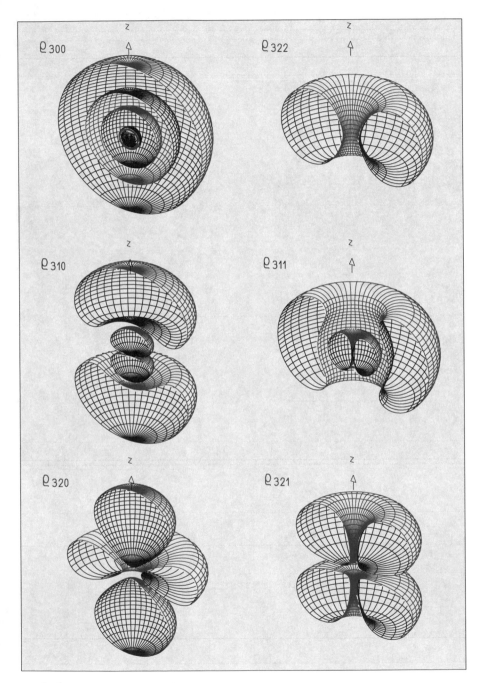

**Fig. 13.21. Surfaces of constant probability density** $\rho_{3\ell m} = 0.0002$ **in the half-space** $x > 0$.

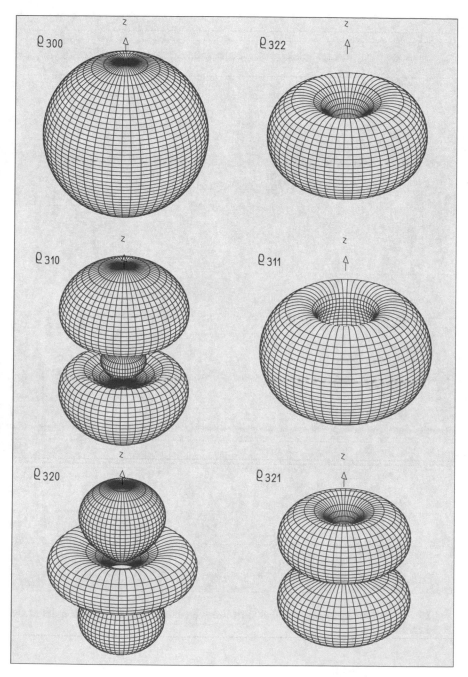

**Fig. 13.22. Surfaces of constant probability density** $\rho_{3\ell m} = 0.0002$ **in full** $x, y, z$ **space.**

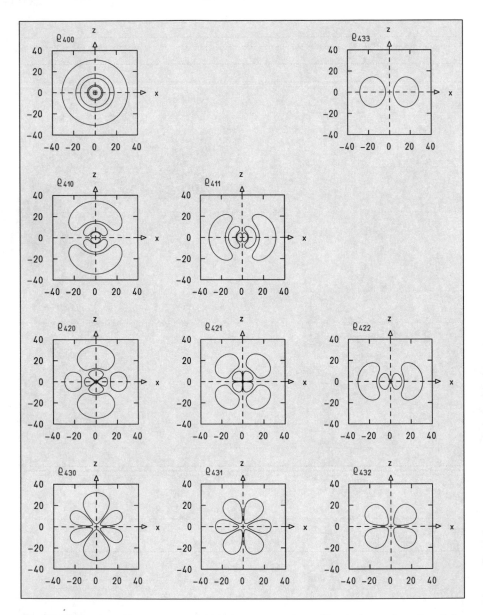

**Fig. 13.23. Contour lines** $\rho_{4\ell m} = 0.00002$ **in the** $x, z$ **plane. Numbers are in units of the Bohr radius.**

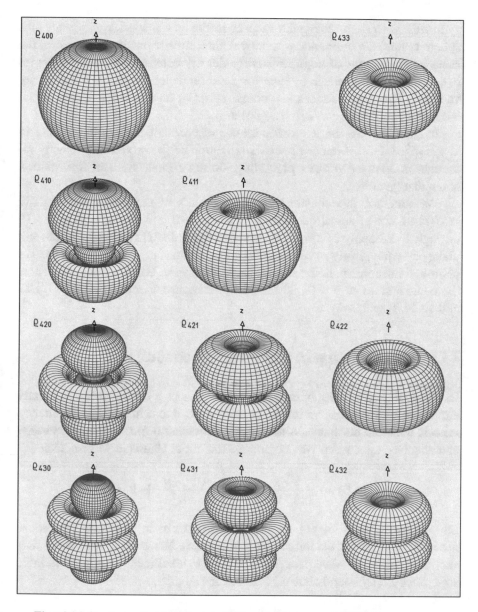

**Fig. 13.24. Surfaces of constant probability density** $\rho_{4\ell m} = 0.00002$ **in** $x, y, z$ **space.**

half-space $x > 0$, i.e., they are cut open to reveal a possible inner structure. Obviously, the lines generated by the cut are the contour lines of Figure 13.17. Regions of high probability density ($\rho > 0.02$) are enclosed by the surfaces. Outside the surfaces the probability density is small ($\rho < 0.02$). The region of high probability density is a kind of torus for $\rho_{211}$, it consists of two lobes symmetrical to each other with respect to the $x, y$ plane for $\rho_{210}$, and of a sphere around the origin and a spherical shell further outside for $\rho_{200}$. In the cases where regions of high probability density consist of separate volumes in space they are separated by *node surfaces* on which the probability density vanishes. These are surfaces $r = \text{const}$ (spheres) or $\vartheta = \text{const}$ (cones about the $z$ axis and, for $\vartheta = \pi/2$, the $x, y$ plane).

In summary we can say that surfaces of constant probability density in $x, y, z$ space allow a rather direct visualization of the probability density although, in contrast to other plots, they do not contain the full information about that quantity.

We conclude this chapter by showing plots of the probability density $\rho_{n\ell m}(\mathbf{r})$ for the principal quantum numbers $n = 3$ and $n = 4$. Figures 13.19 through 13.22 apply to $n = 3$ and contain the probability density in the $x, z$ plane or – equivalently – $r, \vartheta$ plane, the contour lines $\rho_{3\ell m} = 0.0002$ in the $x, z$ plane and the corresponding surfaces in $x, y, z$ space. For $n = 4$ we only show the contour lines $\rho_{4\ell m} = 0.00002$ and the contour surfaces in Figures 13.23 and 13.24, respectively.

## 13.5 Kepler Motion in Quantum Mechanics

In classical mechanics motion under the action of a central force is greatly simplified by the conservation of angular momentum in addition to energy. Also in quantum mechanics time-independent central forces conserve energy and angular momentum. We decompose the initial Gaussian wave packet

$$\psi(\mathbf{r}, 0) = \frac{1}{(2\pi)^{3/4}\sigma^{3/2}} \exp\left\{ -\frac{(\mathbf{r} - \mathbf{r}_0)^2}{4\sigma^2} + i\mathbf{k}_0 \cdot \mathbf{r} \right\}$$

into a linear superposition of eigenfunctions of the hydrogen atom. We choose the initial position $\mathbf{r}_0$, the initial momentum $\mathbf{p}_0 = \hbar \mathbf{k}_0$, and the spatial width $\sigma$ such that the initial wave packet $\psi(\mathbf{r}, 0)$ can to a sufficient approximation be superimposed by bound-state eigenfunctions $\varphi_{n\ell m}(\mathbf{r})$,

$$\psi(\mathbf{r}, 0) = \sum_{n=1}^{\infty}\sum_{\ell=0}^{n-1}\sum_{m=-\ell}^{\ell} b_{n\ell m}\varphi_{n\ell m}(\mathbf{r}) \quad .$$

The $\varphi_{n\ell m}(\mathbf{r})$ are eigenfunctions of the Hamiltonian containing the Coulomb potential to the eigenvalue (cf. Section 13.4)

$$E_n = -\frac{E_1}{n^2} \quad , \qquad n = 1, 2, 3, \ldots \quad .$$

Here

$$E_1 = \frac{1}{2}Mc^2\alpha^2 = 13.61\,\text{eV}$$

is the modulus of the ground-state energy.

The time-dependent wave packet can be obtained from $\psi(\mathbf{r},0)$ by multiplying the members of the sum with the time-dependent phase factors $\exp(-i\omega_n t)$ with the angular frequencies

$$\omega_n = \frac{E_n}{\hbar} \quad .$$

The wave packet at time $t$ is given by

$$\psi(\mathbf{r},t) = \sum_{n=1}^{\infty}\sum_{\ell=0}^{n-1}\sum_{m=-\ell}^{\ell} b_{n\ell m}\mathrm{e}^{-i\omega_n t}\varphi_{n\ell m}(\mathbf{r}) \quad .$$

As discussed in Section 10.10 the angular-momentum content of a Gaussian wave packet is described by the probability $W_{\ell m}$ for the total angular momentum $\ell$ and the $z$ component $m$ in the direction of classical angular momentum $\mathbf{L}_{\mathrm{cl}} = \mathbf{r}_0 \times \mathbf{p}_0$. In terms of the coefficients $b_{n\ell m}$ the probabilities $W_{\ell m}$ of angular momentum are given by the sum $W_{\ell m} = \sum_{n=1}^{\infty} |b_{n\ell m}|^2$.

In Figure 13.25 the plot of the probability $W_{\ell m}$ for the Gaussian wave packet chosen is shown. As already remarked in Section 10.10 for fixed $\ell$ the probabilities $W_{\ell\ell}$ are at maximum if the quantization axis $\mathbf{n}$ is chosen in the direction of classical angular momentum. The marginal distribution $W_\ell$ has its maximum close to the value $\ell_{\mathrm{cl}} = L_{\mathrm{cl}}/\hbar$ given by the classical angular momentum.

The distribution of the principal quantum number $n$ and total angular momentum $\ell$ is given by the probabilities

$$P_{n\ell} = \sum_{m=-\ell}^{\ell} |b_{n\ell m}|^2 \quad .$$

Additional summation over $\ell$ yields the marginal distribution

$$P_n = \sum_{\ell=0}^{n-1} P_{n\ell}$$

of the energy eigenvalues, while by summing over $n$ we obtain the marginal distribution $W_\ell$,

$$W_\ell = \sum_{n=\ell+1}^{\infty} P_{n\ell} \quad .$$

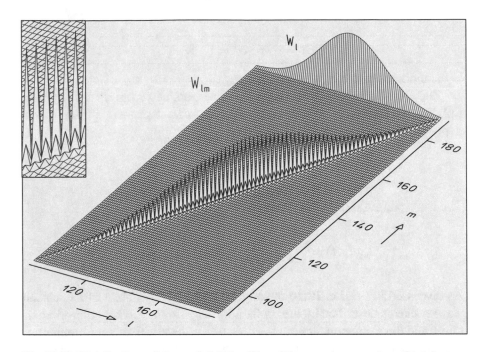

**Fig. 13.25. Distributions of the probabilities** $W_{\ell m}$ **of the quantum number of total angular momentum** $\ell$ **and its** $z$ **component** $m$ **for the wave packet shown in Figures 13.27 through 13.30. The quantities are defined for integer values of** $\ell, m$ **only. Their graphical representations are connected by straight lines. Also shown is the marginal distribution** $W_\ell$. **The inlay is a magnification of the central part of the figure.**

Figure 13.26 exhibits the distribution of $P_{n\ell}$ for the same wave packet together with the two marginal distributions $P_n$ and $W_\ell$. The maximum of $P_n$ is close to the value

$$n_{\text{cl}} = \frac{E_{\text{cl}}}{E_1}$$

given by the classical energy

$$E_{\text{cl}} = \frac{\mathbf{p}_0^2}{2M} - \frac{\alpha \hbar c}{r_0}$$

in terms of the initial expectation values $\mathbf{p}_0$ and $r_0 = |\mathbf{r}_0|$ of momentum and position of the Gaussian wave packet.

Figure 13.27 presents the first revolution of the wave packet at the time instants $t = 0, \frac{1}{3}T_K, \frac{2}{3}T_K$, and $T_K$, where $T_K$ is the classical Kepler period. The solid line represents the classical elliptical Kepler orbit for the initial conditions $\mathbf{r}_0, \mathbf{p}_0$. The dot on the ellipse marks at time $t$ the position of the classical particle with the initial conditions $\mathbf{r}_0, \mathbf{p}_0$. The function shown is the spatial probability density

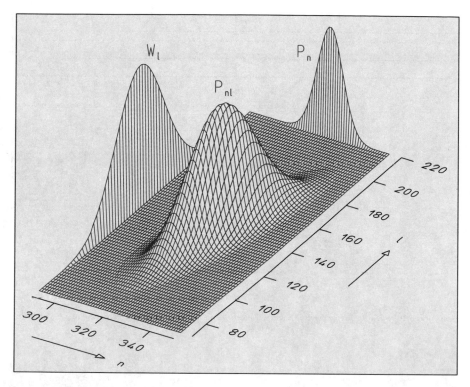

**Fig. 13.26. Probability distribution** $P_{n\ell}$ **of the principal quantum number** $n$ **and the angular-momentum quantum number** $\ell$ **of the wave packet shown in Figures 13.27 through 13.30 together with the marginal distributions** $W_\ell$ **and** $P_n$. **The probability is set equal to zero if** $P_{n\ell} < 10^{-5}$.

$$\rho(\mathbf{r},t) = |\psi(\mathbf{r},t)|^2$$

over the plane $\mathbf{r} = (x,y,0)$ of the classical orbit.

We compare the behavior of the quantum-mechanical wave packet with the time development of a classical phase-space distribution which has at $t = 0$ the form

$$\rho^{\mathrm{cl}}(\mathbf{r},\mathbf{p}) = \frac{1}{(2\pi)^3 \sigma^{3/2} \sigma_p^{3/2}} \exp\left\{ -\frac{(\mathbf{r}-\mathbf{r}_0)^2}{2\sigma^2} - \frac{(\mathbf{p}-\mathbf{p}_0)^2}{2\sigma_p^2} \right\}$$

of the product of two Gaussian distributions in space and momentum. The momentum width $\sigma_p$ is chosen according to Heisenberg's uncertainty relation for a Gaussian wave packet,

$$\sigma_p = \hbar/(2\sigma) \quad .$$

The time evolution $\rho^{\mathrm{cl}}(\mathbf{r},\mathbf{p},t)$ of the phase-space distribution is calculated with Newton's laws. The classical probability distribution in space is the

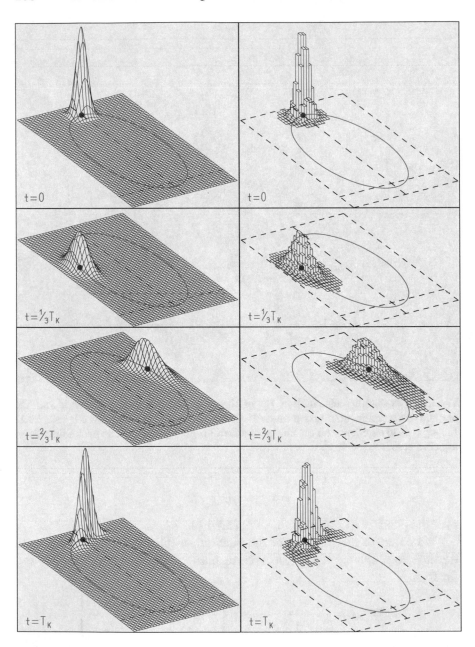

**Fig. 13.27. Plots in the left column show the time evolution of an initially Gaussian wave packet in the plane $z = 0$, i.e., the plane of the classical Kepler orbit indicated by the ellipse. The full dot represents the corresponding position of the classical particle. The temporal instants shown are the thirds of the first Kepler period $T_K$. Plots in the right column show the time evolution of the spatial probability density of the corresponding classical phase-space distribution.**

marginal distribution

$$\rho_{\mathbf{r}}^{\mathrm{cl}}(\mathbf{r},t) = \int \rho^{\mathrm{cl}}(\mathbf{r},\mathbf{p},t)\,\mathrm{d}^3\mathbf{p}$$

obtained by integration over all values of the momentum.

The time development of the classical spatial probability distribution is shown also in Figure 13.27. For computational reasons the plots are not smooth but look as composed of columns. They are called histograms. The height of each column is proportional to the probability per rectangular region in $x$ and $y$. The histogram shows only columns which correspond at least to a certain minimum probability. The main features of the classical probability distribution are very similar to the quantum-mechanical distribution.

The deformation of the Gaussian wave packet showing at $t = T_{\mathrm{K}}/3$ for the first time is thus seen to be a purely classical effect. It is in particular due to the distribution of momentum in classical phase space initially of Gaussian shape about the point $\mathbf{p}_0$. For example, the distribution at any given point $\mathbf{r}$ in the $x, y$ plane also contains momenta $\mathbf{p}_0 + \mathbf{p}_1$, where $\mathbf{p}_1$ is pointing toward the center of force. The time of revolution for a particle with this momentum is shorter than for one with $\mathbf{p}_0$. Therefore the spatial distribution will have a forerunning part inside the classical ellipse. Analogous arguments show that momenta $\mathbf{p}_0 - \mathbf{p}_1$ are responsible for a delayed tail located outside the classical orbit. These features catch the eye at first glance in the plots of Figure 13.27.

As time elapses the distributions widen and finally wind around the orbit. As soon as the bow of the quantum-mechanical wave function overlaps with its stern we expect and observe interference phenomena, Figure 13.28. An earlier example of this phenomenon we have met in Section 6.2 of a wave packet in a deep square well. There, the interference leads to a revival of the wave packet at the revival time $t = T_{\mathrm{rev}}$ and to fractional revivals at times $t = \frac{k}{\ell} T_{\mathrm{rev}}$, $k$, $\ell$ integer. The same phenomenon of the revival of the wave packet exists for the wave packet on a Kepler orbit with the revival time given by

$$T_{\mathrm{rev}} = \frac{n_{\mathrm{cl}}}{3} T_{\mathrm{K}} \quad ,$$

where $n_{\mathrm{cl}} = E_{\mathrm{cl}}/E_1$ is the classical value of the principal quantum number. The existence of a revival time in the quantum-mechanical Kepler problem was pointed out by Parker and Stroud (Physical Review Letters 56, 716 (1986)). Figure 13.29 exhibits the distribution $\rho(x, y, 0, t)$ at the times $t = (n_{\mathrm{cl}}/3 - \frac{1}{2})T_{\mathrm{K}}$, $n_{\mathrm{cl}}T_{\mathrm{K}}/3$, $(n_{\mathrm{cl}}/3 + \frac{1}{2})T_{\mathrm{K}}$. It is surprising that there is a time difference between the positions of the revived wave packet and the classical particle. A more detailed discussion of the origin of the quasiperiodicity in the quantum-mechanical hydrogen atom as carried out by Averbukh and Perelman in Physics Letters A 139, 449 (1989) reveals the reason.

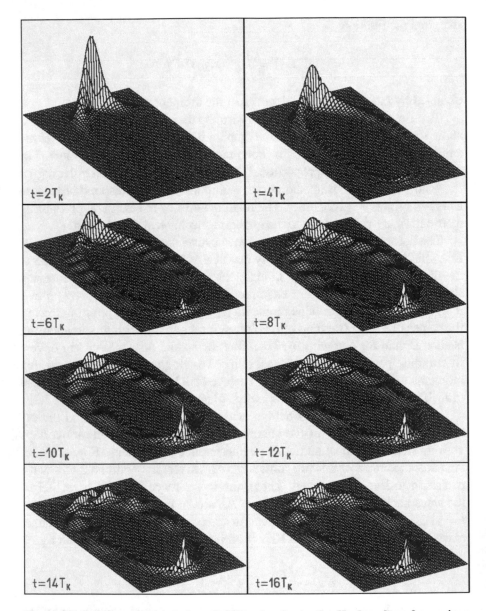

**Fig. 13.28. Quantum-mechanical probability density in the Kepler plane for various multiples of the Kepler period. The wave packet widens from period to period. Once both ends of the wave packet overlap, interference sets in.**

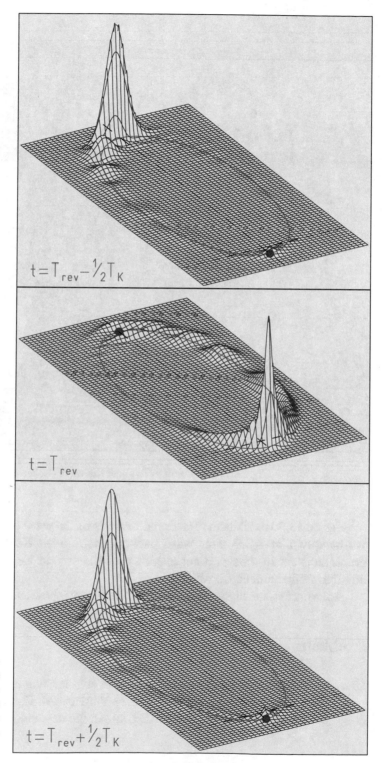

$$t = T_{rev} - \tfrac{1}{2} T_K$$

$$t = T_{rev}$$

**Fig. 13.29.** Revival of the wave packet for the times $t = T_{rev} - T_K/2$, $T_{rev}$, $T_{rev} + T_K/2$.

$$t = T_{rev} + \tfrac{1}{2} T_K$$

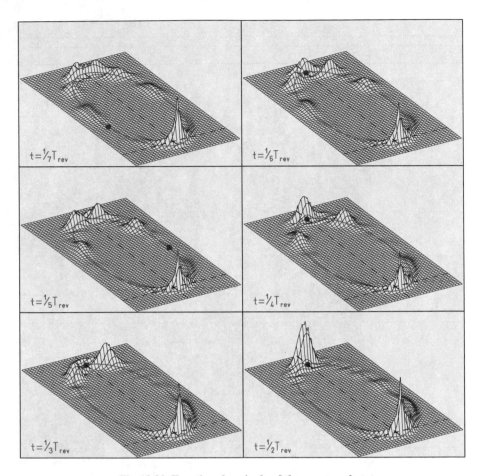

**Fig. 13.30. Fractional revivals of the wave packet.**

Figure 13.30 exhibits the fractional revivals of the wave packet. At $T_{\text{rev}}/2$ we have two, at $T_{\text{rev}}/3$ three wave packets, etc., on the Kepler orbit. They are equidistant in time as a consequence of the second Kepler law, i.e., of angular-momentum conservation.

As expected, the classical spatial distribution $\rho^{\text{cl}}$ shows no such revivals.

## Problems

13.1. Calculate the energies $E_n$ of the states of angular momentum zero for an infinitely deep potential well in three dimensions. Compare this spectrum with the one in Figure 13.1. Explain the deviations.

13.2. Why are the energies of the same quantum numbers $n$ for $\ell = 1, 2$ in Figure 13.1 larger than those for $\ell = 0$?

13.3. Why does the energy of the lowest (in general, the $n$th) state decrease with increasing width of the spherical square-well potential of the same depth (Figure 13.2)?

13.4. Why does the difference $E_{1\ell} - V_0$ of the state of lowest energy for a given angular momentum $\ell$ increase as the potential well deepens?

13.5. Explain the structure of the product function $\varphi'_{210}(x_1, x_2, x_3)$, as plotted in Figure 13.5, in terms of the structures of the harmonic-oscillator functions in one dimension, as plotted in Figure 6.4.

13.6. Why does the average probability density in a spherical shell of unit volume, given by $r^2 R_{n\ell}^2(r)$ as plotted in Figure 13.8, increase toward the harmonic-oscillator wall?

13.7. What do you expect the harmonic-oscillator probability densities for $n = 5, \ell = 0, 1, 2, 3$ to roughly look like? Describe their nodes in $r$ and $\vartheta$ in analogy to those in Figures 13.9 and 13.10.

13.8. Verify by explicit calculation that the angular-momentum eigenstate $\varphi_{220}(\mathbf{r})$ of the harmonic oscillator can be decomposed as was done in Figure 13.11.

13.9. Let us consider time-dependent motion in a rotationally symmetric harmonic oscillator. The wave function of the initial state $\psi(\mathbf{r}, 0)$ at $t = 0$ is given explicitly in terms of a decomposition into eigenfunctions $\varphi'_{n_1 n_2 n_3}(x_1, x_2, x_3)$,

$$\psi(\mathbf{r}, 0) = \sum_{n_1, n_2, n_3} a_{n_1 n_2 n_3} \varphi'_{n_1 n_2 n_3}(x_1, x_2, x_3) \quad,$$

corresponding to the eigenvalues $E_n = (n + \frac{3}{2})\hbar\omega$, $n = n_1 + n_2 + n_3$, of the harmonic oscillator as described in Section 13.2. The $a_{n_1 n_2 n_3}$ are the spectral coefficients of the initial state in the harmonic-oscillator base $\varphi'_{n_1 n_2 n_3}$. Show that the time-dependent wave function $(n = n_1 + n_2 + n_3)$

$$\psi(\mathbf{r}, t) = \sum_{n_1, n_2, n_3} e^{-iE_n t/\hbar} a_{n_1 n_2 n_3} \varphi'_{n_1 n_2 n_3}(x_1, x_2, x_3)$$

is a solution of the time-dependent Schrödinger equation

$$i\hbar \frac{\partial}{\partial t}\psi(\mathbf{r}, t) = \left(-\frac{\hbar^2}{2M}\nabla^2 + \frac{k}{2}r^2\right)\psi(\mathbf{r}, t)$$

and fulfills the initial condition.

13.10. Analyze the behavior of the three-dimensional wave packet under the influence of a harmonic force, as plotted in Figure 13.12, in terms of the behavior of three independent one-dimensional oscillators, as plotted in Figures 6.6 and 6.8. Describe the initial conditions of these independent oscillators in terms of classical mechanics.

13.11. Show that the general solution $\psi(\mathbf{r},t)$ for the motion in a harmonic oscillator,

$$\psi(\mathbf{r},t) = \sum_{n_1,n_2,n_3} \exp\left[-\frac{i}{\hbar}E_n t\right] a_{n_1 n_2 n_3} \varphi'_{n_1 n_2 n_3}(x_1, x_2, x_3) \quad,$$

with the energy of the state $\varphi'_{n_1 n_2 n_3}$,

$$E_n = (n + \frac{3}{2})\hbar\omega \quad, \qquad n = n_1 + n_2 + n_3 \quad,$$

possesses the following periodicity property:

$$\psi\left(\mathbf{r}, t + m\frac{2\pi}{\omega}\right) = e^{-im\pi}\psi(\mathbf{r},t), \qquad m = 1, 2, 3, \dots \quad .$$

The periodicity property implies that

$$\left|\psi\left(\mathbf{r}, t + m\frac{2\pi}{\omega}\right)\right|^2 = |\psi(\mathbf{r},t)|^2 \quad .$$

This result can be read off Figure 13.12.

13.12. Calculate the minimum of the effective potential, $V_{2,\min}^{\text{eff}}$, for $\ell = 2$ of the hydrogen atom,

$$V_2^{\text{eff}}(r) = \frac{\hbar^2}{2M}\frac{6}{r^2} - \hbar c\frac{\alpha}{r} \quad .$$

Find the differences between the eigenvalues $E_n$ of the electron in the hydrogen atom and this minimum, $E_n - V_{2,\min}^{\text{eff}}$. Explain why only states with $n \geq 3$ exist for angular momentum $\ell = 2$ in the hydrogen atom.

13.13. Show that the Bohr radius $a$ as given in Section 13.4 is at the position of the maximum of $r^2 R_{10}^2(r)$, that is, show that, at $r = a$,

$$\frac{d}{dr}[r R_{10}(r)] = 0 \quad .$$

13.14. The energy of the ground state of a two-particle system bound by a Coulomb force is

$$E_1 = -\frac{1}{2}\mu c^2 \alpha^2 \quad,$$

where $\mu = M_1 M_2/(M_1 + M_2)$ is the reduced mass of the system of two particles of masses $M_1$ and $M_2$. For $M_1 \ll M_2$, $\mu$ tends toward $M_1$. Using this formula, calculate the ground-state energy $E_1$ for the hydrogen atom and for a positronium, which is a system of an electron and a positron, that is, an electron of positive charge.

13.15. Muons are particles similar to the electron but possessing a mass

$$m_\mu = 105.6\,\text{MeV}/c^2 \quad.$$

The Bohr radius, that of the innermost orbit, of a system made up of a positive and negative charge is

$$a = \frac{\hbar}{\alpha \mu c} \quad,$$

where $\mu$ is the reduced mass of the system, as given in the preceding problem. Calculate the Bohr radii for a hydrogen atom; for a muonic hydrogen atom, whose electron has been replaced by a muon; and for positronium, a hydrogen-like system in which the proton has been replaced by a positron.

13.16. The Bohr radius $a$ of the innermost orbit of a nucleus with atomic number $Z$ is

$$a = \frac{\hbar}{Z\alpha \mu c} \quad,$$

where $\mu$ is the reduced mass of the system. For the uranium nucleus, $Z = 92$, the reduced mass can safely be taken as the mass of the particle in the innermost orbit. Calculate the Bohr radius for a muonic uranium atom and compare the result with the radius of the uranium nucleus, $r_0 \approx 6 \times 10^{-15}\,\text{m}$.

# 14. Hybridization

## 14.1 Introduction

In Section 13.4 we studied in detail the stationary wave functions

$$\varphi_{n\ell m}(\mathbf{r}) = \varphi_{n\ell m}(r,\vartheta,\phi) = R_{n\ell}(r)Y_{\ell m}(\vartheta,\phi)$$

of the electron in the hydrogen atom and the corresponding probability densities

$$\rho(\mathbf{r}) = \rho_{n\ell m}(r,\vartheta) = |\varphi_{n\ell m}(r,\vartheta,\phi)|^2 \quad .$$

Here $r,\vartheta,\phi$ are the spherical coordinates, $n$ is the principal quantum number, $\ell$ the quantum number of angular momentum, and $m$ that of its $z$ component; $R_{n,\ell} = R_{n,\ell}(r)$ is the radial wave function, which depends on the radial position $r$, and $Y_{\ell,m} = Y_{\ell,m}(\vartheta,\phi)$ the spherical harmonic (Section 10.3), which depends on the angles $\vartheta$ and $\phi$. The energy eigenvalues $E_n$ depend on $n$ only. Accordingly, in general, there are several eigenstates of the same eigenvalue (which are called *degenerate*) and therefore linear combinations of these are again eigenstates.

Figures 13.16 to 13.24 show that the $\rho_{n\ell m}$ are either spherically symmetric with respect to the origin or rotationally symmetric around the $z$ axis. Common to all is a mirror symmetry with respect to the $xy$ plane. In a particular model of chemical bonds, the *hybridization model*, wave functions are constructed for which the probability density has a single preferred direction, pointing from the nucleus of one atom towards the nucleus of a partner atom to which it is bound. In this section we show how such wave functions can be constructed starting out from the stationary states $\varphi_{n\ell m}$.

We consider exclusively the superposition of a state $\ell = 0, m = 0$ (called an *s state*) with a state $\ell = 1, m = 0$ (a *p state*) and introduce the notation

$$s_n = \psi_{n,0,0} = R_{n,0}Y_{0,0} = \frac{1}{\sqrt{4\pi}}R_{n,0} \quad ,$$

$$p_n = \psi_{n,1,0} = R_{n,1}Y_{1,0} = \sqrt{\frac{3}{4\pi}}\cos\vartheta\, R_{n,1} \quad .$$

The wave functions $s_n$ and $p_n$ are real; $s_n$ has no angular dependence whereas $p_n$ is proportional to $\cos\vartheta$, i.e., it is antisymmetric with respect to the equato-

S. Brandt and H.D. Dahmen, *The Picture Book of Quantum Mechanics*, 294
DOI 10.1007/978-1-4614-3951-6_14, © Springer Science+Business Media New York 2012

rial plane $\vartheta = \pi/2$. There is no $\phi$ dependence. Because of the orthonormality of the $R_{n,\ell}$ and the $Y_{\ell,m}$ the states $s_n$ and $p_n$ are orthonormal, i.e.,

$$\int |s_n|^2 \, dV = 1 \quad , \qquad \int |p_n|^2 \, dV = 1 \quad , \qquad \int s_n p_n \, dV = 0 \quad ;$$

the integration is performed over all space.

A normalized linear combination of $s_n$ and $p_n$,

$$h_n = \frac{1}{\sqrt{1+\lambda^2}}(s_n + \lambda p_n) \quad ,$$

is called a *hybrid state* with the *hybridization parameter* $\lambda$.

Figures 14.1 and 14.2 show, in a plane containing the $z$ axis, the functions $s_2$, $p_2$, $h_2$, and $|h_2|^2$ for a particular value of $\lambda$. Due to the different symmetry properties of $s_2$ and $p_2$ the hybrid $h_2$ is asymmetric with respect to $\vartheta = \pi/2$. The square $|h_2|^2$, which is a probability density, extends much farther along the negative $z$ axis than along the positive one. This feature of hybrid states is used to explain some type of chemical bonds: An electron in a hybrid state in one atom reaches out to its partner in another atom along that preferred direction.

The hybrid state is rotationally symmetric around the $z$ axis, which we call the *orientation axis*. We can denote it by the unit vector $\hat{\mathbf{a}}$; here $\hat{\mathbf{a}} = \mathbf{e}_z$. In general, the orientation axis is given by a unit vector $\hat{\mathbf{a}}$, characterized by a polar angle $\vartheta_a$ and an azimuthal angle $\phi_a$ with respect to the $x, y, z$ coordinate frame.

The general form of a hybrid state is therefore

$$h_n(\lambda; \vartheta_a, \phi_a) = \frac{1}{\sqrt{1+\lambda^2}}(s_n + \lambda p_n(\vartheta_a, \phi_a)) \quad ,$$

where $p_n(\vartheta_a, \phi_a)$ is a $p$ state of principal quantum number $n$ with the symmetry axis

$$\hat{\mathbf{a}} = \sin\vartheta_a \cos\phi_a \, \mathbf{e}_x + \sin\vartheta_a \sin\phi_a \, \mathbf{e}_y + \cos\vartheta_a \, \mathbf{e}_z \quad .$$

In the form of $p_n$ which we wrote down first, for which the symmetry axis is the $z$ axis, the angular dependence showed up as $\cos\vartheta$ or, written as scalar product, as $\hat{\mathbf{r}} \cdot \mathbf{e}_z$. Its generalization reads $\hat{\mathbf{r}} \cdot \hat{\mathbf{a}}$. Therefore the general form of $p_n$ is

$$p_n(\vartheta_a, \phi_a) = \sqrt{\frac{3}{4\pi}}(\hat{\mathbf{r}} \cdot \hat{\mathbf{a}}) R_{n,1} \quad .$$

A good impression of the spatial orientation is given by a contour-surface plot. In Figure 14.3 such a plot is shown for the hybrid probability density $|h_2(\lambda = 1.73; \vartheta_a = 0, \phi_a = 0)|^2$. It is obtained by rotating one of the contour lines in the bottom right of Figure 14.2 about the $z$ axis.

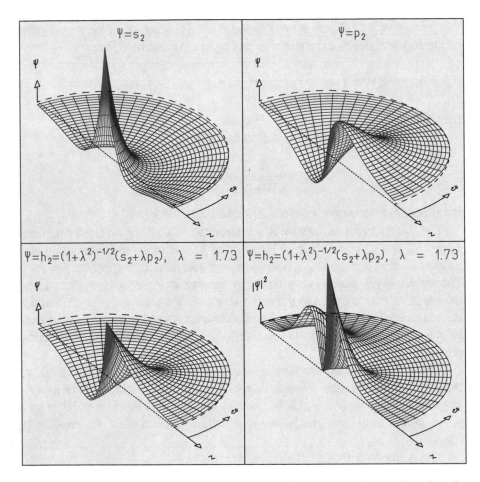

**Fig. 14.1.** The wave function $\varphi = s_2$ is symmetric with respect to the $xy$ plane, i.e., the plane perpendicular to the $z$ axis. The wave function $\varphi = p_2$ is antisymmetric. A superposition of both, the hybrid $\varphi = h_2$, displays neither symmetry nor antisymmetry: it is unsymmetric. Its absolute square, the probability density $|\varphi|^2 = |h_2|^2$, is markedly more extended along the negative $z$ direction compared to the positive $z$ direction.

Atoms other than hydrogen have more than one electron. Here we do not attempt a quantitative discussion of their properties. Nevertheless, the study of the simple hybrid states $h_n(\lambda; \vartheta_a, \phi_a)$ allows some understanding of hybrid bonds. In Section 14.2 we summarize the assumptions made in the simplest hybridization model and try to justify them by qualitative arguments. In Section 14.3 we shall compute the hybridization parameter and the orientations for some hybrid states in situations of particularly high symmetry.

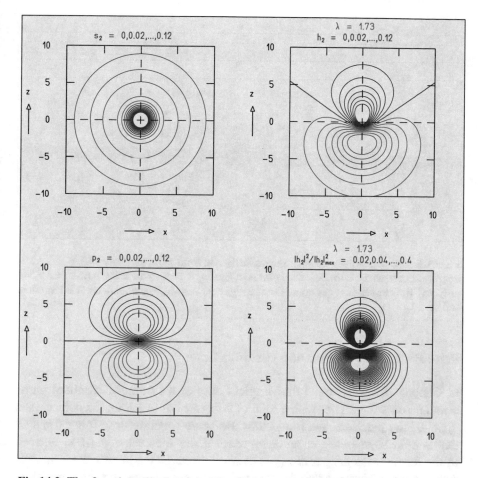

**Fig. 14.2.** The functions displayed in Fig. 14.1 are shown here in the form of contour plots in the $xz$ plane. Function values are positive on blue lines, negative on magenta lines, and vanish on red lines. The unit length used for the scales in $x$ and $z$ is the Bohr radius.

## 14.2 The Hybridization Model

In our qualitative discussion of the model we first study the structure of atoms in the first few periods of the Periodic Table and shall see that the wave functions of the valence electrons, responsible for the binding of atoms, retain some similarity to the hydrogen wave functions. We then consider the binding energy which becomes available when atoms approach each other and the role it plays in bringing about hybrid states.

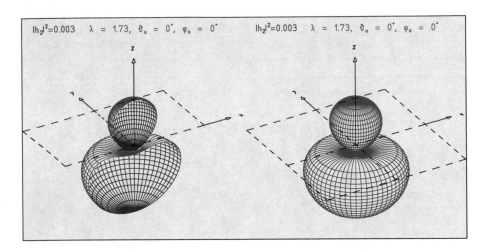

$|h_2|^2 = 0.003$   $\lambda = 1.73$, $\vartheta_a = 0°$, $\varphi_a = 0°$      $|h_2|^2 = 0.003$   $\lambda = 1.73$, $\vartheta_a = 0°$, $\varphi_a = 0°$

**Fig. 14.3. The probability density shown in the bottom right of Figs. 14.1 and 14.2 is displayed here as a contour-surface plot. On the left it is shown for the half-space $y > 0$ only and thus appears as cut open. On the right the surface is closed since it is shown in full space.**

### Pauli Exclusion Principle and Periodic Table

In Chapter 9 we discussed the simplest case of a system of identical particles, i.e., only 2 particles with a total of 2 degrees of freedom, the coordinates $x_1, x_2$ of the particles. We found that the wave function describing the system is either symmetric or antisymmetric under the exchange of $x_1$ and $x_2$. In the former case the particles are called *bosons*, in the latter case *fermions*. Antisymmetry signifies that the two fermions cannot have the same position. In a realistic case of more particles, each with more degrees of freedom, antisymmetry means that there cannot be two identical fermions in the same quantum-mechanical state. This statement is the *Pauli exclusion principle*.

Particles with half-integer spin ($\hbar/2, 3\hbar/2, \ldots$) are fermions, those with integer spin ($0, \hbar, 2\hbar, \ldots$) are bosons. Electrons carry spin $\hbar/2$ and thus are fermions. An atom with atomic number $Z$ consists of a nucleus with electric charge $Ze$ and $Z$ electrons, each of charge $-e$. Because of the Pauli principle there cannot be two electrons in an atom with exactly the same set of quantum numbers $n, \ell, m, s_z$. For a given principal quantum number $n$ there are $n$ different values of the angular-momentum quantum number $\ell$ ($\ell = 0, \ldots, n-1$); for each value of $\ell$ there are $2\ell + 1$ different values of the quantum number $m$ of the $z$ component of angular momentum ($m = -\ell, -\ell+1, \ldots, \ell$); there are two possible values, $s_z = \pm 1/2$, of the quantum number of the $z$ component of spin for every electron. In all, for a given value of $n$ there are $2n^2$ different sets of quantum numbers $\ell, m, s_z$.

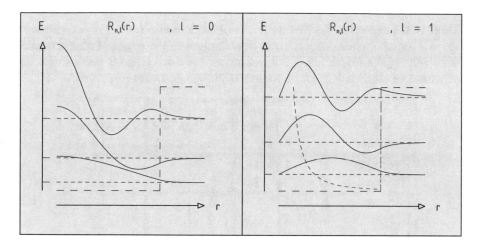

**Fig. 14.4. Radial wave functions and their eigenvalues in a 3D square-well potential. The eigenvalues are systematically higher for states with $\ell = 1$ (right) compared to the corresponding states with $\ell = 0$ (left).**

For $n = 1$ there are 2 states. One is occupied in the hydrogen atom ($Z = 1$), both are filled in the helium atom ($Z = 2$). In the case of the lithium atom ($Z = 3$) there is no free place for an electron in the inner *shell* of lowest energy. The third electron necessarily has the principal quantum number $n = 2$. Its probability density is shifted further away from the nucleus. This electron "sees" a potential similar to that of the hydrogen nucleus, because the nuclear charge $3e$ is "shielded" by the charge $-2e$ of the two electrons in the inner shell. If the shielding were perfect, all states with $n = 2$ were degenerate. Since it is not, the states with $n = 2, \ell = 0$, called $2s$ states, have a somewhat lower energy than the $2p$ states ($n = 2, \ell = 1$). (For an illustration, in which we use a square-well rather than a Coulomb potential, see Figure 14.4.)

As $Z$ increases, more electrons are added and new states are filled, see Table 14.1. For the lithium atom, one of these $2s$ states is filled; in the beryllium atom both are occupied. Boron, in addition to the two $2s$ electrons, has one $2p$ electron; carbon has two, nitrogen three. In all, there are 6 different $2p$ states. They are all filled in neon, which concludes the second period of the Periodic Table. In the third period, the outer shell comprises states with $n = 3$. That shell is begun with sodium and completed with argon. In particular, we note that silicon, situated directly below carbon in the Periodic Table, in its outer shell has two $3s$ and two $3p$ electrons. Its situation is that of carbon but with $n = 3$ instead of $n = 2$.

Only the electrons in the outermost shell of an atom, the *valence electrons*, contribute significantly to chemical binding. For the single valence electron of an alkali atom (H, Li, Na, ... ) the shielding of the nuclear charge by the

**Table 14.1.** Elements in the first 3 periods of the Periodic Table. For each element the number of electrons with allowed combinations of the principal quantum number $n$ and angular-momentum quantum number $\ell$ is listed. The top line contains the conventional abbreviations for such combinations, e.g., $2p$ for $n = 2, \ell = 1$.

| Element | | Z | $1s$ $n = 1$ $\ell = 0$ | $2s$ $n = 2$ $\ell = 0$ | $2p$ $n = 2$ $\ell = 1$ | $3s$ $n = 3$ $\ell = 0$ | $3p$ $n = 3$ $\ell = 1$ |
|---|---|---|---|---|---|---|---|
| Hydrogen | H | 1 | 1 | | | | |
| Helium | He | 2 | 2 | | | | |
| Lithium | Li | 3 | 2 | 1 | | | |
| Beryllium | Be | 4 | 2 | 2 | | | |
| Boron | B | 5 | 2 | 2 | 1 | | |
| Carbon | C | 6 | 2 | 2 | 2 | | |
| Nitrogen | N | 7 | 2 | 2 | 3 | | |
| Oxygen | O | 8 | 2 | 2 | 4 | | |
| Fluorine | F | 9 | 2 | 2 | 5 | | |
| Neon | Ne | 10 | 2 | 2 | 6 | | |
| Sodium | Na | 11 | 2 | 2 | 6 | 1 | |
| Magnesium | Mg | 12 | 2 | 2 | 6 | 2 | |
| Aluminum | Al | 13 | 2 | 2 | 6 | 2 | 1 |
| Silicon | Si | 14 | 2 | 2 | 6 | 2 | 2 |
| Phosphorous | P | 15 | 2 | 2 | 6 | 2 | 3 |
| Sulfur | S | 16 | 2 | 2 | 6 | 2 | 4 |
| Chlorine | Cl | 17 | 2 | 2 | 6 | 2 | 5 |
| Argon | Ar | 18 | 2 | 2 | 6 | 2 | 6 |

remaining electrons is nearly perfect. Nevertheless, in our simplified computations of hybrid wave functions we do assume perfect shielding also for the valence electrons of other atoms: We compute the hydrogen wave functions for given quantum numbers and then superimpose them to obtain the desired hybrid wave function.

### Simple Example

As a first qualitative example we consider a molecule of lithium hydride LiH, formed of an atom of lithium Li and one of hydrogen H. The single electron of hydrogen is in the $1s$ state. Since there is no $1p$ state, there can be no question of hybridization in that atom. The outer electron of lithium is in the $2s$ state, which is nearly degenerate with the $2p$ state, since shielding is nearly perfect.

The electron is assumed to be in a hybrid state formed by a superposition of the states $2s$ and $2p$. There are models in which the binding energy of the LiH molecule is computed from the overlap of this hybrid wave function with the $1s$ wave function in the hydrogen atom. The binding energy outweighs the slight extra energy of the lithium hybrid state compared to its original $2s$ state. The hybrid state is of the form $h_2(\lambda; \vartheta_a, \phi_a)$; the orientation axis $\hat{\mathbf{a}}$ points along the line connecting the hydrogen nucleus with that of the lithium atom. The hybridization parameter $\lambda$ depends on the details of the model used.

### Binding Energy and Promotion. Orientation of $p_n$ States

Here and in Section 14.3 we concentrate on the atoms of carbon C and of silicon Si. Both have four electrons in their outer shell with principal quantum number $n = 2$ for C and $n = 3$ for Si. In a solitary C or Si atom there are two $s$ electrons and two $p$ electrons in the outer shell. In the presence of one or more neighboring atoms, to which they are bound, the electrons can assume "binding" states of lower energy, see Figure 14.5. Part of the binding energy can be used to "promote" one of the two $s$ electrons to a $p$ state; that is possible as long as the binding energy per atom exceeds the promotion energy. With this *promotion* there are one $s$ electron and three $p$ electrons in the outer shell.

The state $s_n$ and and the three states $p_n$ ($n$, as usual, is the principal quantum number) are normalized and all are orthogonal to each other. We align the $p_n$ states along the coordinate directions $z, x, y$, i.e., we write them in the form

$$p_{nz} = p_n(\vartheta_a = 0) = \sqrt{\frac{3}{4\pi}} R_{n,1} \cos\vartheta \quad,$$

$$p_{nx} = p_n(\vartheta_a = 90°, \phi_a = 0) = \sqrt{\frac{3}{4\pi}} R_{n,1} \sin\vartheta \cos\phi \quad,$$

$$p_{ny} = p_n(\vartheta_a = 90°, \phi_a = 90°) = \sqrt{\frac{3}{4\pi}} R_{n,1} \sin\vartheta \sin\phi \quad,$$

see Figure 14.6.

We now consider superpositions of the three $p_n$ states along the Cartesian coordinate axes with coefficients $a_z, a_x, a_y$:

$$a_z p_{nz} + a_x p_{nx} + a_y p_{ny} =$$

$$= \sum_{f=x,y,z} a_f p_{nf} = \sum_{f=x,y,z} a_f \sqrt{\frac{3}{4\pi}} R_{n,1}(\hat{\mathbf{r}} \cdot \mathbf{e}_f)$$

$$= \sqrt{\frac{3}{4\pi}} R_{n,1} \left( \hat{\mathbf{r}} \cdot \sum_{f=x,y,z} a_f \mathbf{e}_f \right) = \sqrt{\frac{3}{4\pi}} R_{n,1}(\hat{\mathbf{r}} \cdot \mathbf{a})$$

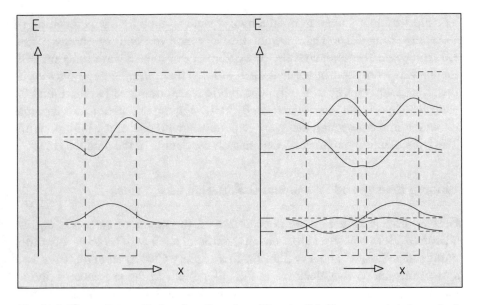

**Fig. 14.5. Eigenvalues and eigenfunctions in a 1D potential. For every state in a single well (left), there are two states in the double well: one with lower, the other with higher energy than in the single well. The state of lower energy is** *binding,* **that of higher energy is** *anti-binding.*

$$= \ |\mathbf{a}|\sqrt{\frac{3}{4\pi}}R_{n,1}(\hat{\mathbf{r}} \cdot \hat{\mathbf{a}}) = |\mathbf{a}|\,p_n(\vartheta_a, \phi_a) \quad,$$

where the coefficients $a_x, a_y, a_z$ have been interpreted as those of an (unnormalized) vector $\mathbf{a}$ defining an orientation axis of a general $p_n$ state. With $|\mathbf{a}| = \sqrt{a_x^2 + a_y^2 + a_z^2}$ and $|\mathbf{a}_\perp| = \sqrt{a_x^2 + a_y^2}$ the relations of these coefficients to the spherical angles $\vartheta_a, \phi_a$, defining the orientation axis, read

$$\cos\vartheta_a = \frac{a_z}{|\mathbf{a}|} \quad, \quad \sin\vartheta_a = \frac{|\mathbf{a}_\perp|}{|\mathbf{a}|} \quad, \quad \cos\phi_a = \frac{a_x}{|\mathbf{a}_\perp|} \quad, \quad \sin\phi_a = \frac{a_y}{|\mathbf{a}_\perp|} \quad.$$

## 14.3 Highly Symmetric Hybrid States

A diamond crystal is a symmetric arrangement of carbon atoms. Every atom is in the center of a tetrahedron with carbon atoms at its four corners. In graphite, a hexagonal planar structure, each C atom has three nearest neighbors to which it is tightly bound; in comparison, the binding between planes is rather loose. In a linear molecule like carbon dioxide $CO_2$ with the structure O–C–O the carbon atom is bound to two oxygen atoms which are situated in exactly opposite positions.

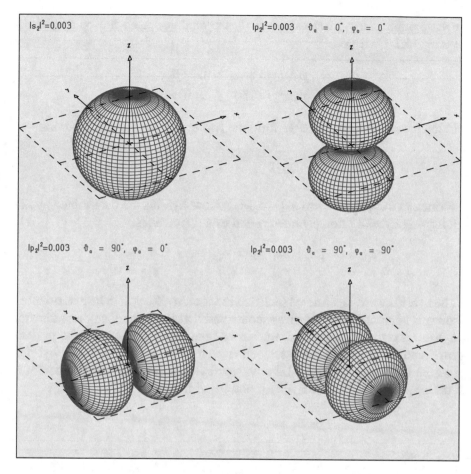

**Fig. 14.6. Contour-surface plots illustrating the four orthogonal states** $s_2, p_{2z}, p_{2x}, p_{2y}$

States resulting from hybridization again have to be normalized and to be orthogonal to each other (and to those states which do not hybridize). With these assumptions the following, particularly symmetric, situations are possible.

***sp* Hybridization** The $s_n$ state and *one* of the $p_n$ states form two hybrid states. We assume that $p_n$ state to be oriented in $z$. The other $p_n$ states stay unchanged. We consider the most symmetric case, in which the two hybrids are identical in form but oriented back to back. This is a picture for the bonds (as far as they are due to hybrids of the carbon atom) in the carbon dioxide molecule $CO_2$.

We first construct two $p_n$ states oriented in $z$ and in $-z$ direction, i.e., with $\vartheta_a = 0$ and $\vartheta_a = 180°$, respectively. (We assign $\phi_a = 0$ to these states even though at the given values of $\vartheta_a$ the angle $\phi_a$ need not be specified.) From (14.2.1) we get

$$
\begin{aligned}
p_{n1} &= p_n(\vartheta_a = 0, \phi_a = 0) = p_{nz} \quad, \\
p_{n2} &= p_n(\vartheta_a = 180°, \phi_a = 0) = -p_{nz} \quad.
\end{aligned}
$$

Using these and the $s_n$ state we form the two orthonormal superpositions

$$
h_{n1} = \frac{1}{\sqrt{2}}(s_n + p_{nz}) \quad, \qquad h_{n2} = \frac{1}{\sqrt{2}}(s_n - p_{nz}) \quad.
$$

Written in the general form $h_n(\lambda; \vartheta_a, \phi_a)$, these hybrids are determined by the following hybridization parameters and orientation axes:

$$
\begin{aligned}
\lambda_1 &= 1 \quad, & \vartheta_{a1} &= 0 \quad, & \phi_{a1} &= 0 \quad, \\
\lambda_2 &= 1 \quad, & \vartheta_{a2} &= 180° \quad, & \phi_{a2} &= 0 \quad.
\end{aligned}
$$

The hybrids were constructed such that the term $p_n(\vartheta_a, \phi_a)$, in turn, is oriented towards each of the neighboring atoms and that the hybridization parameter $\lambda$ ensures orthonormalization. Because of the high symmetry in the present case (and the two cases discussed below), $\lambda$ has the same value for all hybrids (within one case). For $n = 2$ the two $sp$ hybrids are illustrated as contour-surface plots of the functions $|h_{21}|^2$ and $|h_{22}|^2$ in Figure 14.7.

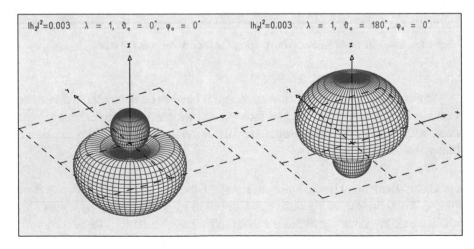

**Fig. 14.7.** $sp$ **hybrids for** $n = 2$: **Contour-surface plots of** $|h_{21}|^2$ **and** $|h_{22}|^2$

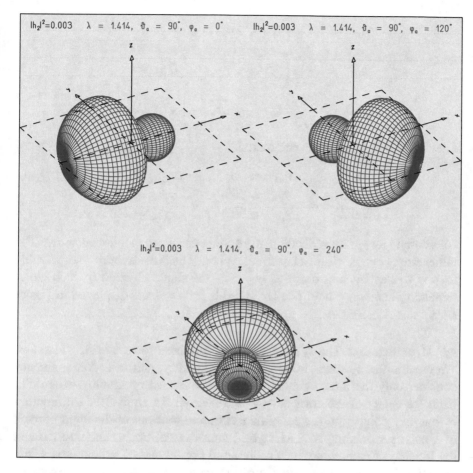

**Fig. 14.8. Contour-surface plots of** $sp^2$ **hybrids for** $n = 2$**.**

$sp^2$ **Hybridization** The $s_n$ state and *two* of the $p_n$ states form three hybrid states. We assume these $p_n$ states to be oriented in $x$ and in $y$. Again, we consider a particularly symmetric case: The hybrid orientation axes lie in the $x, y$ plane, each forming angles of $120°$ with the other two. We begin by constructing three $p_n$ states with these orientations,

$$p_{n1} = p_n(\vartheta_a = 90°, \phi_a = 0) = p_{nx} \quad ,$$

$$p_{n2} = p_n(\vartheta_a = 90°, \phi_a = 120°) = -\frac{1}{2}p_{nx} + \frac{\sqrt{3}}{2}p_{ny} \quad ,$$

$$p_{n3} = p_n(\vartheta_a = 90°, \phi_a = 240°) = -\frac{1}{2}p_{nx} - \frac{\sqrt{3}}{2}p_{ny} \quad .$$

Superposition with $s_n$ and proper normalization yield the three hybrids

$$h_{n1} = \frac{1}{\sqrt{3}}\left(s_n + \sqrt{2}p_{nx}\right) \quad ,$$

$$h_{n2} = \frac{1}{\sqrt{3}}\left(s_n - \frac{1}{2}\sqrt{2}p_{nx} + \frac{1}{2}\sqrt{6}p_{ny}\right) \quad ,$$

$$h_{n3} = \frac{1}{\sqrt{3}}\left(s_n - \frac{1}{2}\sqrt{2}p_{nx} - \frac{1}{2}\sqrt{6}p_{ny}\right) \quad .$$

They are determined by the parameters

$$\begin{aligned}
\lambda_1 &= \sqrt{2} \quad, & \vartheta_{a1} &= 90° \quad, & \phi_{a1} &= 0 \quad, \\
\lambda_2 &= \sqrt{2} \quad, & \vartheta_{a2} &= 90° \quad, & \phi_{a2} &= 120° \quad, \\
\lambda_3 &= \sqrt{2} \quad, & \vartheta_{a3} &= 90° \quad, & \phi_{a3} &= 240° \quad.
\end{aligned}$$

An example for $sp^2$ hybridization is graphite, which we mentioned above. The rather strong bonds within a layer are explained by these hybrids. Responsible for the weaker binding between layers is the single electron in each atom, remaining in the $p_{nz}$ state. The $sp^2$ hybrids for $n = 2$ are displayed in Figure 14.8.

$sp^3$ **Hybridization** The $s_n$ state and *three* $p_n$ states form four hybrid states. We assume the $p_n$ states to be oriented in $x$, in $y$, and in $z$. Yet again, we consider only the most symmetric case: The hybrid orientation axes point from the center to the corners of a tetrahedron. An equivalent formulation is that they point from the coordinate origin to four out of the eight corners of a cube surrounding it. Such hybrid states serve, for instance, to explain the binding of carbon atoms in a diamond crystal and of silicon atoms in a crystal with diamond structure. Construction of the $p_n$ states with the desired orientations yields

$$p_{n1} = p_n(\vartheta_a = 54.74°, \phi_a = 45°) = \frac{1}{\sqrt{3}}(p_{nx} + p_{ny} + p_{nz}) \quad ,$$

$$p_{n2} = p_n(\vartheta_a = 125.26°, \phi_a = 135°) = \frac{1}{\sqrt{3}}(p_{nx} - p_{ny} - p_{nz}) \quad ,$$

$$p_{n3} = p_n(\vartheta_a = 54.74°, \phi_a = 225°) = \frac{1}{\sqrt{3}}(-p_{nx} + p_{ny} - p_{nz}) \quad ,$$

$$p_{n4} = p_n(\vartheta_a = 125.26°, \phi_a = 315°) = \frac{1}{\sqrt{3}}(-p_{nx} - p_{ny} + p_{nz}) \quad .$$

Superposition with the $s_n$ state yields the hybrids

$$h_{n1} = \frac{1}{2}(s_n + p_{nx} + p_{ny} + p_{nz}) \quad , \qquad h_{n2} = \frac{1}{2}(s_n + p_{nx} - p_{ny} - p_{nz}) \quad ,$$

$$h_{n3} = \frac{1}{2}(s_n - p_{nx} + p_{ny} - p_{nz}) \quad , \qquad h_{n4} = \frac{1}{2}(s_n - p_{nx} - p_{ny} + p_{nz})$$

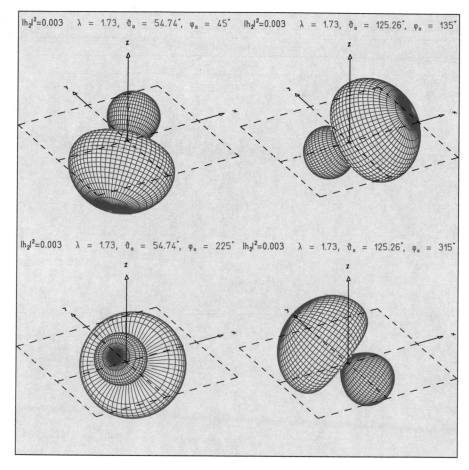

**Fig. 14.9. Contour-surface plots of $sp^3$ hybrids for $n = 2$.**

with the parameters

$$
\begin{aligned}
\lambda_1 &= \sqrt{3} \ , & \vartheta_{a1} &= 54.74° \ , & \phi_{a1} &= 45° \ , \\
\lambda_2 &= \sqrt{3} \ , & \vartheta_{a2} &= 125.26° \ , & \phi_{a2} &= 135° \ , \\
\lambda_3 &= \sqrt{3} \ , & \vartheta_{a3} &= 54.74° \ , & \phi_{a3} &= 225° \ , \\
\lambda_4 &= \sqrt{3} \ , & \vartheta_{a4} &= 125.26° \ , & \phi_{a4} &= 315° \ .
\end{aligned}
$$

An illustration of the $sp^3$ hybrids with $n = 2$ is given in Figure 14.9.

**Hybrids for $n = 3$**   So far, all illustrations showed hybrids for the principal quantum number $n = 2$; they applied to elements in the second period of the Periodic Table and, in particular, to carbon for which we mentioned several examples. If we proceed to the third period, and especially to silicon, the

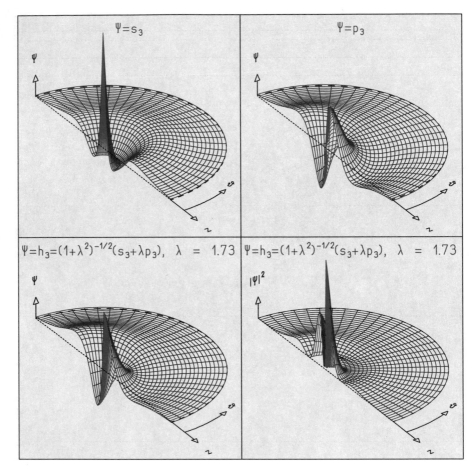

**Fig. 14.10. The wave functions** $\varphi = s_3$, $\varphi = p_3$, **their superposition** $\varphi = h_3$, **and its absolute square, the probability density** $|\varphi|^2 = |h_3|^2$. **In comparison to Fig. 14.1 the scale in the** $r\vartheta$ **plane is different; the functions shown here extend over a wider region.**

only change is in the radial wave functions which now are $R_{30}(r)$ and $R_{31}(r)$ rather than $R_{20}(r)$ and $R_{21}(r)$. These functions display more zeros in the radial coordinate $r$. They also extend to larger values of $r$. After all, they describe electrons in the third shell.

In Figure 14.10 we display the functions $s_3$, $p_3$ as well as the $sp^3$ hybrid wave function $h_3$, constructed from them as a linear combination, and the corresponding spatial probability density $|h_2|^2$. (The corresponding functions for $n = 2$ were shown in Figure 14.1.) The functions $h_3$ and $|h_3|^2$ are depicted as contour-line plots in Figure 14.11. The probability density extends farthest along the negative $z$ axis, appreciably farther than in the case of $n = 2$. Because of the comparatively complicated form of of the radial wave

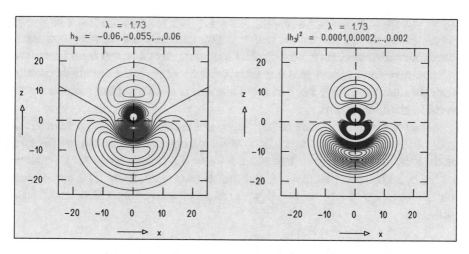

**Fig. 14.11.** The functions $\varphi = h_3$ and $|\varphi|^2 = |h_3|^2$ displayed in Fig. 14.10 are shown here in the form of contour plots in the $xz$ plane. Function values are positive on blue lines, negative on magenta lines, and vanish on red lines. The unit length used for the scales in $x$ and $z$ is the Bohr radius. Note the difference in scale with respect to Fig. 14.2.

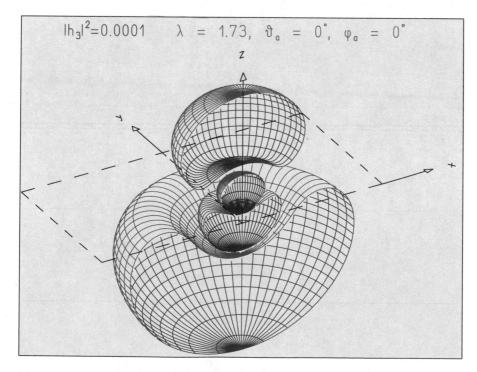

**Fig. 14.12.** The probability density of the $sp^3$ hybrid for $n = 3$ shown as contour-surface plot in the half-space $y > 0$.

functions there is some interesting structure in the inner region where $r$ is small. Since that is the region which is populated by the electrons on inner shells, our assumption of perfect shielding certainly does not hold there. We do not, therefore, expect that our simple calculation yields a realistic picture near the nucleus; but we keep in mind that it is the outer region which matters most in chemical bonds.

The full spatial structure of $|h_3|^2$ becomes apparent in the contour-surface plot of Figure 14.12. As we did above for carbon we can construct $sp$, $sp^2$, and $sp^3$ hybrids for silicon. The same rules as for carbon apply for the values of the hybridization parameter $\lambda$ and the spatial orientations. The parameters chosen for Figs. 14.10 to 14.12 were those of an $sp^3$ hybrid oriented in $z$ direction.

# 15. Three-Dimensional Quantum Mechanics: Resonance Scattering

## 15.1 Scattering by Attractive Potentials

We now return to the discussion of scattering in three dimensions. In Chapter 12 we looked only at scattering by repulsive potentials. Now we shall study the effects of an attractive potential.

In Figure 15.1 the wave function $\varphi_{\mathbf{k}}^{(+)}(\mathbf{r})$ is shown in terms of its real part, imaginary part, and absolute square. The figure is analogous to Figure 12.1, except for the sign of the square-well potential in region I. In comparing Figures 12.1 and 15.1, we observe no striking differences except that in region I where the potential is nonzero the probability density $|\varphi_{\mathbf{k}}^{(+)}(\mathbf{r})|^2$ is appreciably larger for the attractive potential. This larger probability density was to be expected since for the repulsive potential the particle can enter region I only by the tunnel effect.

Figure 15.2 presents the scattered spherical wave $\eta_{\mathbf{k}}(\mathbf{r})$ as defined in Section 12.2. Again we observe that the plot of $|\varphi_{\mathbf{k}}^{(+)}(\mathbf{r})|^2$ has a ripple structure whereas the plot of $|\eta_{\mathbf{k}}(\mathbf{r})|^2$ does not. As discussed at the end of Section 12.1, the ripples of $|\varphi_{\mathbf{k}}^{(+)}(\mathbf{r})|^2$ are caused by the interference of the incident wave $\exp(i\mathbf{k}\cdot\mathbf{r})$ and the scattered spherical wave $\eta_{\mathbf{k}}(\mathbf{r})$. The absolute square of $\eta_{\mathbf{k}}(\mathbf{r})$ shows no such ripples, and for larger $r$ there is only a $|f(\vartheta)|^2/r^2$ falloff.

Comparing the two sets of pictures for the attractive and repulsive potentials (Figures 15.1 and 12.1), we realize that the forward scattering, the scattering into angles $\vartheta$ close to zero, is for the repulsive potential only shadow scattering. In other words, immediately beyond the repulsive square well there is very little probability of finding the particle, whereas there is considerable probability that the particle has traversed the attractive square-well region.

S. Brandt and H.D. Dahmen, *The Picture Book of Quantum Mechanics*, DOI 10.1007/978-1-4614-3951-6_15, © Springer Science+Business Media New York 2012

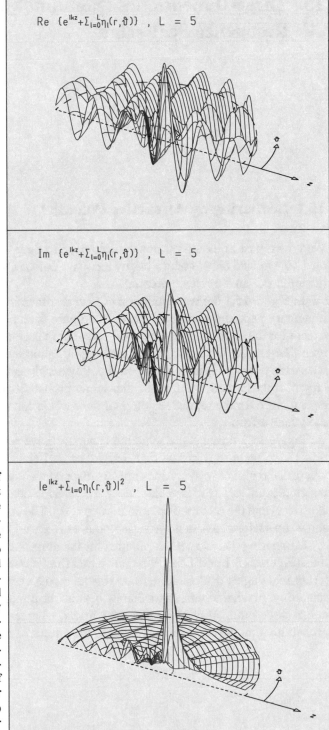

**Fig. 15.1. Scattering of a plane wave incident from the left along the $z$ direction by an attractive potential. The potential is confined to region $r < d$ indicated by the small half-circle. Shown are the real part, the imaginary part, and the absolute square of the wave function $\varphi_{\mathbf{k}}^{(+)}$. The figure corresponds exactly to the situation of Figure 12.1, except for the change $V_0 \rightarrow -V_0$ in the scattering potential.**

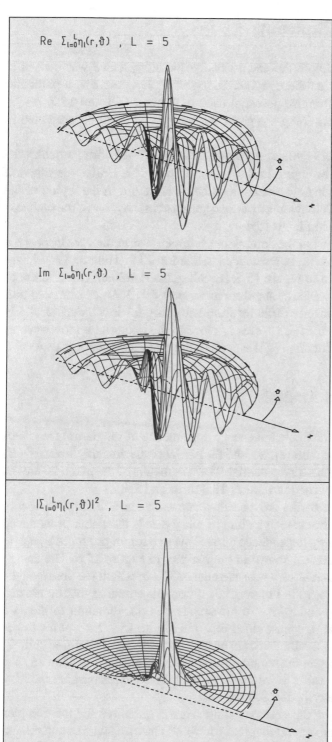

Fig. 15.2. Real part, imaginary part, and absolute square of the scattered spherical wave $\eta_k$ resulting from the scattering of a plane wave by an attractive potential, as shown in Figure 15.1.

## 15.2 Resonance Scattering

In the preceding example the energy of the incoming wave was chosen at random. Let us now consider the scattering of a plane wave at a particular energy $E_{\text{res}}$ by the attractive potential used in Figures 15.1 and 15.2. A systematic way for determining the particular energy $E_{\text{res}}$ will be presented in Section 15.3.

Real and imaginary parts of the wave function $\varphi_{\mathbf{k}}^{(+)}(\mathbf{r})$ with particular energy $E_{\text{res}}$ are plotted in Figure 15.3 together with $|\varphi_{\mathbf{k}}^{(+)}|^2$. Unlike the situation in Figure 15.1, there is now a rather symmetric structure in the region of the attractive potential. This symmetry is also apparent in the plots of the scattered spherical wave $\eta_{\mathbf{k}}(\mathbf{r})$ in Figure 15.4.

To clarify the origin of the dominating symmetric structure, we inspect the scattered partial waves $\eta_\ell$, as introduced in Section 12.2. Their real and imaginary parts are plotted in Figure 15.5, revealing the dominant contribution of the scattered partial wave for angular momentum $\ell = 3$. Since scattered partial waves are significantly different from zero only for low values of $\ell$ – in our example for $\ell = 0, 1, 2, 3$ – clearly close to the potential region wave $\eta_3$ dominates the wave function $\varphi_{\mathbf{k}}^{(+)}$ as well as the scattered spherical wave $\eta_{\mathbf{k}}$.

## 15.3 Phase-Shift Analysis

In this section we investigate the energy dependence of the partial cross sections $\sigma_\ell(E)$, the phase shifts $\delta_\ell(E)$, and the partial scattering amplitudes $f_\ell(E)$ for scattering by an attractive potential. The parameters of the potential are the same as those already used in Figures 15.3 through 15.5.

In Figure 15.6 the partial cross sections are shown as a function of energy for $\ell = 0, 1, \dots, 5$. The striking feature of this figure is the rather pronounced maximum in the energy dependence of $\sigma_3$. This maximum produces a peak in the total cross section $\sigma_{\text{tot}}$, shown in the top plot of Figure 15.6. The energy value of this maximum is very near the energy $E_{\text{res}}$ at which we observed the striking structure in $\eta_3(\mathbf{r})$ in Figure 15.5. It was this structure that dominated the functions $\varphi_{\mathbf{k}}^{(+)}(\mathbf{r})$ and $\eta_{\mathbf{k}}(\mathbf{r})$. To investigate this phenomenon further, we study the behavior of the phase shifts $\delta_\ell(E)$ in Figure 15.7. Except for $\ell = 3$, the phase shifts show a rather smooth energy dependence. The phase shift $\delta_3$, however, rises sharply in the neighborhood of $E_{\text{res}}$ crossing the value $3\pi/2$ at $E_{\text{res}}$. From the phase shifts $\delta_\ell$ we now construct the complex partial scattering amplitudes $f_\ell$, as described in Section 12.3.

Figure 15.8 shows the corresponding Argand diagrams for the complex functions $f_\ell(E)$. The Argand diagram for $f_3(E)$ shows a swift counterclockwise motion of point $f_3$ in the complex plane as the energy passes through

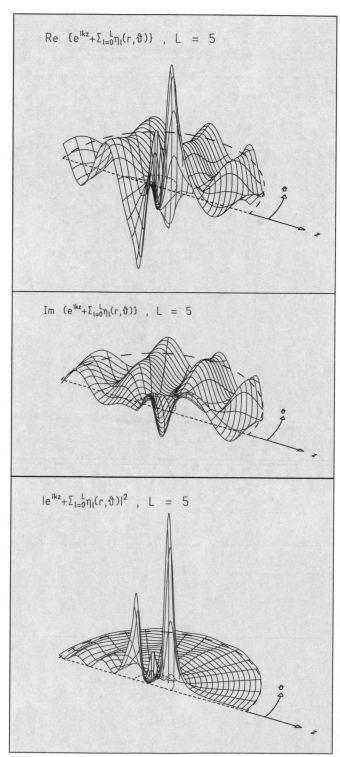

**Fig. 15.3.** Real part, imaginary part, and absolute square of the wave function $\varphi_{\mathbf{k}}^{(+)}$ for the scattering of a plane wave by an attractive potential as given in Figure 15.1, but for a resonance energy $E = E_{\mathrm{res}}$ of the wave.

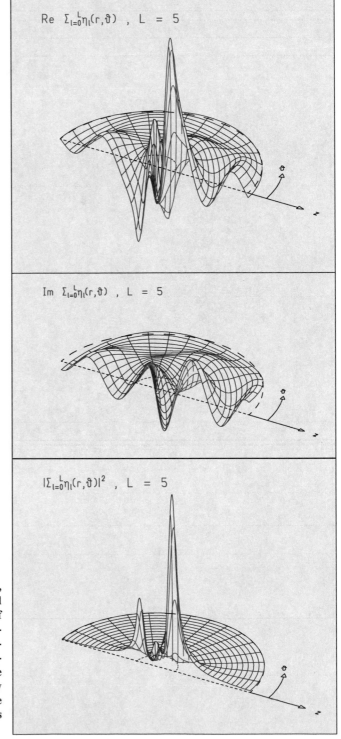

**Fig. 15.4. Real part, imaginary part, and absolute square of the scattered spherical wave $\eta_k$ resulting from the scattering of a plane wave of resonance energy $E = E_{res}$ by the same attractive potential as in Figure 15.3.**

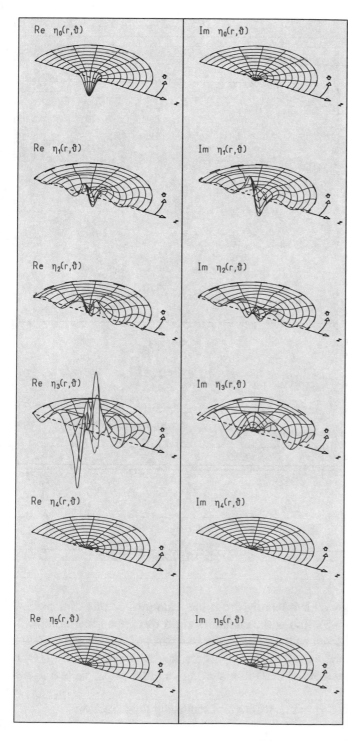

**Fig. 15.5. Real and imaginary parts of the scattered partial waves $\eta_\ell$ resulting from the scattering of a plane wave of resonance energy $E = E_{\mathrm{res}}$ by the attractive potential as in Figures 15.3 and 15.4. The resonance is in the partial wave for $\ell = 3$.**

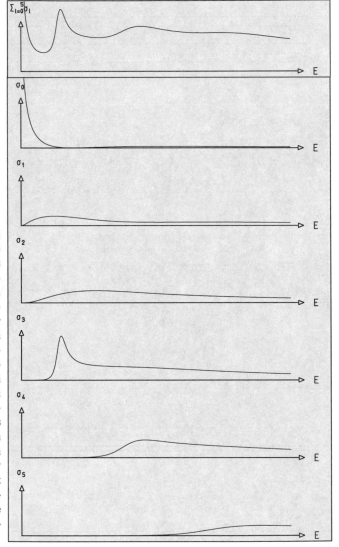

**Fig. 15.6. The partial cross sections $\sigma_\ell(E)$ and the total cross section $\sigma_{\text{tot}}(E)$ approximated by the sum over the first few partial cross sections for the scattering of a plane wave of energy $E$ by the attractive potential used in Figures 15.1 through 15.5. For resonant energy $E = E_{\text{res}}$ there is a sharp maximum in $\sigma_3$ which is reflected in $\sigma_{\text{tot}}$ approximated by the sum over the first six partial cross sections and shown in the top diagram of the figure.**

the energy $E_{\text{res}}$. As we have learned from the examples of one-dimensional scattering (Section 5.5), this is the signature for a resonance-scattering process. As the phase ascends through $\pi/2$, the real part passes through zero in a sharp decrease, whereas the imaginary part reaches its maximum, Im $f_\ell = 1$. Figure 15.8 indicates that none of the scattering amplitudes $f_0$, $f_1$, and $f_2$ has a resonance.

Of particular interest is the differential scattering cross section

$$\frac{\mathrm{d}\sigma}{\mathrm{d}\Omega} = |f(\vartheta)|^2$$

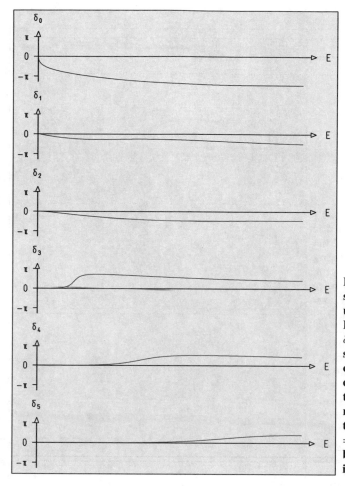

Fig. 15.7. The phase shifts $\delta_\ell(E)$ for the situation of Figure 15.6. For $E = 0$ we put $\delta_\ell(0) = 0$. All phase shifts except $\delta_3$ vary only slowly with energy. Near $E = E_{\text{res}}$ the phase shift $\delta_3(E)$ rises sharply, passing through $\delta_3(E = E_{\text{res}}) = \pi/2$, see also the bottom right diagram in Figure 15.8.

with

$$f(\vartheta) = \frac{1}{k} \sum_{\ell=0}^{\infty} (2\ell + 1) f_\ell(k) P_\ell(\cos \vartheta) \quad .$$

The differential cross section is used to measure the angular momentum of resonances. If near the resonance energy the absolute values of all partial scattering amplitudes $f_\ell$ except the resonant one are small, the differential cross section is determined by the square of the Legendre polynomial corresponding to the angular momentum of the resonance. This is the case for our example. At the resonance energy we expect the differential cross section to be approximately proportional to $[P_3(\cos \vartheta)]^2$. Figure 15.9b shows the differential cross section as a function of $\cos \vartheta$ for various energies. For the resonance energy it is indeed very similar to $(P_3)^2$, as we can see by comparing this figure with Figure 10.3. In Figure 15.9a the intensity of the scattered

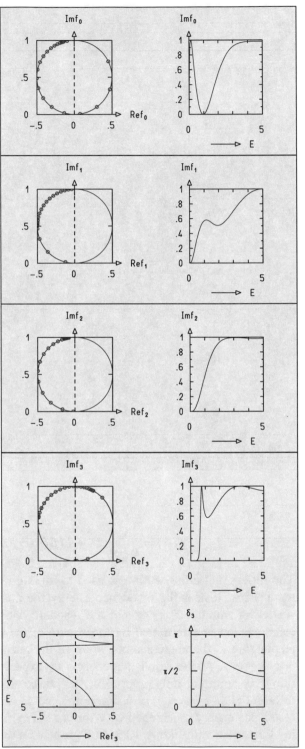

Fig. 15.8. Argand diagrams, that is, diagrams of the energy dependence of the complex partial scattering amplitudes $f_\ell(E)$, for the scattering of a plane wave of energy $E$ by the attractive potential used in Figures 15.1 through 15.7 for $\ell = 0, 1, 2, 3$. The amplitude $f_\ell$ moves on a circle in the complex plane. Small circles are placed on the circle at points equidistant in energy. For the nonresonant partial waves, $\ell = 0, 1, 2$, only the Argand diagram itself and its projection on the $\mathrm{Im}\, f_\ell, E$ plane are shown. The function $\mathrm{Im}\, f_\ell(E)$ is closely related to the partial cross section $\sigma_\ell(E)$. For resonant wave $\ell = 3$ both $\mathrm{Im}\, f_\ell(E)$ and $\mathrm{Re}\, f_\ell(E)$ projections and the phase shift $\delta_3(E)$ are shown. Near resonance energy $E = E_{\mathrm{res}}$ the partial scattering amplitude $f_3(E)$ performs a swift counterclockwise motion through point $(0, 1)$ in the complex plane, giving rise to (1) the pronounced maximum in $\mathrm{Im}\, f_3(E_{\mathrm{res}})$, (2) the steep drop of $\mathrm{Re}\, f_3(E)$ through $\mathrm{Re}\, f_3(E_{\mathrm{res}}) = 0$, and (3) the sharp rise of $\delta_3(E)$ through $\delta_3(E_{\mathrm{res}}) = \pi/2$.

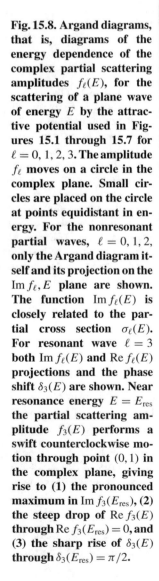

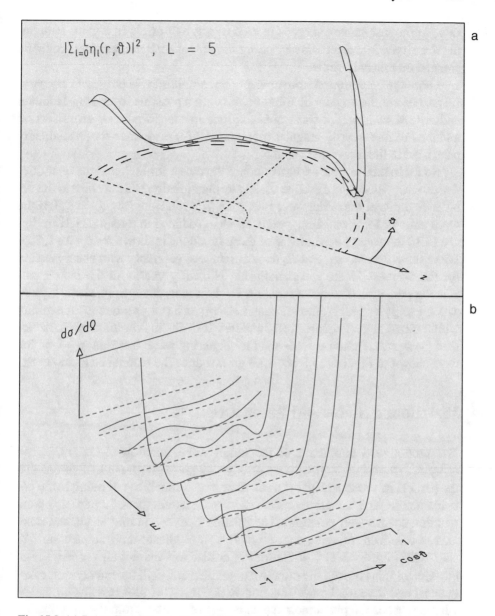

**Fig. 15.9. (a)** Intensity of the scattered spherical wave resulting from the scattering of a plane wave incident in the $z$ direction onto an attractive potential restricted to a small region $r < d$, indicated by the small dashed half-circle. The intensity far outside the potential region is a function of the scattering angle $\vartheta$. The energy of the incident wave is the resonance energy $E = E_{\mathrm{res}}$. **(b)** Energy dependence of the differential scattering cross section $\mathrm{d}\sigma/\mathrm{d}\Omega$ shown over a linear scale in $\cos\vartheta$. The differential cross section is constant in $\cos\vartheta$, indicating isotropic scattering, for $E \approx 0$ (background). At resonance energy $E = E_{\mathrm{res}}$ **(fourth line from the back)** it is given approximately by the square of the Legendre polynomial $P_3(\cos\vartheta)$, since the partial scattering amplitude $f_3$ dominates the cross section.

wave at the resonance energy is plotted over a half-circle in a plane containing the $z$ axis. The detectors measuring the flux of scattered particles could be situated on this half-circle.

Often the background from nonresonant amplitudes is not small. By careful analysis of the angular distribution, it is often possible to separate resonant and nonresonant partial-wave contributions in the differential cross section and thus to measure the angular momentum of a resonance that has already been seen in the total cross section.

So far in this section we have studied the phase shifts $\delta_\ell$ and the quantities derived from them that describe the scattering globally. We now turn to the detailed features of the radial wave functions $R_\ell(k,r)$. Here $k = \sqrt{2ME}/\hbar$ is the wave number for a vanishing potential, as introduced in Section 11.1. In Figure 15.10 the energy dependence of these functions is shown for $\ell = 0, 1, 2, 3$. We observe that $R_0$, $R_1$, and $R_2$ do not change appreciably with energy except for the decrease in the wavelength that is clearly visible in the region outside the potential. The wave function $R_3$, however, changes its shape rapidly as the energy varies. At the resonance energy it has a pronounced maximum within the attractive square-well potential. It is this maximum that characterized the wave function $\varphi_{\mathbf{k}}^{(+)}(\mathbf{r})$ and the scattering wave function $\eta_{\mathbf{k}}(\mathbf{r})$, which were shown in Figures 15.3 and 15.4 to introduce the resonance phenomena.

## 15.4 Bound States and Resonances

The pronounced maximum of the radial wave function $R_3(k_{\text{res}},r)$, $k_{\text{res}} = \sqrt{2ME_{\text{res}}}/\hbar$, in the range of the attractive square-well potential signifies that the particle in a resonant state has a rather large probability of being in the potential range. This situation resembles to some extent that of a particle bound within a square-well potential. The relation between bound states and resonances is indeed intimate, and we shall try to indicate their connection. We start with Figure 15.11d. It shows the attractive square-well potential used throughout this chapter, the effective potential for $\ell = 3$, the energy of the resonance, and the radial wave function $R_3(k_{\text{res}}^{(1)},r)$. This plot reveals the reason why the resonance phenomenon occurs. We remember from the introduction to Chapter 11 that the effective potential is the sum of the potential $V(r)$ and centrifugal potential $\hbar^2\ell(\ell+1)/(2Mr^2)$. Thus the effective potential has a wall just outside the square well, that is, a region where $V_\ell^{\text{eff}}$ is larger than $E_{\text{res}}$. This wall keeps the particle from leaving the potential region except by the tunnel effect. The wall is of finite thickness, however, so that the particle, unlike the particle in the bound state, can also populate the outside region. For this depth of the potential there is no bound state for $\ell = 3$. We now increase the depth of the potential well. There is a continuous decrease of the

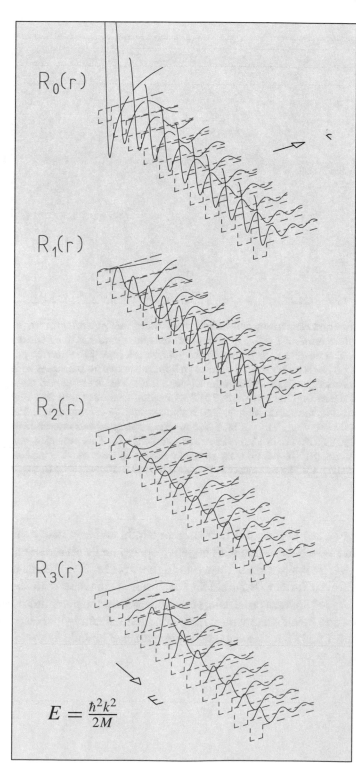

$R_0(r)$

$R_1(r)$

$R_2(r)$

$R_3(r)$

$$E = \frac{\hbar^2 k^2}{2M}$$

**Fig. 15.10. Energy dependence of the radial wave function $R_\ell(k,r)$ for scattering by an attractive square-well potential. The form of the potential is indicated by the long-dash line, the wave energy by the short--dash line, which also serves as zero line for the wave function. Whereas $R_0$, $R_1$, and $R_2$ change very little within the potential region, near the energy $E_{res}$ the wave function $R_3$ of the resonant partial wave develops a very pronounced maximum. Outside the potential region all wave functions show trivial shortening of the wavelength with growing energy.**

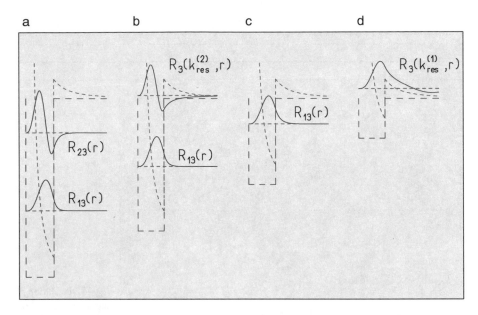

**Fig. 15.11. Bound states and resonances of an attractive square-well potential for angu-lar-momentum quantum number $\ell = 3$. The potential wells have constant fixed widths but different depths. The potential $V(r)$ is shown as a long-dash line. The effective po-tential is also shown. (a) For a rather deep potential well there are two bound states with negative energies indicated by the horizontal short-dash lines. The lower bound state has no radial nodes; the second has one node. (b) A somewhat shallower well has only one bound state but it does have a resonance. The resonance energy corresponds to the horizontal line of positive energy. The radial wave function $R_3(k_{res}^{(2)},r)$ has one node in the potential region, just as the second bound state in part a has. (c) This potential well has only one bound state. (d) The bound state in part c now reappears as a resonance. Its wave function is $R_3(k_{res}^{(1)},r)$. The resonance is the same as that in Figures 15.3 through 15.10.**

resonance energy $E_{res}$ as the potential increases in depth, and eventually the resonant state turns into a bound state with negative energy and with the radial wave function $R_{13}(r)$. This situation is shown in Figure 15.11c. If we increase the potential depth even further (Figure 15.11b), a new resonance with the wave function $R_3(k_{res}^{(2)},r)$ appears possessing one node within the potential re-gion, just as the second bound state would have. When the depth is increased even further (Figure 15.11a), this resonance too becomes a bound state with the wave function $R_{23}(r)$.

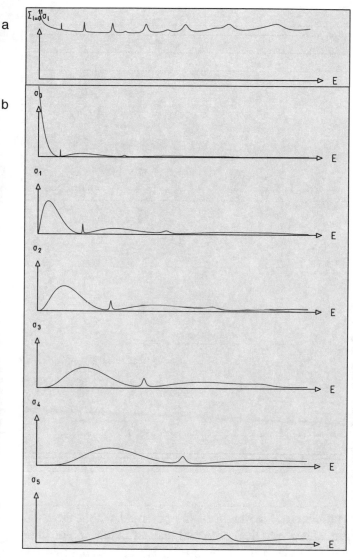

Fig. 15.12. Energy dependence of the partial cross sections $\sigma_\ell(E)$ and of the total cross section $\sigma_{\text{tot}}(E)$, which is approximated by the sum over the first few partial cross sections. Resonances for the different partial waves are visible as maxima in $\sigma_\ell$ and $\sigma_{\text{tot}}$. The maxima are rather sharp for the first resonance and broader for the second. The resonances shift systematically to higher energies as angular-momentum quantum number $\ell$ increases. The energy ranges from $E = 0$ to $E = 2V_0$.

## 15.5 Resonance Scattering by a Repulsive Shell

We have found that resonances occur when there is a repulsive wall in the effective potential. In our example of an attractive square-well potential, this wall originated from the centrifugal force. We can also study resonance scattering on a repulsive shell potential:

$$V(r) = \begin{cases} 0, & 0 \leq r < d_1 \\ V_0, & d_1 \leq r < d_2 \\ 0, & d_2 \leq r < \infty \end{cases} .$$

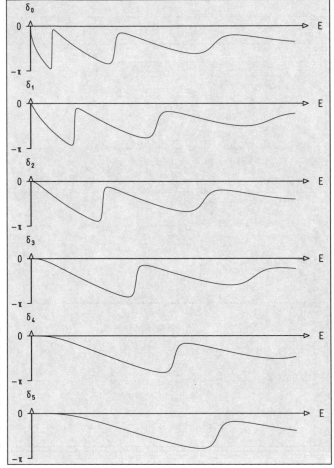

**Fig. 15.13. Energy dependence of the phase shifts** $\delta_\ell(E)$. **At a resonance energy the corresponding phase shift rises steeply and passes through** $-\pi/2$. **See also Figure 15.14. The energy ranges from** $E = 0$ **to** $E = 2V_0$.

Here $V_0$ is positive and denotes the height of the potential within the shell. The shell potential provides a spherical potential wall of height $V_0$, of inner radius $d_1$, and of outer radius $d_2$ around the origin. We can expect that this wall will produce resonances quite independent of a centrifugal force.

Figure 15.12 shows the total cross section $\sigma_{\text{tot}}$ and the partial cross sections $\sigma_\ell$, and Figure 15.13 shows the phase shifts $\delta_\ell$, for $\ell = 0, 1, \ldots, 5$. The resonances are clearly visible as peaks in $\sigma_\ell$ and as jumps in $\delta_\ell$. For $\ell = 0, 1, 2, 3$ there are two resonances at two different energies. We shall refer to them as first and second resonances. For $\ell = 4, 5$ only the first resonance is visible in the energy range of the figure. The second resonance is much wider in energy than the first. The width of both resonances increases with angular momentum $\ell$. There is also a striking regularity between the angular momentum and the energy of the first resonance. In a plane spanned by energy $E$ and angular

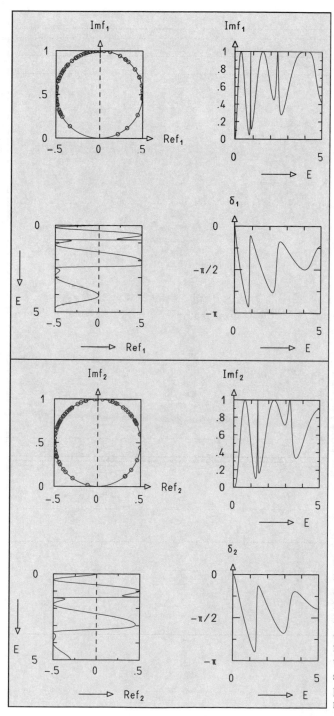

**Fig. 15.14. Argand diagrams for the complex partial scattering amplitudes** $f_1(E)$ **and** $f_2(E)$ **for scattering by a repulsive shell. As in Figures 15.12 and 15.13, the energy ranges from** $E = 0$ **to** $E = 2V_0$. **The resonances have a swift counterclockwise motion of** $f_\ell$ **through the point** $(0, 1)$ **in the complex plane, and the characteristic resonance patterns in** $\mathrm{Im}\, f_\ell(E)$, $\mathrm{Re}\, f_\ell(E)$, **and** $\delta_\ell(E)$ **already familiar from Figure 15.8 (bottom). Because of the shell structure of the potential, there are now more resonances.**

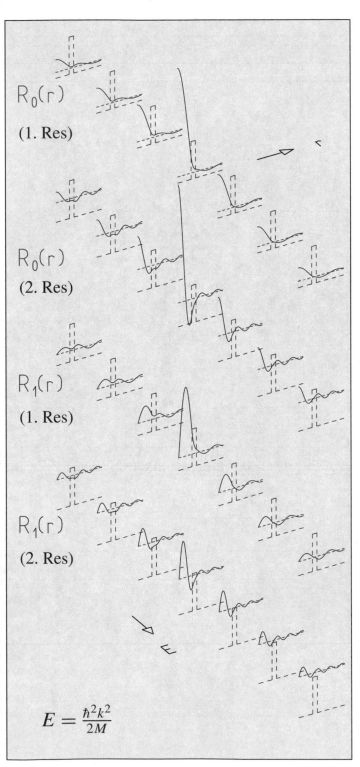

**Fig. 15.15. Energy dependence of the radial wave functions $R_\ell(k,r)$ within restricted energy intervals surrounding the resonances in $\ell = 0$ and in $\ell = 1$ for scattering by a repulsive shell. The form $V(r)$ of the potential is indicated by the long-dash line, the energy $E$ of the wave by the short-dash line. The middle diagram of each series corresponds to the resonance energy. The wave functions $R_\ell(k_{\text{res}}, r)$ shown in these middle diagrams display no node and one node inside the shell for the first and second resonance, respectively.**

$R_0(r)$

(1. Res)

$R_0(r)$

(2. Res)

$R_1(r)$

(1. Res)

$R_1(r)$

(2. Res)

$E = \dfrac{\hbar^2 k^2}{2M}$

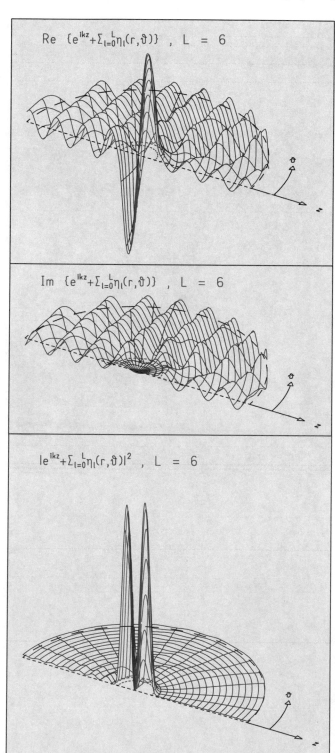

**Fig. 15.16.** Wave functions $\varphi_{\mathbf{k}}^{(+)}$ for the scattering of a plane wave incident along the $z$ direction by a repulsive shell. The energy is that of the first resonance in partial wave $\ell = 1$. The two half-circles near the center indicate the inner and outer boundaries of the spherical potential shell.

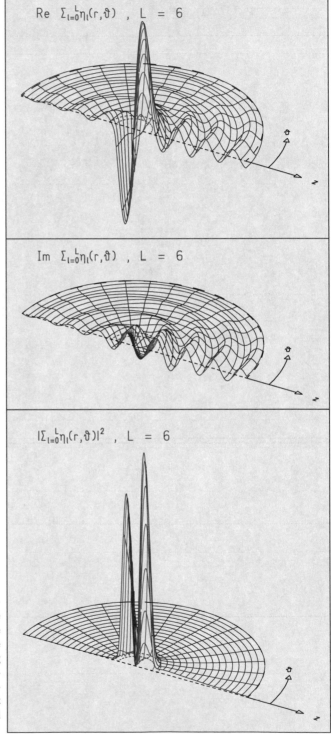

**Fig. 15.17. The scattered spherical wave $\eta_k$ resulting from the scattering of a plane wave incident along the $z$ direction by a repulsive shell. The energy is that of the first resonance in partial wave $\ell = 1$.**

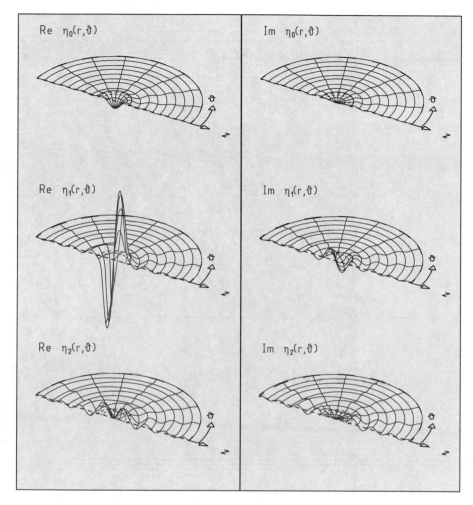

**Fig. 15.18. Scattered partial waves** $\eta_\ell$, $\ell = 0, 1, 2$, **resulting from the scattering of a plane wave incident along the** $z$ **direction by a repulsive shell. The partial wave** $\eta_1$ **has its first resonance at this particular energy of the incident plane wave.**

momentum $\ell$, the first resonances fall on a curved, smooth line called a *Regge trajectory*. There is a similar trajectory for the second resonances.

In the total cross section, also shown in Figure 15.12, the various resonances in different partial waves appear as peaks. When they are sufficiently narrow, they can easily be separated from the smooth background.

Figure 15.14 gives the Argand diagrams for the partial-wave amplitudes $f_1$ and $f_2$. It shows the resonance structure known from the bottom part of Figure 15.8. In both amplitudes there are now two resonances indicated by

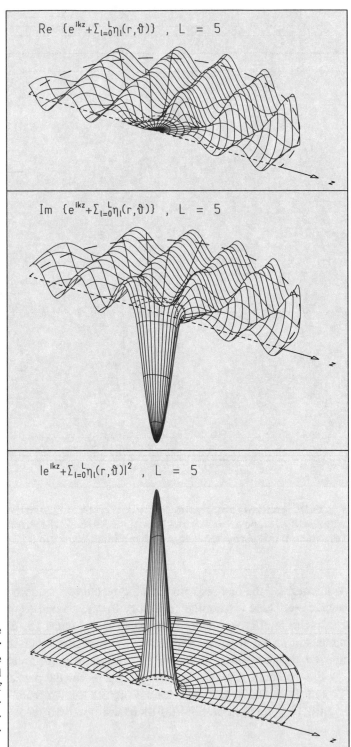

**Fig. 15.19.** Wave function $\varphi_{\mathbf{k}}^{(+)}$, for the first resonance in $\ell = 0$ produced by the scattering of a plane wave incident along the $z$ direction by a repulsive potential shell.

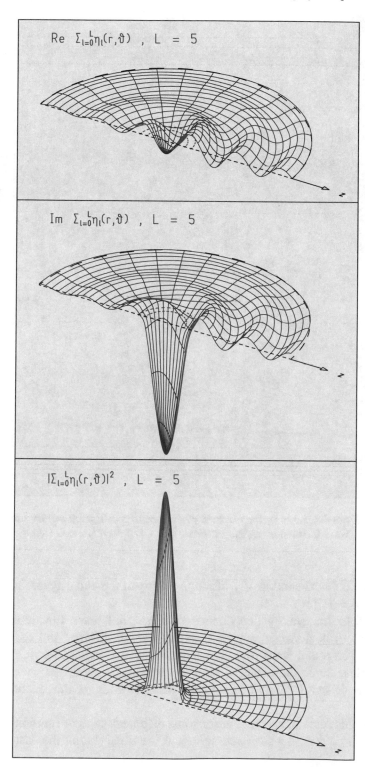

**Fig. 15.20.** Scattered spherical wave $\eta_k$ for the first resonance in $\ell = 0$ produced by the scattering of a plane wave incident along the $z$ direction by a repulsive potential shell.

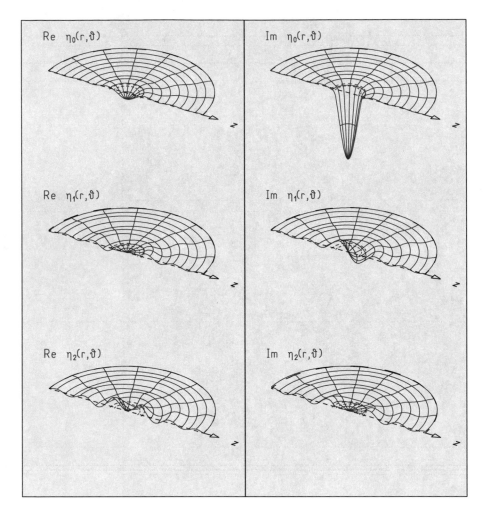

**Fig. 15.21. Scattered partial waves $\eta_\ell$ for the first resonance in $\ell = 0$ produced by the scattering of a plane wave incident along the $z$ direction by a repulsive potential shell.**

the swift counterclockwise motion of $f_\ell$ through the top of the unitarity circle, that is, point Re $f_\ell = 0$, Im $f_\ell = 1$.

In Figure 15.15 the energy dependence of the radial wave functions $R_0(k,r)$ and $R_1(k,r)$ is shown in energy intervals around the first and second resonances. The radial wave functions have the typical enhancement at resonance energies. Since the potential vanishes inside the shell, the wave function has no node in $r$ for the first resonance and one node for the second resonance.

We conclude this section by showing for some resonances the full stationary wave function $\varphi_\mathbf{k}^{(+)}(\mathbf{r})$, the scattered spherical wave $\eta_\mathbf{k}(\mathbf{r})$, and the scattered partial waves $\eta_\ell(\mathbf{r})$ for $\ell = 1, 2, 3$. In each figure the size of the spherical potential shell is indicated by the two half-circles near the origin. They correspond to the inner and outer boundaries of the shell. The first resonance with angular momentum $\ell = 1$ is illustrated in Figures 15.16 through 15.18. In Figure 15.18 we observe that only the partial scattered wave for $\ell = 1$, that $\eta_1$ shows a resonance structure. It has no node in $r$, indicating the first resonance. There is one node in the polar angle $\vartheta$ at $\vartheta = \pi/2$. It is caused by the Legendre polynomial $P_1(\cos\vartheta) = \cos\vartheta$, which determines the $\vartheta$ dependence of $\eta_1$ as discussed at the beginning of Section 12.2. The scattered spherical wave $\eta_\mathbf{k}(\mathbf{r})$ in Figure 15.17 is obtained by summing up the partial scattered waves $\eta_\ell(\mathbf{r})$. Since the dominating term in this sum is $\eta_1(\mathbf{r})$, it is not surprising that the structure of $\eta_\mathbf{k}(\mathbf{r})$ in the central region is that of $\eta_1(\mathbf{r})$, displaying no node in $r$ but one node in $\vartheta$. Even the full stationary wave function $\varphi_\mathbf{k}^{(+)}(\mathbf{r})$, which is shown in Figure 15.16 and is a superposition of the incoming harmonic plane wave and the scattered spherical wave $\eta_\mathbf{k}(\mathbf{r})$, retains much of this structure.

Finally, we turn to the resonance of angular momentum zero. Figures 15.19, 15.20, and 15.21 show the functions $\varphi_\mathbf{k}^{(+)}(\mathbf{r})$, $\eta_\mathbf{k}(\mathbf{r})$, and $\eta_\ell(\mathbf{r})$ for the first resonance. The resonating partial wave is now $\eta_0$. It has no node in $\vartheta$ since the Legendre polynomial $P_0$ does not depend on $\vartheta$. As a first resonance it also has no node in $r$. These simple features of $\eta_0$ are very clearly retained in the scattered spherical wave $\eta_\mathbf{k}(\mathbf{r})$, Figure 15.20, and in the stationary wave function $\varphi_\mathbf{k}^{(+)}(\mathbf{r})$, Figure 15.19.

## Problems

15.1. What is the relation between the wavelengths inside and outside the potential region of Figure 15.1?

15.2. Explain the features of the plots of Figure 15.1 with the help of the plots of Figure 15.2 in terms of the initial plane wave, the scattered spherical wave, and their interference pattern.

15.3. Find the features in the $r$ and $\vartheta$ dependence of the resonant partial wave $\eta_3$ in Figure 15.5 that are characteristic for angular momentum $\ell = 3$.

15.4. Given the form of the resonant partial wave $\eta_3$ in Figure 15.5, is there for $\ell = 3$ a bound state or resonance with lower energy than the one plotted?

15.5. Relate the appearance of the backward peak in the differential scattering cross section at resonance energy in Figure 15.9 to the partial-wave decomposition of the scattered wave in Figure 15.5 and to $|\eta_\mathbf{k}|^2$ in Figure 15.4.

15.6. Describe the behavior, for large values of $r$, of the bound-state wave function of the first excited state in Figure 15.11a and of the resonance wave function in Figure 15.11b.

15.7. Compare the criteria for a resonance, as found in Figure 15.8, with the resonances indicated in the Argand diagrams in Figure 15.14. Which peaks correspond to resonance phenomena and why?

15.8. Are the energies of the resonances in Figure 15.15 higher or lower than the energies of bound states in an infinitely deep potential well?

# 16. Coulomb Scattering

## 16.1 Stationary Solutions

In Section 13.4 we discussed the stationary bound states in the attractive Coulomb potential. The bound-state wave functions were characterized by three quantum numbers, the principal quantum number $n$, total angular-momentum quantum number $\ell$, and the quantum number $m$ of the $z$ component of angular momentum. The energies of the bound states are all negative, $E_n = -Mc^2\alpha/(2n^2)$. The set of the bound states in the hydrogen atom is not complete. Since the Coulomb potential vanishes at infinity there exists also a set of continuum scattering states for all positive energy eigenvalues $E$.

As in Section 13.4 the radial Schrödinger equation reads

$$\left[ -\frac{\hbar^2}{2M}\frac{1}{r}\frac{\mathrm{d}}{\mathrm{d}r}r + V_\ell^{\mathrm{eff}}(r) \right] R_\ell(k,r) = E R_\ell(k,r)$$

with the continuous energy eigenvalue

$$E = \frac{\hbar^2 k^2}{2M}$$

parametrized in terms of the wave number $k$.

The effective potential for angular momentum $\ell$,

$$V_\ell^{\mathrm{eff}}(r) = \frac{\hbar^2}{2M}\frac{\ell(\ell+1)}{r^2} - Z\hbar c\frac{\alpha}{r} \quad ,$$

again consists of the repulsive centrifugal term and the Coulomb term. In contrast to the corresponding formula in Section 13.4 here we have incorporated the charge number $Z$ of the nucleus off which the electron is scattered. If the charge number $Z$ is negative the Coulomb potential is repulsive and the Schrödinger equation describes repulsive Coulomb scattering. Of course, in this case no bound states exist. For instance, the charge number $Z = -1$ would describe the charge of an antiproton with a negative elementary charge. Of course, repulsive scattering also takes place if the scattering center carries

S. Brandt and H.D. Dahmen, *The Picture Book of Quantum Mechanics*,
DOI 10.1007/978-1-4614-3951-6_16, © Springer Science+Business Media New York 2012

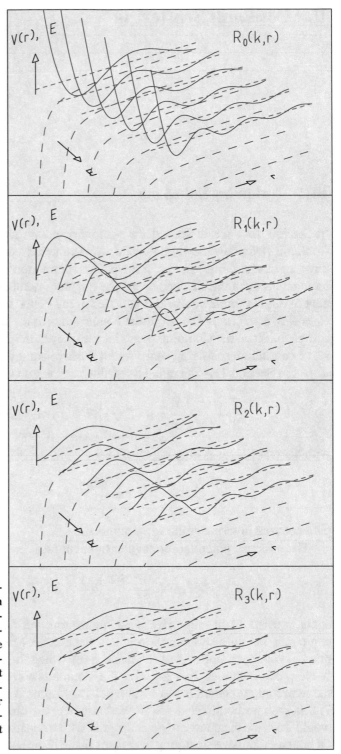

**Fig. 16.1. Radial scattering wave function** $R_\ell(k,r)$ **in an attractive Coulomb potential. In each of the four plots the total energy** $E$ **is varied but the angular-momentum quantum number** $\ell$ **is kept fixed. The latter is varied from plot to plot.**

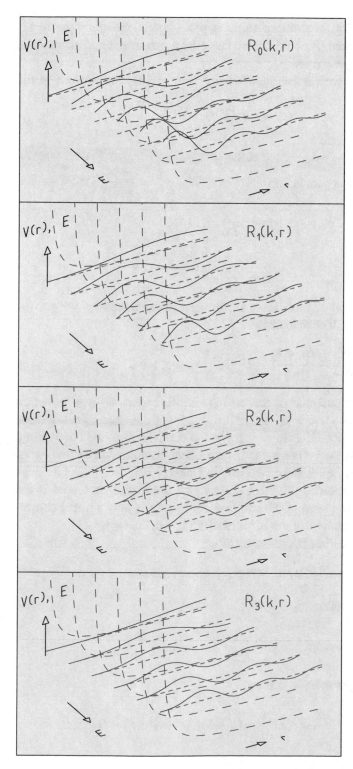

Fig. 16.2. As Figure 16.1 but for a repulsive Coulomb potential.

positive charge, i.e., in an ordinary nucleus, provided the incoming particle is also positively charged. It could for instance be a positron, another nucleus, or a positively charged meson.

We divide the Schrödinger equation by $(-\hbar^2)/(2M)$ and introduce the function

$$w_\ell = r R_\ell(k,r) \quad ,$$

the dimensionless variable

$$\zeta = kr \quad ,$$

and the dimensionless parameter

$$\eta = -\frac{Z}{ka} \quad ,$$

where

$$a = \frac{\hbar}{\alpha M c}$$

is the Bohr radius, cf. Section 13.4. In terms of these quantities we get the differential equation of second order:

$$\left( \frac{\mathrm{d}^2}{\mathrm{d}\zeta^2} - \frac{\ell(\ell+1)}{\zeta^2} - 2\frac{\eta}{\zeta} + 1 \right) w_\ell(\eta,\zeta) = 0 \quad .$$

In the case of a potential of finite range $R_0$ like a square-well potential vanishing for all $r > R_0$ or of a potential falling off faster than $1/r^2$ for large $r > R_0$ the leading potential for $r > R_0$ is the centrifugal term. The leading term in the solution $w_\ell$ for large $r$ is then a linear combination of two exponentials of the form $\exp(\pm ikr)$ as we saw in Chapter 12.

For Coulomb scattering for large distances $r$ the leading potential is the $1/r$ term. The leading term of the solution for large $r$ is now a linear combination of two exponentials of the form $\exp\{\pm i(kr - \eta \ln 2kr)\}$.

We introduce the dimensionless variable

$$z = -2ikr = -2i\zeta \quad , \qquad \zeta = kr \quad ,$$

and factorize the function $w_\ell$,

$$w_\ell(\eta,\zeta) = \mathrm{e}^{-\frac{1}{2}z} \left( \frac{\mathrm{i}}{2}z \right)^{\ell+1} v_\ell(\eta,z) \quad ,$$

to get *Laplace's differential equation*

$$\left[ z\frac{\mathrm{d}^2}{\mathrm{d}z^2} + (2\ell+2 - z)\frac{\mathrm{d}}{\mathrm{d}z} - (\ell+1+\mathrm{i}\eta) \right] v_\ell(\eta,z) = 0 \quad .$$

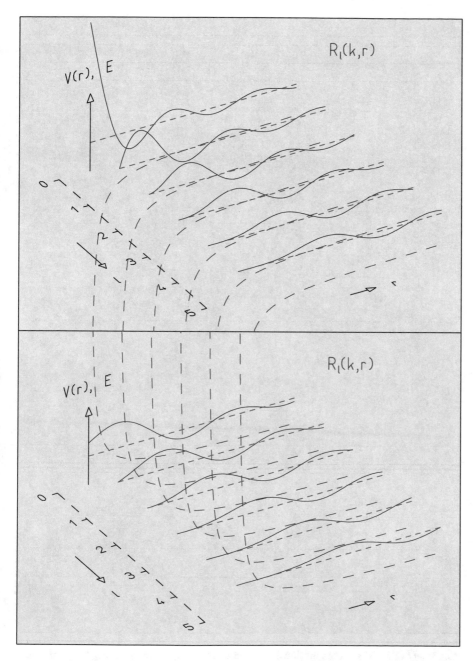

**Fig. 16.3. Radial scattering wave function** $R_\ell(k,r)$ **in an attractive (top) and in a repulsive (bottom) Coulomb potential for different values** $\ell$ **of the angular-momentum quantum number but for fixed energy.**

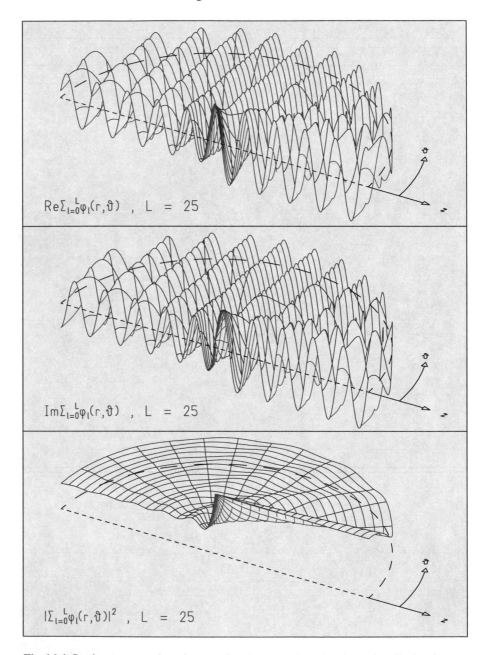

**Fig. 16.4. Stationary wave function $\varphi(\mathbf{r})$ for the scattering of an incoming Coulomb wave by an attractive Coulomb potential. Shown are the real part, the imaginary part, and the absolute square of $\varphi(\mathbf{r})$.**

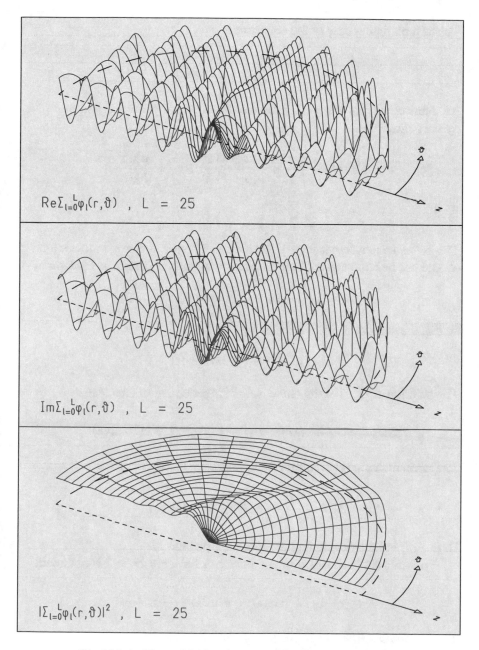

**Fig. 16.5. As Figure 16.4 but for a repulsive Coulomb potential.**

This is a particular case of the equation

$$\left(z\frac{d^2}{dz^2}+(b-z)\frac{d}{dz}-a\right)F(a|b|z)=0 \quad .$$

Its solution, regular at $z=0$, is the *confluent hypergeometric function* given by the series

$$F(a|b|z)=1+\frac{a}{b}\frac{z}{1!}+\frac{a(a+1)}{b(b+1)}\frac{z^2}{2!}+\cdots+\frac{(a)_n}{(b)_n}\frac{z^n}{n!}+\cdots$$

with

$$(a)_n=a(a+1)\cdot\ldots\cdot(a+n-1) \quad .$$

The series is convergent for all complex $z$. The parameter $a$ in this equation should not be confused with the Bohr radius. For our particular case we have

$$a=\ell+1+i\eta \quad , \qquad b=2(\ell+1) \quad ,$$

so that the solution $v_\ell(\eta,z)$ is given by

$$v_\ell(\eta,z)=F(\ell+1+i\eta|2(\ell+1)|z) \quad .$$

The solution $R_\ell(k,r)$ of the radial Schrödinger equation thus reads

$$R_\ell(k,r)=\frac{A_\ell}{r}e^{ikr}(kr)^{\ell+1}F(\ell+1+i\eta|2(\ell+1)|-2ikr) \quad .$$

The normalization is given by

$$A_\ell=\frac{2^\ell}{(2\ell+1)!}e^{-\frac{1}{2}\pi\eta}|\Gamma(\ell+1+i\eta)| \quad .$$

Here $\Gamma(z)$ is Euler's gamma function, cf. Appendix E.
    With this choice the asymptotic behavior for large $kr\gg1$ is obtained:

$$R_\ell(k,r)\xrightarrow[r\to\infty]{}\frac{1}{r}\sin(kr-\eta\ln2kr-\frac{1}{2}\ell\pi+\delta_\ell) \quad .$$

The *Coulomb scattering phase* $\delta_\ell$ is given by

$$e^{i2\delta_\ell}=\frac{\Gamma(\ell+1+i\eta)}{\Gamma(\ell+1-i\eta)} \quad , \qquad \delta_\ell=\frac{1}{2i}\ln\frac{\Gamma(\ell+1+i\eta)}{\Gamma(\ell+1-i\eta)} \quad .$$

The asymptotic form of $R_\ell(k,r)$ differs from the corresponding formulae of Section 12.3 by the additional phase $\eta\ln2kr$ which diverges for $kr\to\infty$. This is a particular feature of the Coulomb interaction.

The solution $\psi_c(\mathbf{r})$ of the three-dimensional Schrödinger equation with Coulomb potential can be reconstructed with the help of the partial-wave decomposition

$$\psi_c(\mathbf{r}) = \sum_{\ell=0}^{\infty} (2\ell + 1) i^\ell e^{i\delta_\ell} R_\ell(k,r) P_\ell(\cos\vartheta) \quad .$$

It is called a Coulomb wave function. The asymptotically leading term for large $|r - z| \to \infty$ has the form

$$\psi_c(\mathbf{r}) \to e^{i(kz + \eta \ln k(r-z))} \quad .$$

This asymptotic Coulomb wave differs from a plane wave because of the appearance of the logarithm in the exponent. This effect is due to the long range of the Coulomb potential $\hbar c \alpha / r$. It is said to have infinite range. Only potentials falling off faster than $r^{-1}$ for large $r$ possess finite range. Scattering solutions in these potentials approach a plane wave $\exp\{ikz\}$ asymptotically.

The total stationary scattering solution is the product of $R_\ell(k,r)$ and the spherical harmonics $Y_{\ell m}(\vartheta, \varphi)$:

$$\varphi_{\ell m}(k,\mathbf{r}) = R_\ell(k,r) Y_{\ell m}(\vartheta, \varphi) \quad .$$

In Figure 16.1 we show the radial wave function $R_\ell(k,r)$ for different total energies and for various fixed values of the angular-momentum quantum number $\ell$. For large $r$ the wave function is an oscillating function of $r$ which can be qualitatively described by an $r$-dependent wavelength. As we would expect the wavelength is large for large distances from the origin, where the kinetic energy is small, and decreases towards the origin. Besides $r$ the wavelength depends on $k$ and therefore on the total energy $E$. It decreases with increasing energy. Near the origin the wave function is suppressed by the centrifugal barrier. This effect becomes particularly evident in Figure 16.3 in which the wave function $R_\ell(k,r)$ is shown at a fixed energy, i.e., at a fixed $k$ value, for different angular-momentum quantum numbers $\ell$.

In Figures 16.4 and 16.5 the real and imaginary part and absolute square of the Coulomb wave function $\varphi(\mathbf{r})$ are shown. It is the solution of the three-dimensional stationary Schrödinger equation for an incoming Coulomb wave scattered by a Coulomb potential. The summation over partial waves $\varphi_{\ell m}(k,\mathbf{r})$ is carried out up to angular momentum $\ell = 25$. For the range shown in $r$ this guarantees sufficient accuracy.

In Figure 16.4 the case of the attractive Coulomb potential is presented. In the region of the singularity of the potential energy the wave function acquires a shorter wavelength due to the increase of the momentum of the particle.

Figure 16.5 exhibits the case of a repulsive Coulomb potential. In the center of the plots the wave function is suppressed since here the repulsive potential dominates over the kinetic energy. Towards the center of repulsion the wavelength increases indicating the loss of momentum of the particle as it climbs the repulsive wall.

## 16.2 Hyperbolic Kepler Motion: Scattering of a Gaussian Wave Packet by a Coulomb Potential

We consider a Gaussian wave packet as described in Section 13.5 as initial wave function of an electron with expectation values of position $\mathbf{r}_0$ and momentum $\mathbf{p}_0$. The spatial width is $\sigma$. The quantities $\mathbf{r}_0$, $\mathbf{p}_0$ and $\sigma$ are chosen in such a way that the wave packet contains negligible contributions from bound states. Thus it can be represented by a superposition of scattering wave functions $\varphi_{\ell m}(k,\mathbf{r})$ only. As quantization axis for angular momentum, i.e., as $z$ axis of the coordinate frame, we choose the direction of classical angular momentum $\mathbf{L}_0 = \hbar \mathbf{r}_0 \times \mathbf{k}_0$.

The decomposition reads

$$\psi(\mathbf{r},0) = \frac{2}{\pi} \sum_{\ell=0}^{\infty} \sum_{m=-\ell}^{\ell} \int_0^{+\infty} b_{\ell m}(k)\varphi_{\ell m}(k,\mathbf{r})k^2 \, \mathrm{d}k \quad .$$

The coefficients $b_{\ell m}(k)$ are probability amplitudes, their absolute squares

$$P_{\ell m}(k) = |b_{\ell m}(k)|^2$$

represent probabilities for angular momentum $\ell, m$ at given $k$ and probability densities in $k$ for given $\ell$, $m$, if we choose as integration measure $(2/\pi)k^2 \, \mathrm{d}k$.

The function

$$P_\ell(k) = \sum_{m=-\ell}^{\ell} |b_{\ell m}(k)|^2$$

is shown in Figure 16.6 for attractive (top) and repulsive (bottom) Coulomb scattering for the same initial wave packet. Its marginal distributions

$$P(k) = \sum_{\ell=0}^{\infty} P_\ell(k)$$

and

$$W_\ell = \frac{2}{\pi} \int_0^{+\infty} P_\ell(k)k^2 \, \mathrm{d}k$$

are also exhibited. The peaks of the distributions are close to the classical values of

$$k_0^2 = \frac{2M}{\hbar^2} E_0 \quad , \qquad E_0 = \frac{\mathbf{p}_0^2}{2M} - Z\hbar c \frac{\alpha}{r_0} \quad .$$

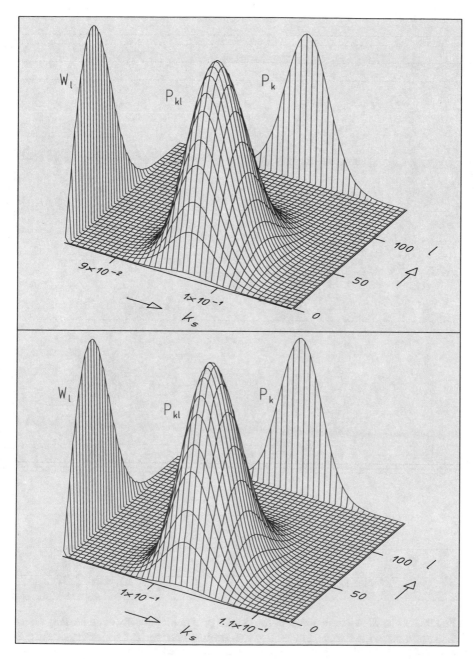

**Fig. 16.6. Probability distributions** $P_\ell(k)$ **and marginal distributions** $W_\ell$ **and** $P(k)$ **for the wave packet used in attractive (top) and repulsive (bottom) Coulomb scattering in Figure 16.7 and 16.8, respectively.**

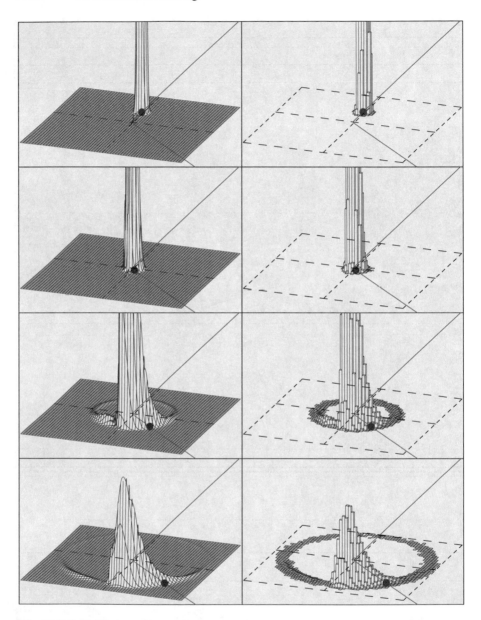

**Fig. 16.7. Left:** Time development of the spatial probability density of an initially Gaussian wave packet undergoing attractive Coulomb scattering. The trajectory of the corresponding classical particle is a hyperbola and indicated by the solid line. The density is shown in the plane of the classical trajectory for four moments in time. The position is marked by a dot. **Right:** The corresponding time development for the spatial probability density in a classical phase-space distribution.

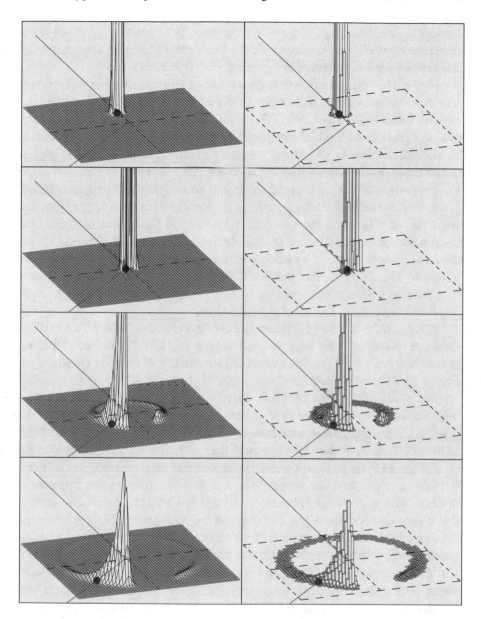

**Fig. 16.8. As Figure 16.7 but for repulsive scattering.**

Depending on the sign of $Z$ we deal with attractive scattering ($Z$ positive) or repulsive scattering ($Z$ negative). As to be expected for repulsive scattering the peak of the $k$ distribution is located at higher $k$ values.

Figure 16.7 presents the plots of the time evolution of an initially Gaussian wave packet with an impact parameter equal to its spatial width under the action of an attractive potential. The solid line indicates the trajectory of the classical particle with initial conditions $\mathbf{r}_0$ and $\mathbf{p}_0 = \hbar\mathbf{k}_0$. The black dot marks its position at time $t$. The circular density distribution with the scatterer as center indicates the scattered spherical wave. Also shown in Figure 16.7 is the time evolution of the corresponding spatial density derived from a classical phase-space distribution which initially is identical to the one of the Gaussian wave packet. The main features are the same as in the quantum-mechanical distribution. In particular, we realize that in both cases the position of the classical particle does not coincide with the maximum of the probability distribution. This is due to the fact that the angle of deflection of the trajectory of a particle with smaller impact parameter is much larger than for a larger impact parameter, cf. Figure 16.9 (top) below.

Analogously, Figure 16.8 presents the corresponding plots for a repulsive Coulomb potential. The large deflection of a particle at very small impact parameters causes the gap in the probability density in forward direction, cf. Figure 16.9 (bottom) below.

In the case of the elliptical orbits the most striking feature of the quantum-mechanical probability density in contrast to the classical distribution is the revival of the wave packet at the time $t = T_{\mathrm{rev}} = (n_{\mathrm{cl}}/3)T_K$. It is a consequence of the interference of the widening wave packet with itself on the closed orbit.

Figure 16.9 (top) shows classical trajectories in an attractive Coulomb field for a series of initial conditions which differ only in the impact parameter. The trajectories intersect in a region situated behind the scattering center as seen from the initial position. We expect this region to be the one where quantum-mechanical interference is important. The quantum-mechanical probability density shown for this region in Figure 16.10 (top) verifies this expectation. For Figure 16.10 the initial spatial width of the wave packet was increased and consequently the initial width in momentum was decreased. In this way the spread in de Broglie wavelength is reduced for the wave packet and the interference pattern characterized by half the de Broglie wavelength becomes more apparent.

Finally, we look for interference in repulsive scattering. An inspection of classical trajectories with different impact parameters, Figure 16.9 (bottom), shows that they interact behind and sideways of the scattering center as seen from the initial position. In Figure 16.10 which shows the quantum-mechanical probability density we indeed observe interference just there.

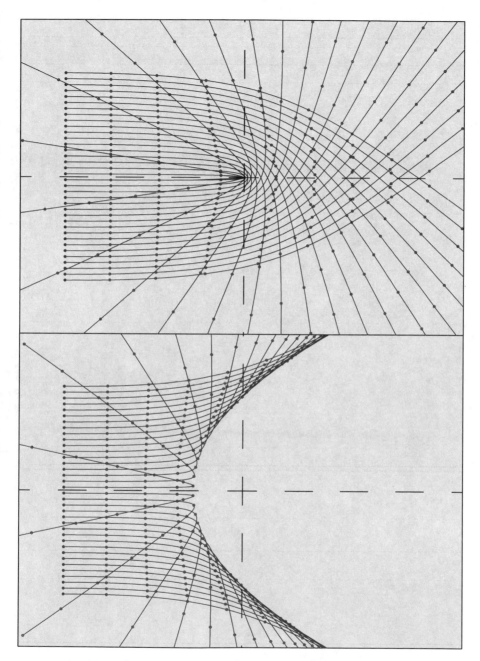

**Fig. 16.9. Classical trajectories in an attractive Coulomb potential (top) and in a repulsive Coulomb potential (bottom). The individual trajectories beginning on the far left and containing marks corresponding to equal time intervals differ only by their impact parameter.**

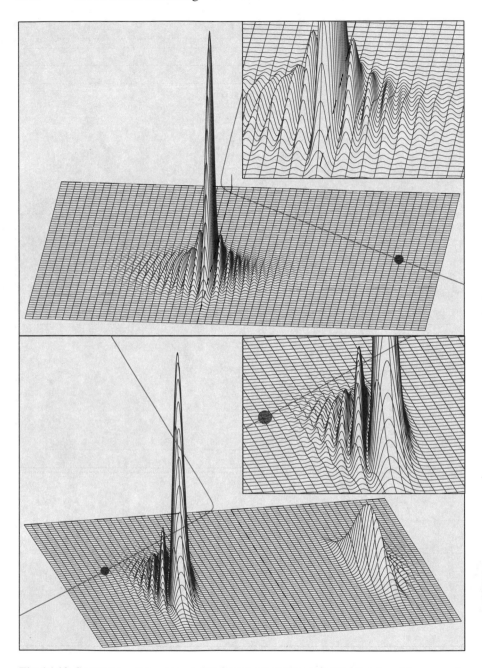

**Fig. 16.10. Scattering of a wave packet by an attractive potential (top) and by a repulsive potential (bottom). All physics parameters of the wave packet are the same as in Figure 16.7 and Figure 16.8, respectively, except for the initial spatial width which is 2.5 times higher, and the momentum width which is 2.5 times lower.**

# 17. Spin

## 17.1 Spin States, Operators and Eigenvalues

In Chapter 10 we introduced orbital angular momentum

$$\hat{\mathbf{L}} = \mathbf{r} \times \hat{\mathbf{p}}$$

in terms of the position operator $\mathbf{r} = (x, y, z)$ and momentum operator $\hat{\mathbf{p}} = (\hbar/\mathrm{i})(\partial/\partial x, \partial/\partial y, \partial/\partial z)$ of a particle moving in space. In a magnetic field electrons on atomic orbits with orbital angular momentum $\hbar\ell$ exhibit magnetic moments

$$\mu = -\frac{e}{2M}\hbar\ell$$

with $-e$ being the charge and $M$ the mass of the electron. The Stern–Gerlach experiment shows, however, that in addition electrons possess an *intrinsic magnetic moment* $\mu_s$. This led George Uhlenbeck and Samuel Goudsmit to postulate an *intrinsic angular momentum* or *spin* of the electron. The intrinsic magnetic moment is then related to the spin quantum number $s$ by

$$\mu_s = -g_s\frac{e}{2M}\hbar s \quad .$$

The coefficient $g_s$ is called *gyromagnetic factor*, its value for the electron is (nearly exactly) 2.

It can be shown that the *spin states* cannot be represented as wave functions of the space coordinates $x$, $y$, $z$. We use the concepts introduced in Appendix B concerning the quantum mechanics in a two-level system and describe the spin state of a particle in a two-dimensional space with the base vectors

$$\eta_1 = \begin{pmatrix} 1 \\ 0 \end{pmatrix} \quad , \quad \eta_{-1} = \begin{pmatrix} 0 \\ 1 \end{pmatrix} \quad .$$

For later use we already introduce here the notation

$$\eta_1^+ = (1, 0) \quad , \quad \eta_{-1}^+ = (0, 1) \quad ,$$

S. Brandt and H.D. Dahmen, *The Picture Book of Quantum Mechanics*, 353
DOI 10.1007/978-1-4614-3951-6_17, © Springer Science+Business Media New York 2012

for the adjoint vectors. The three Pauli matrices $\sigma_1$, $\sigma_2$, $\sigma_3$ together with the unit matrix $\sigma_0$ form a basis for all Hermitean $2 \times 2$ matrices in this space. We introduce the matrices

$$S_x = \frac{\hbar}{2}\sigma_1 \quad , \qquad S_y = \frac{\hbar}{2}\sigma_2 \quad , \qquad S_z = \frac{\hbar}{2}\sigma_3 \quad .$$

The commutation relations for the components $S_x$, $S_y$, $S_z$ and for the sum

$$\mathbf{S}^2 = S_x^2 + S_y^2 + S_z^2$$

of their squares are the same as for the components $\hat{L}_i$ and the square $\hat{\mathbf{L}}^2$ of the orbital angular-momentum operator $\hat{\mathbf{L}} = (\hat{L}_x, \hat{L}_y, \hat{L}_z)$, i.e.,

$$[S_x, S_y] = i\hbar S_z \quad , \qquad [S_y, S_z] = i\hbar S_x \quad , \qquad [S_z, S_x] = i\hbar S_y \quad ,$$

$$[\mathbf{S}^2, S_a] = 0 \quad , \qquad a = x, y, z \quad .$$

Therefore, we interpret the components of the vector

$$\mathbf{S} = (S_x, S_y, S_z)$$

as the *operator of the spin* of the electron. Because of the non-commutativity of these components there are no common eigenstates of the three components. As for orbital angular momentum, we choose the third component $S_z$ and the square $\mathbf{S}^2$ for the definition of a base of common eigenvector equations

$$\mathbf{S}^2 \boldsymbol{\eta}_r = \frac{3}{4}\hbar^2 \boldsymbol{\eta}_r \quad , \qquad S_z \boldsymbol{\eta}_r = \frac{1}{2}r\hbar \boldsymbol{\eta}_r \quad , \qquad r = 1, -1 \quad .$$

The eigenvalues of $S_z$ are $m_s\hbar$, $m_s = \pm 1/2$, the one of $\mathbf{S}^2$ is $s(s+1)\hbar^2 = (3/4)\hbar^2$ with $s = 1/2$ corresponding to the situation for orbital angular momentum $\mathbf{L}$.

We conclude the electron possesses half-integer intrinsic angular momentum, $s\hbar = \frac{1}{2}\hbar$. In analogy to the integer quantum numbers $\ell$ and $m$ of orbital angular momentum ($m = \ell, \ell - 1, \ldots, -\ell$) we introduce the *spin quantum numbers* $s = \frac{1}{2}$ and $m_s = \frac{1}{2}, -\frac{1}{2}$.

The expectation values of the operators $S_x$, $S_y$, $S_z$ for the states $\boldsymbol{\eta}_1$, $\boldsymbol{\eta}_{-1}$ are

$$\boldsymbol{\eta}_r^+ S_x \boldsymbol{\eta}_r = 0 \quad , \qquad \boldsymbol{\eta}_r^+ S_y \boldsymbol{\eta}_r = 0 \quad , \qquad \boldsymbol{\eta}_r^+ S_z \boldsymbol{\eta}_r = \frac{r}{2}\hbar \quad .$$

The three equations are equivalent to one vectorial equation:

$$\langle \mathbf{S} \rangle_r = \boldsymbol{\eta}_r^+ \mathbf{S} \boldsymbol{\eta}_r = \frac{r}{2}\hbar \mathbf{e}_z \quad , \qquad r = 1, -1 \quad .$$

Since the two states $\boldsymbol{\eta}_1$, $\boldsymbol{\eta}_{-1}$ are eigenstates of $S_z$ the variance of $S_z$ vanishes,

$$(\Delta S_z)^2 = \boldsymbol{\eta}_r^+ (S_z^2 - \langle S_z \rangle_r^2) \boldsymbol{\eta}_r = 0 \quad .$$

For the two other components we get

$$(\Delta S_x)^2 = (\Delta S_y)^2 = \frac{1}{4}\hbar^2 \quad .$$

## 17.2  Directional Distribution of Spin

The general spin state is a linear superposition of the two basic states $\eta_1$ and $\eta_{-1}$,

$$
\begin{aligned}
\chi &= \chi_1 \eta_1 + \chi_{-1} \eta_{-1} \\
&= e^{-i\Phi/2} \cos\frac{\Theta}{2} \eta_1 + e^{i\Phi/2} \sin\frac{\Theta}{2}\eta_{-1} \quad .
\end{aligned}
$$

The expectation value for the spin vector of the state $\chi$ is

$$
\langle S \rangle_\chi = \chi^+(\Theta,\Phi) S \chi(\Theta,\Phi) = \frac{\hbar}{2}\mathbf{n}(\Theta,\Phi) \quad .
$$

Here $\mathbf{n}(\Theta,\Phi)$ is the unit vector in the direction given by the polar angle $\Theta$ and the azimuthal angle $\Phi$ in the $x,y,z$ coordinate system

$$
\mathbf{n}(\Theta,\Phi) = \mathbf{e}_x \sin\Theta\cos\Phi + \mathbf{e}_y \sin\Theta\sin\Phi + \mathbf{e}_z\cos\Theta
$$

and

$$
\begin{aligned}
\chi^+ &= \chi_1^* \eta_1^+ + \chi_{-1}^* \eta_{-1}^+ \\
&= e^{i\Phi/2} \cos\frac{\Theta}{2}\eta_1^+ + e^{-i\Phi/2}\sin\frac{\Theta}{2}\eta_{-1}^+ \quad .
\end{aligned}
$$

The general spin state $\chi(\Theta,\Phi)$ is an eigenstate of the $\mathbf{n}$ component $\mathbf{n}\cdot\mathbf{S}$ of the spin vector $\mathbf{S}$,

$$
(\mathbf{n}\cdot\mathbf{S})\chi(\Theta,\Phi) = \frac{\hbar}{2}\chi(\Theta,\Phi) \quad .
$$

The two basic states $\eta_1$ and $\eta_{-1}$ may also be considered as eigenstates to $\mathbf{e}_z\cdot\mathbf{S}$ and $-\mathbf{e}_z\cdot\mathbf{S}$, respectively.

In Section 10.5 we have used the angular-momentum state $Y_{\ell\ell}(\vartheta,\phi,\mathbf{n})$ to analyze another angular-momentum state $Y_{\ell m}(\vartheta,\phi,e_z)$. Likewise we now use the state $\chi(\Theta,\Phi)$ to analyze our basic states $\eta_1$ and $\eta_{-1}$. The scalar product

$$
\chi^+(\Theta,\Phi)\cdot\eta_r = D^{(1/2)}_{r/2,1/2}(\Phi,\Theta,0) \quad , \qquad r = 1,-1 \quad ,
$$

is the probability amplitude for detecting the spin angular momentum $\frac{\hbar}{2}\mathbf{n}$ in the state $\eta_r$. Here

$$
\begin{aligned}
D^{(1/2)}_{1/2,1/2}(\Phi,\Theta,0) &= e^{-i\Phi/2}\cos\frac{\Theta}{2} \quad , \\
D^{(1/2)}_{-1/2,1/2}(\Phi,\Theta,0) &= e^{i\Phi/2}\sin\frac{\Theta}{2}
\end{aligned}
$$

are the *Wigner functions* for spin 1/2.

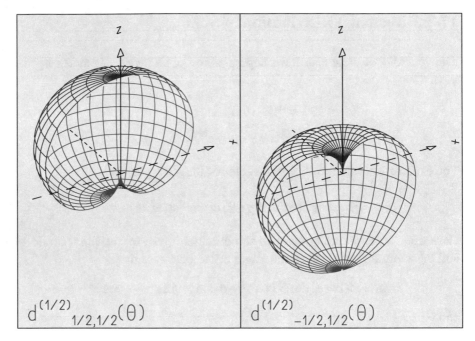

$$d^{(1/2)}_{1/2,1/2}(\theta) \qquad\qquad d^{(1/2)}_{-1/2,1/2}(\theta)$$

**Fig. 17.1. Polar diagrams of the Wigner functions $d^{(1/2)}_{m_s,1/2}(\Theta)$.**

The absolute square of this amplitude is

$$\left|\chi^+(\Theta,\Phi)\cdot\boldsymbol{\eta}_r\right|^2 = \left[d^{(1/2)}_{r/2,1/2}(\Theta)\right]^2 \quad , \qquad r = 1,-1 \quad ,$$

where the functions

$$d^{(1/2)}_{1/2,1/2}(\Theta) = \cos\frac{\Theta}{2} \quad ,$$

$$d^{(1/2)}_{-1/2,1/2}(\Theta) = \sin\frac{\Theta}{2}$$

are also called Wigner functions for spin $1/2$. In order to obtain a directional distribution we normalize like in Section 10.5 by a factor $(2s+1)(s+1)/(4\pi s) = 3/(2\pi)$,

$$f_{1/2,m_s}(\Theta,\Phi) = \frac{3}{2\pi}\left[d^{(1/2)}_{m_s,1/2}(\Theta)\right]^2 \quad , \qquad m_s = \frac{1}{2},-\frac{1}{2} \quad .$$

Figure 17.1 shows the Wigner function, and Figure 17.2 the directional distribution for spin $\pm1/2$. For $f_{1/2,1/2}$ the probability is largest for the direction $\Theta = 0$. In contrast to the distributions $f_{\ell\ell}$ for integer values of $\ell$, cf. Figure 10.11, the distribution for the electron spin is not sharply peaked in the direction $\Theta = 0$. The distribution of $f_{1/2,-1/2}$ is the mirror image of $f_{1/2,1/2}$ under a reflection $\Theta \to \pi - \Theta$.

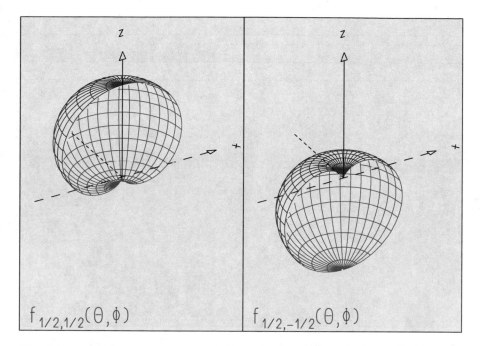

$f_{1/2,1/2}(\theta,\phi)$    $f_{1/2,-1/2}(\theta,\phi)$

**Fig. 17.2. Polar diagrams of the directional distributions** $f_{1/2,1/2}(\Theta,\Phi)$ **and** $f_{1/2,-1/2}(\Theta,\Phi)$.

In complete analogy to our discussion in Section 10.5 we can also construct angular distributions

$$f_{1/2,m_s,\Theta}(\Theta) = 2\pi f_{1/2,m_s}(\Theta,0)\sin\Theta$$

or, explicitly,

$$f_{1/2,1/2,\Theta}(\Theta) = 3\cos^3\frac{\Theta}{2}\sin\frac{\Theta}{2} \quad ,$$

$$f_{1/2,-1/2,\Theta}(\Theta) = 3\sin^3\frac{\Theta}{2}\cos\frac{\Theta}{2} \quad .$$

They are shown in Figure 17.3. These functions have their maxima at the angles given by

$$\cos\Theta_{1/2,1/2} = \frac{1}{2} \quad , \qquad \cos\Theta_{1/2,-1/2} = -\frac{1}{2} \quad ,$$

i.e., $\Theta_{1/2,1/2} = 60°$, $\Theta_{1/2,-1/2} = 120°$.

For the corresponding semiclassical angular-momentum vectors $\mathbf{L}^{sc}_{1/2,m_s}$ of length $\sqrt{1/2(1/2+1)}\hbar = (\sqrt{3}/2)\hbar$ and $z$ component $\pm(1/2)\hbar$ the angles are given by

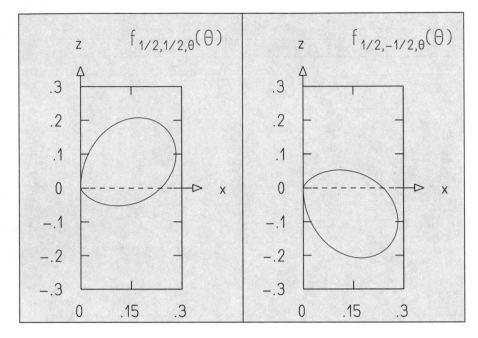

**Fig. 17.3. Polar diagrams of the angular distributions** $f_{1/2,1/2,\Theta}(\Theta)$ **and** $f_{1/2,-1/2,\Theta}(\Theta)$.

$$\cos\Theta^{\mathrm{sc}}_{1/2,1/2} = \frac{1}{\sqrt{3}} \quad , \qquad \cos\Theta^{\mathrm{sc}}_{1/2,1/2} = -\frac{1}{\sqrt{3}} \quad ,$$

i.e.,

$$\Theta_{1/2,1/2} \approx 55° \quad , \qquad \Theta_{1/2,-1/2} \approx 125° \quad .$$

In Figure 17.4 we show the angular distributions $f_{1/2,\pm1/2,\Theta}(\Theta)$, the direction given by the most probable angles $\Theta_{1/2,\pm1/2}$ and compare them with the semiclassical angular-momentum vectors $\mathbf{L}^{\mathrm{sc}}_{1/2,\pm1/2}$.

The spin state $\chi(\Theta_0,\Phi_0)$ for a polar angle $\Theta_0$ and the azimuth $\Phi_0$ is eigenstate to the spin projection $\mathbf{n}_0 \cdot \mathbf{S}$,

$$(\mathbf{n}_0 \cdot \mathbf{S})\chi(\Theta_0,\Phi_0) = \frac{\hbar}{2}\chi(\Theta_0,\Phi_0) \quad ,$$

onto the direction $\mathbf{n}_0$ characterized by $\Theta_0$ and $\Phi_0$.

The directional distribution describing – after dividing by 3 – the probability to find in $\chi(\Theta_0,\Phi_0)$ the spin $s = \frac{1}{2}$ in the direction $\mathbf{n}(\Theta,\Phi)$ is given by

$$
\begin{aligned}
f_{1/2,1/2}(\Theta,\Phi,\Theta_0,\Phi_0) &= \frac{3}{2\pi}\left(\frac{\mathbf{n}(\Theta,\Phi)+\mathbf{n}(\Theta_0,\Phi_0)}{2}\right)^2 \\
&= \frac{3}{4\pi}(1+\mathbf{n}(\Theta,\Phi)\cdot\mathbf{n}(\Theta_0,\Phi_0)) \quad .
\end{aligned}
$$

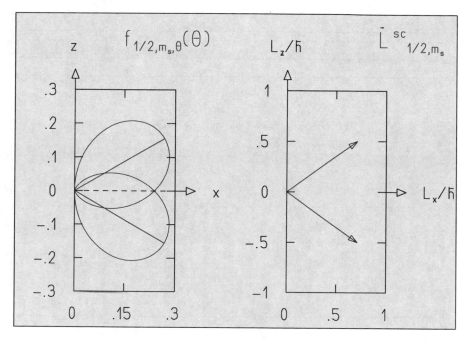

**Fig. 17.4. The left-hand plot contains polar diagrams of the angular distributions** $f_{1/2,m_s,\Theta}(\Theta)$ **for** $m_s = \pm 1/2$. **Also shown are lines from the origin to the points** $f_{1/2,m_s,\Theta}(\Theta_{1/2,m_s})$ **where** $\Theta_{1/2,m_s}$ **is the angle for which** $f_{1/2,m_s,\Theta}$ **has its maximum. The right-hand diagram shows the semiclassical angular-momentum vectors** $\mathbf{L}^{sc}_{1/2,m_s}$.

Plots of this distribution are shown in Figure 17.5. They exhibit the same apple-shaped form as those in Figure 17.2 but now with $\mathbf{n}_0$ as the symmetry axis rather than the $z$ axis.

## 17.3 Motion of Magnetic Moments in a Magnetic Field. Pauli Equation

With the help of the spin operator $\mathbf{S}$ the *magnetic-moment operator* of the electron can now be expressed as the negative product of the *gyromagnetic ratio*

$$\gamma = g_s \frac{e}{2M}$$

and the spin vector $\mathbf{S}$:

$$\mu = -\gamma \mathbf{S} \quad .$$

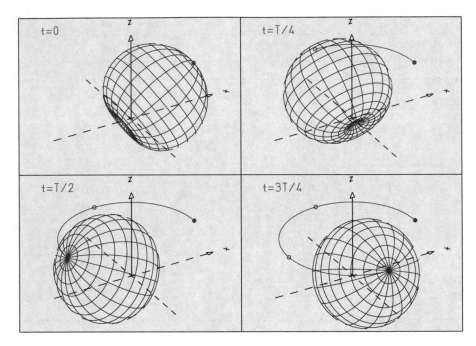

**Fig. 17.5. Polar diagrams of the distribution** $f_{1/2,1/2}(\Theta, \Phi, t)$ **for the direction of the spin of an electron precessing around the** $z$ **axis which is the direction of a homogeneous field of magnetic induction. The tip of the vector of the spin expectation value moves with angular frequency** $\Omega$ **on a circle around the** $z$ **axis. The plots show the directional distribution for** $t = 0, T/4, T/2, 3T/4$ **with** $T = 2\pi/\Omega$ **being the precession period.**

The potential energy of the static magnetic moment $\boldsymbol{\mu}$ in a field of magnetic induction **B** is given by

$$H = -\boldsymbol{\mu} \cdot \mathbf{B} = \gamma \mathbf{B} \cdot \mathbf{S} = \gamma \hbar \mathbf{B} \cdot \frac{\boldsymbol{\sigma}}{2} \quad .$$

The factor in front of the Pauli matrices can be expressed by a *precession frequency*

$$\Omega = \gamma B \quad .$$

We choose the $z$ axis of a coordinate frame in the direction of the $B$ field:

$$\mathbf{B} = B\mathbf{e}_z \quad .$$

The Hamiltonian now takes the form

$$H = -\boldsymbol{\mu} \cdot \mathbf{B} = \frac{\hbar}{2}\Omega \mathbf{e}_z \cdot \boldsymbol{\sigma} = \frac{\hbar}{2}\Omega \sigma_3 \quad .$$

A homogeneous field of magnetic induction exerts no force on the electron but it does exert a torque on its magnetic moment. In analogy to the Schrödinger equation we may write down the *Pauli equation* for the temporal evolution of the spin state $\boldsymbol{\chi}(t)$ of the electron,

$$i\hbar \frac{d}{dt}\chi(t) = H\chi(t) = \hbar \frac{\Omega}{2}\sigma_3 \chi(t) \quad .$$

Its solution is

$$\chi(t) = \left( \cos\frac{\Omega}{2}t - i\sigma_3 \sin\frac{\Omega}{2}t \right)\chi_0 \quad ,$$

where $\chi_0$ is the initial spin state at time $t = 0$. For the two base states $\eta_1$, $\eta_{-1}$ taken as initial states we find in particular

$$\chi_{\pm 1}(t) = \left( \cos\frac{\Omega}{2}t \mp i\sin\frac{\Omega}{2}t \right)\eta_{\pm 1} = e^{\mp i\frac{\Omega}{2}t}\eta_{\pm 1} \quad .$$

As a particular case we study the motion of the magnetic moment initially described by a state

$$\chi_0 = \chi(\Theta_0, \Phi_0) = \exp\left\{ -i\frac{\Phi_0}{2} \right\} \cos\frac{\Theta_0}{2}\eta_1 + \exp\left\{ i\frac{\Phi_0}{2} \right\} \sin\frac{\Theta_0}{2}\eta_{-1}$$

in a homogeneous field of magnetic induction along the $z$ direction.

The time-dependent state is given by

$$\chi(t) = \chi_1(t)\eta_1 + \chi_{-1}(t)\eta_{-1}$$

with the complex coefficients

$$\chi_1(t) = \exp\left\{ -i\frac{\Omega t + \Phi_0}{2} \right\}\cos\frac{\Theta_0}{2} \quad , \quad \chi_{-1}(t) = \exp\left\{ i\frac{\Omega t + \Phi_0}{2} \right\}\sin\frac{\Theta_0}{2} \quad .$$

For the expectation values of the components of the spin vector we find

$$\langle S \rangle_\chi = \chi^+(t)S\chi(t) = \frac{\hbar}{2}[\mathbf{e}_x \sin\Theta_0 \cos(\Omega t + \Phi_0)$$

$$+ \mathbf{e}_y \sin\Theta_0 \sin(\Omega t + \Phi_0) + \mathbf{e}_z \cos\Theta_0] = \frac{\hbar}{2}\mathbf{n}(t) \quad .$$

We differentiate the expectation value $\langle S \rangle_\chi$ with respect to time,

$$\frac{d}{dt}\langle S \rangle_\chi = \frac{\hbar}{2}\frac{d\mathbf{n}}{dt} = \frac{\hbar}{2}\Omega[-\mathbf{e}_x \sin\Theta_0 \sin(\Omega t + \Phi_0)$$

$$+ \mathbf{e}_y \sin\Theta_0 \cos(\Omega t + \Phi_0)] \quad .$$

Using $\Omega = \gamma B$ we recognize the right-hand side as the vector product of the magnetic-induction field $\mathbf{B} = B\mathbf{e}_z$ with the expectation value of the spin vector $\langle S \rangle_\chi$:

$$\gamma \mathbf{B} \times \langle \mathbf{S} \rangle_\chi = \frac{\hbar}{2} \Omega \left[ \mathbf{e}_z \times \mathbf{n}(t) \right] \quad .$$

Thus, the expectation value $\langle \mathbf{S} \rangle_\chi$ of the spin-vector operator obeys the equation of motion

$$\frac{d}{dt} \langle \mathbf{S} \rangle_\chi = \gamma \mathbf{B} \times \langle \mathbf{S} \rangle_\chi \quad .$$

Introducing $\boldsymbol{\mu} = -\gamma \mathbf{S}$ we obtain for the expectation value $\langle \boldsymbol{\mu} \rangle$ of the magnetic moment

$$\frac{d}{dt} \langle \boldsymbol{\mu} \rangle = \gamma \mathbf{B} \times \langle \boldsymbol{\mu} \rangle \quad .$$

These results correspond to a rotation of the vector $\langle \mathbf{S} \rangle_{\chi(t)}$ of expectation values about the $\mathbf{e}_z$ axis with angular frequency $\Omega = g_s \frac{e}{2m} B$. The expectation value of the spin vector and therefore of the magnetic moment exhibit *Larmor precession* about the $z$ axis. The time dependence of the expectation value is identical to the result obtained in classical physics for the motion of the magnetic moment in a homogeneous magnetic field.

Of course, the expectation values and their motion present only part of the quantum-mechanical information contained in the time-dependent state $\chi(t)$. The probability for finding the spin $\frac{1}{2}$ in the direction $\mathbf{n}(\Theta, \Phi)$ is given by

$$
\begin{aligned}
\frac{1}{3} f_{\frac{1}{2},\frac{1}{2}}(\Theta, \Phi, t) &= \frac{1}{2\pi} \left| \chi^+(\Theta, \Phi) \chi(t) \right|^2 \\
&= \frac{1}{2\pi} \left| \chi_1(t) D^{(1/2)}_{\frac{1}{2}\frac{1}{2}}(\Phi, \Theta, 0) + \chi_{-1}(t) D^{(1/2)}_{-\frac{1}{2}\frac{1}{2}}(\Phi, \Theta, 0) \right|^2 \\
&= \frac{1}{2\pi} \left| d^{(1/2)}_{\frac{1}{2}\frac{1}{2}}(\Theta_n) \right|^2 \quad ,
\end{aligned}
$$

where $\Theta_n$ is the polar angle relating to the time-dependent direction

$$
\mathbf{n}(t) = \begin{pmatrix} \sin \Theta_0 \cos(\Omega t + \Phi_0) \\ \sin \Theta_0 \sin(\Omega t + \Phi_0) \\ \cos \Theta_0 \end{pmatrix} \quad .
$$

In Figure 17.5 plots of $f_{\frac{1}{2},\frac{1}{2}}(\Theta, \Phi, t)$ for the time instants $t = 0, \frac{1}{4}T, \frac{1}{2}T,$ $\frac{3}{4}T$ of one period $T = 2\pi/\Omega$ are shown. The initial distribution $f_{\frac{1}{2},\frac{1}{2}}(\Theta, \Phi, 0)$ is centered about the initial axis

$$
\mathbf{n}(0) = \begin{pmatrix} \sin \Theta_0 \\ 0 \\ \cos \Theta_0 \end{pmatrix} \quad .
$$

For later times it moves as a rigid structure with a time-dependent axis $\mathbf{n}(t)$ rotating with constant angular frequency $\Omega$ on a cone about the $z$ axis with an opening angle $\Theta_0$.

## 17.4 Magnetic Resonance. Rabi's Formula

We study the motion of the spin of a particle under the influence of a time-independent magnetic-induction field $\mathbf{B}_0 = B_0\mathbf{e}_z$ in $z$ direction and a time-dependent field perpendicular to the $z$ direction,

$$\mathbf{B}_1(t) = B_1(\cos\omega t\,\mathbf{e}_x + \sin\omega t\,\mathbf{e}_y) \quad,$$

rotating with angular frequency $\omega$ about the $z$ axis. In the total field

$$\mathbf{B}(t) = \mathbf{B}_0 + \mathbf{B}_1(t)$$

the magnetic moment

$$\boldsymbol{\mu} = -\gamma\mathbf{S}$$

moves under the action of forces described by the time-dependent potential energy

$$H(t) = -\boldsymbol{\mu}\cdot\mathbf{B}(t) = -\frac{\mu}{2}\mathbf{B}(t)\cdot\boldsymbol{\sigma} \quad.$$

We introduce the two frequencies,

$$\Omega_0 = -\frac{\mu}{\hbar}B_0 \quad, \qquad \Omega_1 = -\frac{\mu}{\hbar}B_1 \quad,$$

related to the components $\mathbf{B}_0$ and $\mathbf{B}_1$ of the magnetic-induction field. This leads to a reformulation of the Hamiltonian

$$H(t) = \frac{\hbar}{2}[\Omega_0\sigma_3 + \Omega_1(\sigma_1\cos\omega t + \sigma_2\sin\omega t)] \quad,$$

or in matrix form

$$H(t) = \frac{\hbar}{2}\begin{pmatrix} \Omega_0 & \Omega_1 e^{-i\omega t} \\ \Omega_1 e^{i\omega t} & -\Omega_0 \end{pmatrix} \quad.$$

The Pauli equation for this situation determines the motion of the spin state $\chi(t)$,

$$i\hbar\frac{d}{dt}\chi(t) = H(t)\chi(t) \quad.$$

The decomposition of the spin state $\chi(t)$ into the basic spinors, $\eta_1$, $\eta_{-1}$,

$$\chi(t) = \chi_1(t)\eta_1 + \chi_{-1}(t)\eta_{-1} \quad,$$

leads to the two coupled equations

$$i\frac{d\chi_1}{dt} = \frac{\Omega_0}{2}\chi_1(t) + \frac{\Omega_1}{2}e^{-i\omega t}\chi_{-1}(t) \quad,$$

$$i\frac{d\chi_{-1}}{dt} = \frac{\Omega_1}{2}e^{i\omega t}\chi_1(t) - \frac{\Omega_0}{2}\chi_{-1}(t)$$

for the time-dependent coefficients $\chi_1(t)$, $\chi_{-1}(t)$. The explicit time dependence of the coefficients can easily be removed by the introduction of a rotating coordinate frame described by the time-dependent spin states $\tilde{\eta}_1(t)$, $\tilde{\eta}_{-1}(t)$,

$$\eta_1 = \exp\left\{i\frac{\omega}{2}t\right\}\tilde{\eta}_1(t) \quad , \qquad \eta_{-1} = \exp\left\{-i\frac{\omega}{2}t\right\}\tilde{\eta}_{-1}(t) \quad ,$$

i.e., for the components

$$\chi_1(t) = \exp\left\{-i\frac{\omega}{2}t\right\}\tilde{\chi}_1(t) \quad , \qquad \chi_{-1}(t) = \exp\left\{i\frac{\omega}{2}t\right\}\tilde{\chi}_{-1}(t) \quad .$$

This leads to the decomposition

$$\chi(t) = \tilde{\chi}_1(t)\tilde{\eta}_1(t) + \tilde{\chi}_{-1}(t)\tilde{\eta}_{-1}(t)$$

of the spin state $\chi(t)$ and to the differential equations

$$i\frac{d}{dt}\tilde{\chi}_1(t) = -\frac{\Delta}{2}\tilde{\chi}_1(t) + \frac{\Omega_1}{2}\tilde{\chi}_{-1}(t) \quad ,$$

$$i\frac{d}{dt}\tilde{\chi}_{-1}(t) = \frac{\Omega_1}{2}\tilde{\chi}_1(t) + \frac{\Delta}{2}\tilde{\chi}_{-1}(t) \quad ,$$

with $\Delta = (\omega - \Omega_0)$. In terms of the initial state

$$\chi_0 = \chi(0) = \chi_1^{(0)}\eta_1 + \chi_{-1}^{(0)}\eta_{-1} = \chi_1^{(0)}\tilde{\eta}_1(0) + \chi_{-1}^{(0)}\tilde{\eta}_{-1}(0)$$

and its components $\chi_1(0) = \chi_1^{(0)}$, $\chi_{-1}(0) = \chi_{-1}^{(0)}$ we find as solution for the components in the rotating frame

$$\tilde{\chi}_1(t) = \chi_1^{(0)}\cos\frac{\Omega}{2}t - i(\omega_1\chi_{-1}^{(0)} - \omega_3\chi_1^{(0)})\sin\frac{\Omega}{2}t \quad ,$$

$$\tilde{\chi}_{-1}(t) = \chi_{-1}^{(0)}\cos\frac{\Omega}{2}t - i(\omega_1\chi_1^{(0)} + \omega_3\chi_{-1}^{(0)})\sin\frac{\Omega}{2}t \quad .$$

Here we used the following notation:

$$\Omega^2 = \Omega_1^2 + \Delta^2 \quad , \qquad \omega_1 = \frac{\Omega_1}{\Omega} \quad , \qquad \omega_3 = \frac{\Delta}{\Omega} \quad .$$

We choose the coordinate frame in such a way that the initial state coincides with the basic spinor

$$\chi_0 = \eta_1 \quad , \qquad \text{i.e.,} \qquad \chi_1^{(0)} = 1 \quad , \qquad \chi_{-1}^{(0)} = 0 \quad .$$

Then, we get as the solution for the components relative to the time-independent coordinate frame $\eta_1$, $\eta_{-1}$,

$$\chi_1(t) = \exp\left\{-i\frac{\omega}{2}t\right\}\left(\cos\frac{\Omega}{2}t + i\frac{\Delta}{\Omega}\sin\frac{\Omega}{2}t\right) \quad,$$

$$\chi_{-1}(t) = -i\frac{\Omega_1}{\Omega}\exp\left\{i\frac{\omega}{2}t\right\}\sin\frac{\Omega}{2}t \quad.$$

Also the time-dependent spinor

$$\chi(t) = \chi_1(t)\boldsymbol{\eta}_1 + \chi_{-1}(t)\boldsymbol{\eta}_{-1}$$

has length one,

$$\chi^+(t)\chi(t) = |\chi_1(t)|^2 + |\chi_{-1}(t)|^2 = 1 \quad.$$

For the frequency $\omega$ of the time-dependent component of the external magnetic-induction field $B_1$ equal to the precession frequency $\Omega_0 = \mu B_0$ of the time-independent field component $B_0$ the difference frequency $\Delta = \frac{1}{2}(\omega - \Omega_0)$ vanishes. In this case the motion of the spin state turns out to be particularly simple:

$$\chi_1(t) = \exp\left\{-i\frac{\omega}{2}t\right\}\cos\frac{\Omega}{2}t \quad, \qquad \chi_{-1}(t) = -i\exp\left\{i\frac{\omega}{2}t\right\}\sin\frac{\Omega}{2}t \quad.$$

The expectation values of the spin vector $\mathbf{S} = \frac{\hbar}{2}\boldsymbol{\sigma}$ are given by

$$\langle S_x\rangle_{\chi(t)} = \frac{\hbar}{2}\left(\frac{\Omega_1}{\Omega}\sin\omega t\sin\Omega t + \frac{\Omega_1}{\Omega}\frac{\Delta}{\Omega}\cos\omega t\cos\Omega t\right.$$
$$\left.-\frac{\Omega_1}{\Omega}\frac{\Delta}{\Omega}\cos\omega t\right) \quad,$$

$$\langle S_y\rangle_{\chi(t)} = -\frac{\hbar}{2}\left(\frac{\Omega_1}{\Omega}\cos\omega t\sin\Omega t - \frac{\Omega_1}{\Omega}\frac{\Delta}{\Omega}\sin\omega t\cos\Omega t\right.$$
$$\left.+\frac{\Omega_1}{\Omega}\frac{\Delta}{\Omega}\sin\omega t\right) \quad,$$

$$\langle S_z\rangle_{\chi(t)} = \frac{\hbar}{2}\left(\frac{\Delta^2}{\Omega^2} + \frac{\Omega_1^2}{\Omega^2}\cos\Omega t\right) \quad.$$

Figure 17.6 exhibits the orbit of the tip of the vector $\langle\mathbf{S}\rangle_{\chi(t)}$ on the sphere of radius $\hbar/2$ for the first period

$$T = 2\pi/\Omega$$

for different values of the ratio $\omega/\Omega_0$ of the frequency $\omega$ of the rotating magnetic field $\mathbf{B}_1(t)$ and the Larmor frequency $\Omega_0$ of the time-independent field $\mathbf{B}_0 = B_0\mathbf{e}_z$ vertical to the plane of rotation of the field vector $\mathbf{B}_1(t)$. At $t = 0$ the vector $\langle\mathbf{S}\rangle_{\chi(t)}$ starts from the $z$ direction spiraling about the $z$ axis up to maximal opening angle $\Theta_{max}$ given by

$$\cos\Theta_{max} = (\Delta^2 - \Omega_1^2)/\Omega^2 \quad.$$

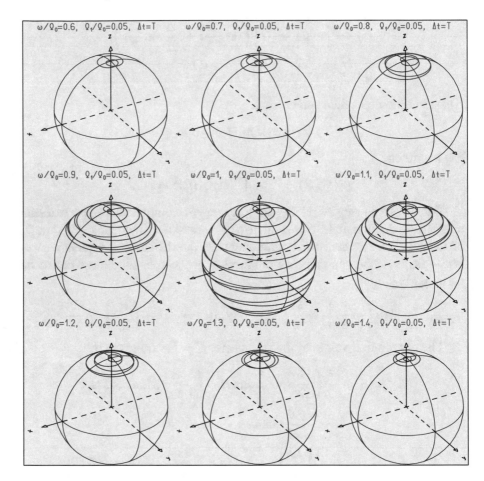

**Fig. 17.6. Magnetic resonance. Trajectory of the tip of the expectation value of the spin vector within one period $T$. The value of $\omega$ is varied from plot to plot whereas $\Omega_0$ and $\Omega_1$ are kept constant. The plot in the middle of the figure corresponds to exact resonance frequency $\omega = \Omega_0$.**

Thus its $z$ component oscillates with the angular frequency $\Omega$ within the range

$$\frac{\hbar}{2}\frac{\Delta^2 - \Omega_1^2}{\Omega^2} \leq \langle S_z \rangle \leq \frac{\hbar}{2} \quad .$$

If the frequency $\omega$ of the time-dependent magnetic-field component $B_1$ coincides with the Larmor frequency $\Omega_0$ corresponding to the time-independent z component $B_0$, i.e., $\Delta = 0$, $\Omega_1 = \Omega$, one observes the phenomenon of *magnetic resonance*. The expectation value of the spin vector **S** becomes simply

$$\langle \mathbf{S} \rangle_{\chi(t)} = \frac{\hbar}{2}\left(\mathbf{e}_x \sin\omega t \sin\Omega t - \mathbf{e}_y \cos\omega t \sin\Omega t + \mathbf{e}_z \cos\Omega t\right)$$
$$= \frac{\hbar}{2}\left(\mathbf{e}_x \sin\Omega t \cos(\omega t - \tfrac{\pi}{2}) + \mathbf{e}_y \sin\Omega t \sin(\omega t - \tfrac{\pi}{2}) + \mathbf{e}_z \cos\Omega t\right) \quad .$$

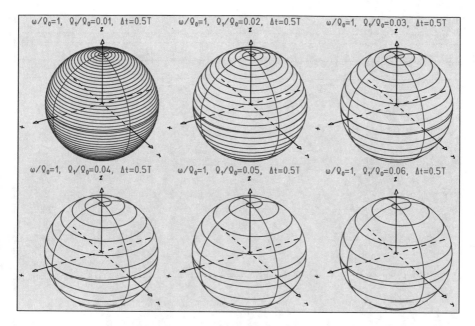

**Fig. 17.7. As Figure 17.6 but for** $\omega$ **fixed to the resonance frequency** $\omega = \Omega_0$ **and for various values of** $\Omega_1$. **In all plots the trajectory is shown for one half of a period** $T$.

The tip of the vector $\langle \mathbf{S} \rangle_\chi$ moves on the surface of a sphere with radius $\hbar/2$ periodically from the $z$ direction to the negative $z$ direction. Polar and azimuthal angle follow the time dependence

$$\Theta(t) = \Omega t \quad , \qquad \varphi(t) = (\omega t - \frac{\pi}{2}) \quad .$$

The $z$ component of the spin-vector expectation value exhausts the full range

$$-\frac{\hbar}{2} \le \langle S_z \rangle_\chi \le \frac{\hbar}{2} \quad .$$

Figure 17.7 presents a set of graphs of the orbits during half a period $T/2 = \pi/\Omega$ for different values of the ratio $\Omega_1/\Omega_0 = B_1/B_0$ of the Larmor frequencies $\Omega_1$, $\Omega_0$ or, equivalently, the field strengths of the rotating field $B_1$ and the constant field $B_0$. For values $\Omega_1 \ll \Omega_0$ the orbit forms a spiral with dense winding on the sphere. The distance of the windings grows with growing ratio $\Omega_1/\Omega_0$.

The directional distribution for the spin direction is, of course, of the same form

$$f_{1/2,1/2}(\Theta, \Phi, t) = \frac{3}{2\pi} \left| d^{(1/2)}_{1/2,1/2}(\vartheta_n) \right|^2$$

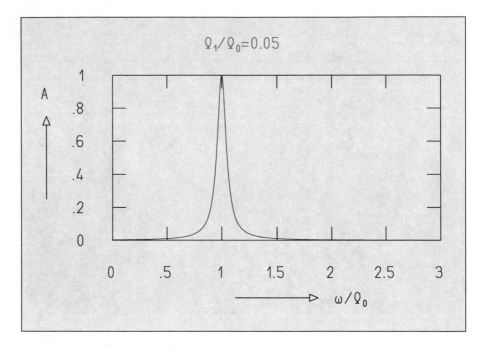

**Fig. 17.8. Amplitude $A$ as a function of $\omega$ for a fixed value of $\Omega_1$. For smaller values of $\Omega_1$ the resonance becomes sharper. For larger values it becomes broader.**

found in Section 17.3 and Figure 17.5. But $\vartheta_n$ is now the polar angle with respect to the expectation value of the spin vector, so that the whole distribution moves along with that vector.

In experiments with atomic or molecular beams Isidor Rabi used magnetic resonance for the measurement of the magnetic moments of protons and nuclei. These can be directly determined from the resonance frequency $\omega = \Omega_0$, since the Larmor frequency $\Omega_0 = \mu B_0$ is directly proportional to the magnetic moment $\mu$.

Finally, we quote the *Rabi formula* from Isidor Rabi's celebrated paper *"space quantization in a gyrating magnetic field"* published in 1937. Starting initially ($t = 0$) from the state $\eta_1$, it gives the probability $P_{-\frac{1}{2}}(t)$ of finding at time $t$ the state $\eta_{-1}$ if the initial state was $\eta_1$,

$$P_{-\frac{1}{2}}(t) = |\eta_1^+ \chi(t)|^2 = |\chi_{-1}(t)|^2 = \frac{\Omega_1^2}{\Omega^2} \sin^2 \frac{\Omega}{2} t \quad .$$

The probability $P_{-\frac{1}{2}}(t)$ is at maximum for odd multiples of the time

$$\frac{T}{2} = \frac{\pi}{\Omega} = \frac{\pi}{\sqrt{\Delta^2 + \Omega_1^2}} \quad , \qquad \Delta = (\omega - \Omega_0) \quad .$$

At these instants the probability has the maximum value

$$A = P_{-\frac{1}{2}}\left(\frac{T}{2}\right) = \frac{\Omega_1^2}{\Omega^2} = \frac{\Omega_1^2}{\Omega_1^2 + (\omega - \Omega_0)^2}$$

$$= \frac{(\Omega_1/\Omega_0)^2}{(\Omega_1/\Omega_0)^2 + (1 - \omega/\Omega_0)^2} \quad .$$

It reaches the value one for the resonance frequency $\omega = \Omega_0$. A plot of $A$ as a function of the ratio $\omega/\Omega_0$ for fixed value $\Omega_1/\Omega_0$ is shown in Figure 17.8. It is of the typical resonance form.

## 17.5  Magnetic Resonance in a Rotating Frame of Reference

At the end of Section 17.3 we obtained the equation of motion

$$\frac{\mathrm{d}}{\mathrm{d}t}\langle\mathbf{S}\rangle = \gamma\mathbf{B}\times\langle\mathbf{S}\rangle$$

for the expectation value $\langle\mathbf{S}\rangle$ of the spin vector in a constant induction field $\mathbf{B} = B\mathbf{e}_z$. The equation described the precession of the vector $\langle\mathbf{S}\rangle$ around the direction of $\mathbf{B}$ with the angular velocity $\Omega = \gamma B$.

For the time-dependent field used in magnetic-resonance experiments,

$$\mathbf{B}(t) = \mathbf{B}_0 + \mathbf{B}_1(t) = B_1\cos(\omega t)\mathbf{e}_x + B_1\sin(\omega t)\mathbf{e}_y + B_0\mathbf{e}_z \quad ,$$

both vectors on the right-hand side of the equation of motion become time dependent.

The discussion simplifies if one considers a *rotating reference frame* $\mathbf{e}'_x, \mathbf{e}'_y, \mathbf{e}'_z = \mathbf{e}_z$ the $x'$ axis of which always coincides with the direction of the rotating field $B_1(t)$,

$$\mathbf{B}_1 = B_1\mathbf{e}'_x(t) = B_1\cos(\omega t)\mathbf{e}_x + B_1\sin(\omega t)\mathbf{e}_y \quad .$$

This implies

$$\begin{aligned}
\mathbf{e}'_x(t) &= \mathbf{e}_x\cos\omega t + \mathbf{e}_y\sin\omega t \quad , \\
\mathbf{e}'_y(t) &= -\mathbf{e}_x\sin\omega t + \mathbf{e}_y\cos\omega t \quad .
\end{aligned}$$

The rotation of the field $\mathbf{B}_1(t)$ and of the vectors $\mathbf{e}'_x(t)$, $\mathbf{e}'_y(t)$ is described by the vector

$$\boldsymbol{\omega} = \omega\mathbf{e}_z$$

of angular velocity. The time derivatives of an arbitrary vector $\langle\mathbf{S}\rangle$ in the laboratory frame and in the rotating frame are connected by

$$\frac{\mathrm{d}\langle\mathbf{S}\rangle}{\mathrm{d}t} = \frac{\mathrm{d}'\langle\mathbf{S}\rangle}{\mathrm{d}t} + \boldsymbol{\omega}\times\langle\mathbf{S}\rangle \quad .$$

We rewrite the relation in the form

$$\frac{d'\langle \mathbf{S}\rangle}{dt} = \frac{d\langle \mathbf{S}\rangle}{dt} - \boldsymbol{\omega} \times \langle \mathbf{S}\rangle$$

and introduce the equation of motion and obtain

$$\frac{d'\langle \mathbf{S}\rangle}{dt} = (\gamma \mathbf{B} - \boldsymbol{\omega}) \times \langle \mathbf{S}\rangle = \gamma \mathbf{B}_{\text{eff}} \times \langle \mathbf{S}\rangle$$

with the *effective field*

$$\mathbf{B}_{\text{eff}} = \mathbf{B} - \frac{\boldsymbol{\omega}}{\gamma} = \left( B_0 - \frac{\omega}{\gamma}\right)\mathbf{e}'_z + B_1 \mathbf{e}'_x \quad ,$$

which is, of course, constant in the rotating frame. The equation of motion describes the precession of the vector $\langle \mathbf{S}\rangle$ about the direction of the field $\mathbf{B}_{\text{eff}}$ in the rotating frame. The frequency of precession is

$$\Omega = \gamma B_{\text{eff}} = \sqrt{(\gamma B_0 - \omega)^2 + \gamma B_1^2} = \sqrt{(\Omega_0 - \omega)^2 + \Omega_1^2} \quad .$$

Since experiments are always performed for $B_1 \ll B_0$, the effective field is practically parallel or antiparallel to the $z$ axis except for frequencies $\omega$ near the Larmor frequency $\Omega_0$ in the static field $\mathbf{B}_0$,

$$\omega = \Omega_0 = \gamma B_0 \quad .$$

If initially the vector $\langle \mathbf{S}\rangle$ is parallel to the $z$ axis then for $\omega$ appreciably different from the resonance frequency it will deviate only very little from the $z$ direction since it precesses around the direction of $\mathbf{B}_{\text{eff}}$ which is nearly parallel (or antiparallel) to the $z$ axis. At resonance, however, $\langle \mathbf{S}\rangle$ precesses about the $x'$ axis since at resonance $\mathbf{B}_{\text{eff}} = \mathbf{B}_1 = B_1 \mathbf{e}'_x$ and the polar angle of $\langle \mathbf{S}\rangle$ with the $z$ axis changes periodically between 0 and $\pi$ with the angular frequency $\Omega_1 = \gamma B_1$.

The situation is illustrated in Figure 17.9. This figure corresponds in all parameters to Figure 17.6 but it shows the trajectory of the tip of $\langle \mathbf{S}\rangle$ in the rotating reference frame $\mathbf{e}'_x, \mathbf{e}'_y, \mathbf{e}'_z$ rather than in the laboratory frame $\mathbf{e}_x, \mathbf{e}_y, \mathbf{e}_z$.

It is interesting to note that in most experiments the time-dependent field $\mathbf{B}_1$ is not realized as a rotating field but as a field oscillating in the $x$ direction,

$$\mathbf{B}_1^{(\text{exp})} = 2 B_1 \cos \omega t\, \mathbf{e}_x \quad .$$

It can, however, be interpreted as a sum

$$\mathbf{B}_1^{(\text{exp})} = \mathbf{B}_{1+}(t) + \mathbf{B}_{1-}(t)$$

with the two fields

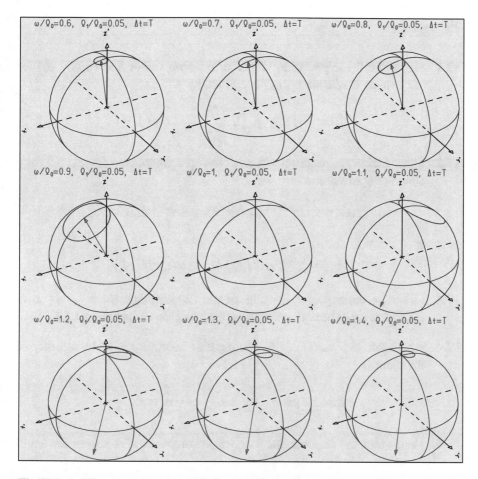

**Fig. 17.9. As Figure 17.6 but presented in the rotating frame of reference. The arrow shown in the $x', z'$ plane is the direction of the effective field $\mathbf{B}_{\mathrm{eff}}$. The tip of the expectation value of the spin vector moves on a circle around that direction. Its initial position is on the $z'$ axis.**

$$\mathbf{B}_{1\pm} = B_1 \cos(\omega t)\mathbf{e}_x \pm B_1 \sin(\omega t)\mathbf{e}_y$$

rotating in opposite directions. The vector of angular velocity of these fields is $\omega\mathbf{e}_z$ and $-\omega\mathbf{e}_z$, respectively. In the frame rotating with $\mathbf{B}_{1+}$ resonance occurs since in the effective field the $z$ component $B_0 - \omega/\gamma$ vanishes for $\omega = \gamma B_0$. In this frame of reference $\mathbf{B}_{1-}$ varies very rapidly with time so that its influence on $\langle \mathbf{S} \rangle$ averages out and can be neglected. In a frame rotating with $\mathbf{B}_{1-}$ the $z$ component of the effective field is $B_0 + \omega/\gamma$ and no cancellation takes place.

## Problems

17.1. Show that the expectation values of the spin vector $\mathbf{S} = (S_x, S_y, S_z)$ for the two basis spinors $\eta_1$, $\eta_{-1}$ are given by

$$\langle \mathbf{S} \rangle_a = \eta_a^+ \mathbf{S} \eta_a = \frac{a}{2} \hbar \, \mathbf{e}_z \quad , \qquad a = 1, -1 \quad .$$

17.2. Show that the expectation value of the spin vector $\mathbf{S} = (S_x, S_y, S_z)$ for a coherent spin state

$$\chi(\Theta, \Phi) = e^{-i\Phi/2} \cos(\Theta/2) \eta_1 + e^{i\Phi/2} \sin(\Theta/2) \eta_{-1}$$

is

$$\langle \mathbf{S} \rangle_\chi = \chi^+(\Theta, \Phi) \mathbf{S} \chi(\Theta, \Phi) = \frac{\hbar}{2} \mathbf{n}(\Theta, \Phi) \quad .$$

17.3. Calculate the expectation value of the Hamiltonian $H = -\boldsymbol{\mu} \cdot \mathbf{B}$, $\boldsymbol{\mu} = g_s e \mathbf{S}/(2m)$ for the coherent state $\chi(\Theta, \Phi)$ given in problem 17.2.

17.4. By Taylor expansion of the exponential function show the validity of the identity

$$\exp\{-i\Omega(\mathbf{n} \cdot \boldsymbol{\sigma})t\} = \cos \frac{\Omega}{2} t - i\mathbf{n} \cdot \boldsymbol{\sigma} \sin \frac{\Omega}{2} t \quad .$$

17.5. Verify that the exponential function of problem 17.4 solves the Pauli equation

$$i\hbar \frac{d}{dt} \exp\{-i\Omega(\mathbf{n} \cdot \boldsymbol{\sigma})t\} = H \exp\{-i\Omega(\mathbf{n} \cdot \boldsymbol{\sigma})t\}$$

with

$$H = \frac{1}{2} \hbar \Omega (\mathbf{n} \cdot \boldsymbol{\sigma}) \quad .$$

# 18. Examples from Experiment

So far we have investigated mechanical systems using the description and tools of quantum mechanics. In this final chapter we look at actual systems as they occur in nature. We shall discuss scattering phenomena, bound systems, and metastable states as they play a role in rather different fields of science.

Before discussing the results of actual experiments, we need to spend a little time on the *units* in which the data are given. The velocities of several of the particles studied are not much slower than the speed of light. To describe them we therefore have to use Einstein's theory of relativity. It states that, if $E$ is the total energy and $p$ the magnitude of the momentum of a particle, the quantity

$$E^2 - p^2 c^2 = m^2 c^4$$

has the same value in any frame of reference in which $E$ and $p$ are measured. Here $c = 3 \times 10^8$ m/s is the speed of light in vacuum. In the particular frame of reference in which the particle is at rest, $p = 0$, we have

$$E = mc^2 \quad .$$

Therefore the constant $m$ is called the *rest mass* of the particle. The quantity $mc^2$ is the *rest energy* of the particle. In a frame of reference in which the particle is not at rest, $p \neq 0$, the total energy is larger:

$$E = \sqrt{m^2 c^4 + p^2 c^2} = mc^2 + E_{\text{kin}} \quad .$$

The additional term is called the *kinetic energy* of the particle.

In the experiments discussed in this section, the particles are characterized by their momentum $p$, their total energy $E$, or their kinetic energy $E_{\text{kin}}$. The energies are measured in *electron volts* (eV). A particle that carries the *elementary charge*

$$e = 1.602 \times 10^{-19} \text{C}$$

and that has traversed an accelerating potential difference of 1 V has gained the kinetic energy

$$1 \text{eV} = 1.602 \times 10^{-19} \text{W s} = 1.602 \times 10^{-19} \text{J} \quad .$$

S. Brandt and H.D. Dahmen, *The Picture Book of Quantum Mechanics*, 373
DOI 10.1007/978-1-4614-3951-6_18, © Springer Science+Business Media New York 2012

A convenient notation for higher energies is $1\,\mathrm{keV} = 10^3\,\mathrm{eV}$, $1\,\mathrm{MeV} = 10^6\,\mathrm{eV}$, $1\,\mathrm{GeV} = 10^9\,\mathrm{eV}$. Since $mc^2$ is an energy, masses can be measured in electron volts per $c^2$:

$$1\frac{\mathrm{eV}}{c^2} = \frac{1.602 \times 10^{-19}}{(3 \times 10^8)^2}\,\mathrm{kg} = 1.782 \times 10^{-36}\,\mathrm{kg} \quad .$$

The rest mass of the electron is

$$m_{\mathrm{e}} = 511\,\mathrm{keV}/c^2 \quad .$$

The rest masses of the proton and the neutron are nearly 2000 times larger,

$$m_{\mathrm{p}} = 938.3\,\mathrm{MeV}/c^2 \quad , \qquad m_{\mathrm{n}} = 939.6\,\mathrm{MeV}/c^2 \quad .$$

It is important to remember that a proton with kinetic energy of $E_{\mathrm{kin}} = 10\,\mathrm{MeV}$ has a total energy of $E = m_{\mathrm{p}}c^2 + E_{\mathrm{kin}} = 948.3\,\mathrm{MeV}$. Often the momentum $p$ is easiest to measure. Since the product $pc$ is an energy, the momentum is measured in electron volts per $c$:

$$1\frac{\mathrm{eV}}{c} = \frac{1.602 \times 10^{-19}}{3 \times 10^8}\,\mathrm{kg\,m/s} = 5.3 \times 10^{-28}\,\mathrm{kg\,m/s} \quad .$$

Once the momentum $p$ and the rest mass $m$ of a particle are known, its total energy $E$ and its kinetic energy $E_{\mathrm{kin}}$ are easily computed.

## 18.1  Scattering of Atoms, Electrons, Neutrons, and Pions

In Chapters 12, 15, and 16 we have discussed the scattering of a particle incident on a spherically symmetric potential, which was assumed to be fixed in space. In actual experiments *projectile* particles scatter on *target* particles. In "colliding beam" experiments the projectile and target particles both move in opposite directions within a storage ring and scatter on each other in a head-on collision. An example for such an arrangement is given at the end of Section 18.4, where the production of the elementary particles $J/\psi$ and $\Upsilon$ in colliding beams of electrons and positrons is discussed. In "fixed target" experiments the target particles are at rest before the scattering process; however, they move after the collision. As in classical mechanics, this two-body process can be reduced to a one-body problem if the coordinate **r** in the wave function is taken to be the distance vector between the two particles, and the mass appearing in the one-body Schrödinger equation is taken to be the reduced mass $M = m_1 m_2/(m_1 + m_2)$ of the two bodies. It is customary to present results of scattering experiments as differential cross sections $\mathrm{d}\sigma/\mathrm{d}\vartheta^*$

with respect to the scattering angle $\vartheta^*$ in the center-of-mass system (CMS). In this reference frame target and projectile have initially equal and opposite momenta (Figure 18.1a).

Figures 18.1b through e show results obtained in scattering experiments in entirely different fields of physics using completely different experimental techniques. Figure 18.1b shows the differential cross section for the scattering of sodium atoms by mercury atoms. The kinetic energy in the laboratory frame is only a fraction of an electron volt. The momentum is of the order of $100\,keV/c$ corresponding to a de Broglie wavelength of about $10^{-11}\,m$, which is one order of magnitude below the atomic radius. Scattering experiments such as this one provide information about the electric potential acting between atoms. Such investigations are helpful in studying problems of chemical bonds.

Nuclear forces can be investigated by using neutrons, which carry no electric charge, as projectiles incident on nuclei. The differential cross section for the scattering of neutrons on lead nuclei is given in Figure 18.1c for two energies $E_{\text{kin}}^{\text{lab}} = 7\,MeV$ and $14.5\,MeV$. The corresponding momenta are $p^{\text{lab}} = 110\,MeV/c$ and $160\,MeV/c$. They in turn correspond to de Broglie wavelengths of roughly $11 \times 10^{-15}\,m$ and $7.6 \times 10^{-15}\,m$. These wavelengths are of the same order of magnitude as the radius of the lead nucleus, which is roughly $7 \times 10^{-15}\,m$. As expected, there are more minima in the differential cross sections for the higher energy, that is, for the shorter wavelength, of the incoming particles (see Figure 15.9b).

To investigate the electric potential of nuclei, we choose electrons as projectiles because they are not affected by the nuclear forces of the nucleus. Figure 18.1d shows the differential cross section of electrons with a laboratory energy of $420\,MeV$ scattered by oxygen nuclei. Because the electron mass is very light, the electron momentum is $420\,MeV/c$, its corresponding wavelength about $3 \times 10^{-15}\,m$. From such experiments the electrical charge distribution of the oxygen nucleus was found to have a characteristic radius of about $3 \times 10^{-15}\,m$.

The nuclei are composed of protons and neutrons, often referred to by the collective term *nucleons*. If we want to study the internal structure of protons or neutrons, we can perform scattering experiments using either electrons or particles with nuclear interaction as projectiles. Results of an experiment using particles with nuclear interaction as projectiles are shown in Figure 18.1e. Here the projectiles are $\pi$ mesons. These particles exert nuclear forces and have a mass of about $140\,MeV/c^2$. The experiment was performed with laboratory energies of $10\,GeV$, that is, a momentum of $10\,GeV/c$ corresponding to a wavelength of $0.1 \times 10^{-15}\,m$, which is one order of magnitude below the proton radius.

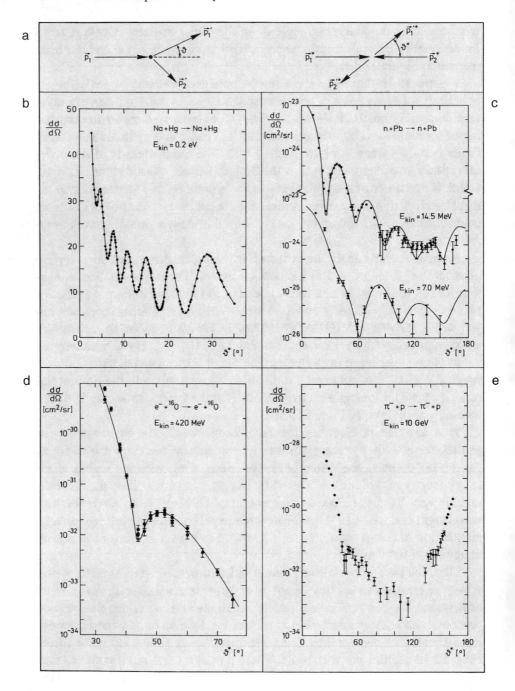

**Fig. 18.1.**

The results given in Figure 18.1 bear a qualitative resemblance to the differential scattering cross sections shown in Figures 12.4b and 15.9b. No quantitative comparison is justified, for the forces acting in the collisions in Figure 18.1 cannot be described by simple square-well potentials. Moreover, effects attributable to the spin of the target and the projectiles were not taken into account in the calculations of Chapters 12 and 15.

## 18.2 Spectra of Bound States in Atoms, Nuclei, and Crystals

The first striking success of quantum mechanics was the explanation of the *hydrogen spectrum*. Sufficiently heated atomic hydrogen emits light with a characteristic wavelength spectrum consisting of discrete wavelengths. In Section 13.4 we found that the energy levels of the electron bound in the hydrogen atom are

$$E_n = -\frac{1}{2}Mc^2\frac{\alpha}{n^2} \quad , \qquad n = 1, 2, \ldots \quad .$$

Here $M$ is the electron mass, $c$ is the speed of light, and $\alpha = 1/137$ is the fine-structure constant. A transition from one level to another is effected by the emission or absorption of the energy difference

$$\Delta E = E_{n_1} - E_{n_2} = -\frac{1}{2}Mc^2\alpha^2\left(\frac{1}{n_1^2} - \frac{1}{n_2^2}\right)$$

in the form of a light quantum of frequency $\nu$ corresponding to

$$\Delta E = h\nu$$

**Fig. 18.1. (a) Scattering of a projectile particle 1 on a target particle 2. In the laboratory the target particle is initially at rest, $p_2 = 0$. In the center-of-mass system (CMS) the particles have initially equal and opposite momenta, $p_1^* = -p_2^*$. For elastic scattering, considered here, the momenta are also equal and opposite after the scattering process, $p_1'^* = -p_2'^*$. (b) Sodium atoms scattered on mercury atoms, (c) neutrons on lead nuclei, (d) electrons on oxygen nuclei, and (e) $\pi$ mesons on protons. The differential cross section $d\sigma/d\Omega$ for the elastic scattering of two particles is given as a function of the CMS scattering angle $\vartheta^*$. The laboratory kinetic energy $E_{\text{kin}}$ of the projectile is given on each figure. For part b the ordinate is a linear scale given in arbitrary units. For parts c, d, and e it is a logarithmic scale given in square centimeters per steradian.** *Sources:* (b) From U. Buck and H. Pauly, *Zeitschrift für Naturforschung* **23a** (1968) 475, copyright © 1968 by Verlag der Zeitschrift für Naturforschung, Tübingen, reprinted by permission. (c) From F. Perey and B. Buck, *Nuclear Physics* **32** (1962) 352, copyright © 1962 by North-Holland Publishing Company, Amsterdam, reprinted by permission. (d) From R. Hofstadter, Nuclear and Nucleon Scattering of Electrons at High Energies, reproduced with permission from the *Annual Review of Nuclear and Particle Science*, Volume 7, copyright © 1957 by Annual Reviews Inc. (e) From a conference contribution by J. Orear et al. as reported by G. Belletini, Intermediate and High Energy Collisions, in *Proceedings of the 14th International Conference on High Energy Physics at Vienna* (J. Prentki and J. Steinberger, editors), copyright © 1968 by CERN, Geneva, reprinted by permission.

or to the wavelength

$$\lambda = \frac{c}{\nu} = \frac{hc}{\Delta E} \quad .$$

A set of transitions for a fixed value of $n_1$ but variable $n_2$ is called a *spectral series* (see Figure 18.2a). In particular, the one with $n_1 = 2$ and $n_2 > 2$ is called the *Balmer series*. Its wavelengths are in the region of visible light and can be easily measured with a prism spectrograph. The spectral lines of the Balmer series are observed in the light emitted by electric discharges in hydrogen gas but also in the light emitted by some stars, proving that there is hot hydrogen in the atmospheres of such stars. Outside the region of some stars that emit light, the hydrogen gas is cold. Then we observe dark lines in the spectrograph for the wavelengths of the Balmer series, indicating that hydrogen atoms of the cold gas have absorbed light. Stellar spectra showing the Balmer series in emission and absorption are given in Figure 18.2b. The energy spectrum shown in Figure 18.2a has already been obtained in Section 13.4. It is characteristic of the Coulomb potential acting between the nucleus of the hydrogen atom and its electron. It possesses an infinite number of levels accumulating at the upper end of the spectrum, $E = 0$. The spectra of more complicated atoms which contain more electrons in the atomic shell become more involved but retain these general features.

Transitions between different energy levels effected by the absorption or emission of photons are also observed in *atomic nuclei*. A typical energy scale for these photons is $1\,\text{MeV}$, compared to $1\,\text{eV}$ in atoms. Nuclear spectra are complex because the nucleus usually consists of a large number of protons and neutrons bound together by nuclear forces. Some of the low-lying levels of nuclei can be explained by the following model. Every nucleon moves in the nuclear potential owing to the presence of all the other nucleons in the nucleus. Since nucleons are fermions and obey the Pauli exclusion principle, they fill up the lowest states in a common nuclear potential, forming the

**Fig. 18.2.** (a) **The energy levels that the electron of the hydrogen atom can take are indicated by horizontal lines and enumerated by the principal quantum number** $n$. **Vertical lines indicate the energies at which transitions between different energy levels take place. Transitions to or from the same lower energy level form a series. For example, transitions to or from energy level** $n = 1$ **make up the Lyman series. Those to or from energy level** $n = 2$ **make up the Balmer series. Transitions to a lower level consist of the emission of a light quantum corresponding to the transition energy. Those to a higher level consist of the absorption of a light quantum.** (b) **Wavelength spectra of light from different stars show the Balmer series in emission (top) and absorption (bottom). The stars are** $\alpha$ **Cassiopeiae and** $\beta$ **Cygni.** From R. W. Pohl, *Optik und Atomphysik*, ninth edition, copyright © 1954 by Springer-Verlag, Berlin, Göttingen, Heidelberg, reprinted by permission. (c) **The different energy levels of the carbon nucleus** $^{12}$C. **The ground state of the nucleus has been chosen to be the zero point of the energy scale. Some of the observed transitions between energy levels are indicated. These transitions, like those for the hydrogen atom in part a, consist of the emission or absorption of a photon.**

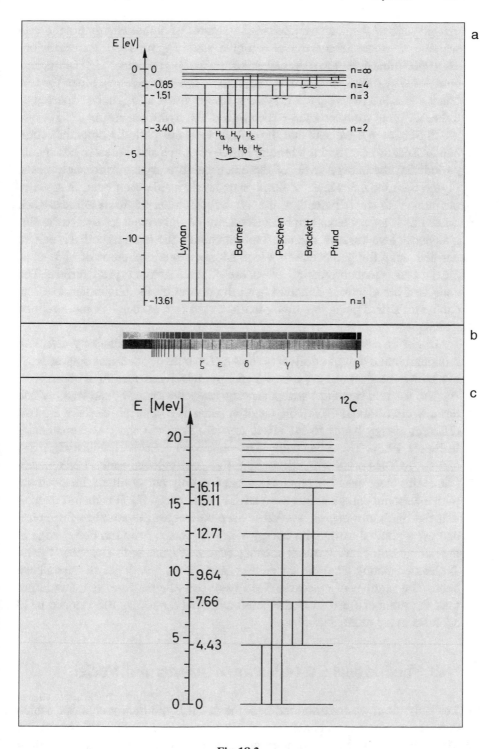

**Fig. 18.2.**

ground state of the nucleus. The simplest states of higher energy are those in which a single nucleon occupies a higher state. Figure 18.2c shows the energy spectrum of the low-lying states of the carbon nucleus $^{12}$C. The nucleus contains six protons and six neutrons, that is, twelve nucleons. Since the carbon nucleus is a twelve-particle system, its spectrum, as might be expected, is rather different from the energy spectrum of the hydrogen atom.

In Section 6.8 we saw that the energy levels of periodic potentials form bands. Because a crystal is a regular lattice of atoms and therefore has spatial periodicity, the energy levels of the electrons in a crystal form such bands. Figure 6.16 indicates that the number of levels inside each band is equal to the number of single potentials, that is, to the number of atoms in the crystal. Since this is a very large number indeed, we do not expect to resolve the single energy levels within a band. Experimentally, the band hypothesis can be verified using the photoelectric effect. Monoenergetic photons of high energies, that is, monochromatic X-rays, are directed onto a crystal surface. The energy of the electrons liberated from the crystal by the photoelectric effect can be measured using the principle illustrated in Figure 1.1 or more refined techniques.

In Figure 18.3a the energy spectrum of electrons obtained by directing monochromatic X-rays on silver is shown. The bulk of photoelectrons appears in the low-energy range between $W_1$ and $W_2$, which has a width of about $5\,\text{eV}$. A small fraction is emitted with an energy range between $W_2$ and $W_{\text{max}}$, which has a width of about $4\,\text{eV}$. This result is taken as evidence that there are two different energy bands in the silver crystal. They are shown schematically in Figure 18.3b. These bands are the *conduction band* with edges $E_{C1}$, $E_{C2}$ and the *valence band* with edges $E_{V1}$, $E_{V2}$. The valence band is completely filled with electrons. The conduction band is only partly filled; the electrons with maximum energy in this band have Fermi energy $E_F$. It is therefore clear that the minimum energy needed to free an electron is equal to the Fermi energy; a photoelectron with energy $W_{\text{max}}$ originates from the Fermi edge in the conduction band. We now identify photoelectrons with energies $W_2$ and $W_1$, as originating from the upper, $E_{V2}$, and lower, $E_{V1}$ edges of the valence band. The number of electrons freed from the valence band is much larger than the number freed from the conduction band because the valence band contains many more electrons.

## 18.3 Shell-Model Classification of Atoms and Nuclei

The only atom we have studied in some detail is the hydrogen atom, which consists of a proton of charge $+e$ as nucleus and an electron of charge $-e$. The heavier atoms have $Z$ protons and additional uncharged neutrons in their

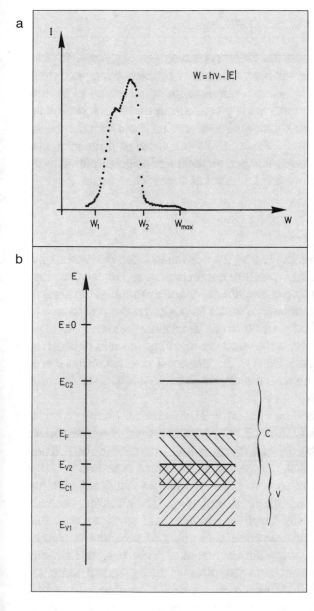

Fig. 18.3. (a) Current $I$ of photoelectrons emitted by a silver crystal, which has been irradiated by monochromatic X-rays, as a function of the kinetic energy $W$ of the photoelectrons. By energy conservation we have $W = h\nu - |E|$, where $h\nu$ is the energy of the X-ray photon and $E$ the energy with which the electron was originally bound in the crystal. Adapted from K. H. Hellwege, *Einführung in die Festkörperphysik*, copyright © 1976 by Springer-Verlag, Berlin, Heidelberg, New York, reprinted by permission. (b) Energy bands of electrons in the silver crystal shown schematically. The conduction band C is only partly filled with electrons, indicated by the hatched area. The valence band V, which is completely filled, partly overlaps with the conduction band. Photoelectrons with the highest energy originate from the region of highest energy in the conduction band, that is, $W_{max} = h\nu - |E_F|$.

nucleus, and $Z$ electrons in their hull. This $Z$ number representing the positive charge of the nucleus of the atom of an element is its *atomic number*. The potential energy of a single electron in the electric field of the nucleus of a heavier atom is

$$V(r) = -Z\alpha\hbar c \frac{1}{r} \quad .$$

Consequently, the energy levels are

$$E_n = -\frac{1}{2}Mc^2\frac{\alpha^2}{n^2}Z^2 \quad .$$

Here the forces acting between the electrons have been neglected. In Section 9.1 we learned that fermions obey the Pauli exclusion principle, which says that two identical fermions cannot populate the same state. Let us now count the number of different states for a given value $n$ of the principal quantum number. The angular-momentum quantum number $\ell$ can take the values $\ell = 0, 1, \ldots, n-1$. For a given $\ell$ there are $2\ell+1$ states of different quantum number $m$, which measures the $z$ component of angular momentum, $m = -\ell, -\ell+1, \ldots, \ell$. Thus the total number of states for a given $n$ is

$$\sum_{\ell=0}^{n-1}(2\ell+1) = n^2 \quad .$$

This number still has to be multiplied by 2 since the electron possesses spin. An electron with given "orbital" quantum numbers $n$, $\ell$, and $m$ can therefore still exist in the two different spin states, characterized by the quantum number $m_s = \frac{1}{2}, -\frac{1}{2}$, so that the total number of states for a given $n$ is equal to $2n^2$.

In our simplified description, all electrons in an atom that have the same principal quantum number $n$ have the same energy. They are said to be in the same *shell*. There can be two electrons in the innermost shell which has $n = 1$, eight electrons in the next shell with $n = 2$, and so on. In this way the *Periodic Table of elements* is easily explained.

For hydrogen ($Z = 1$) and helium ($Z = 2$) the electrons have principal quantum numbers $n = 1$. For lithium ($Z = 3$) both states with $n = 1$ are filled; therefore the third electron has to be in state $n = 2$ and in a second shell. When all the states with $n = 2$ are filled, the element is the noble gas neon ($Z = 10 = 2 \times 1^2 + 2 \times 2^2$). The element sodium ($Z = 11$) has an additional electron with $n = 3$, which goes in the third shell, and so on. The electrons in shells that are filled up are chemically inactive, which is seen from the chemical inertia of the noble gases helium and neon. Elements with the same number of electrons in an unfilled shell possess similar chemical properties, for example, lithium, sodium, and so on. The consecutive filling of the $n = 3$ shell continues only until the $\ell = 0$ and $\ell = 1$ states are all occupied. The element is argon ($Z = 18$), which again has the chemical properties of a noble gas. After argon the shell with $n = 4$ and $\ell = 0$ begins filling, forming potassium ($Z = 19$) and calcium ($Z = 20$). Only then are the so-far vacant states with $n = 3$ and $\ell = 2$ filled. The reason for this irregularity is that the states with $n = 4, \ell = 0$ are situated at lower energy than the states $n = 3, \ell = 2$. This situation is in contrast to our simple scheme, in which we have totally neglected the forces between the electrons in an atom.

Because of the forces acting between electrons, the energy levels of atoms with more than one electron are not simply the energy levels of a hydrogen-like atom with $Z$ protons in its nucleus and with its lowest states filled with $Z$ electrons. In fact, the actual calculation of the levels of many-electron atoms is complicated and can be carried out only with simplifying approximations. The levels least influenced by interactions between the electrons are the innermost levels for $n = 1$ and $n = 2$. Their $Z$ dependence is given by the formula

$$E_n = -\frac{1}{2}Mc^2\alpha^2\frac{Z^2}{n^2}.$$

The difference between the energy of the state with the principal quantum number $n_2$ and that of the ground state with $n_1 = 1$ for an atom with atomic number $Z$ is then

$$\Delta E = -\frac{1}{2}Mc^2\alpha^2\left(\frac{1}{n_2^2} - \frac{1}{n_1^2}\right)Z^2 = \frac{1}{2}Mc^2\alpha^2\left(1 - \frac{1}{n_2^2}\right)Z^2.$$

This difference can be measured in an experiment in which electrons accelerated to some $10\,\text{keV}$ knock an electron out of the ground state of an atom with atomic number $Z$. The unoccupied state ($n_1 = 1$) can be filled by an electron jumping from state $n_2 = 2$, $n_2 = 3$, and so on in the atom to the ground state.

The energy difference between the two states is radiated off as an X-ray quantum of frequency

$$\nu = \frac{1}{h}\Delta E.$$

With this formula for $\Delta E$, we find a linear relation between the atomic number $Z$ and the square root of the frequency, $\sqrt{\nu}$, of the emitted X-ray:

$$\sqrt{\nu} = \sqrt{Mc^2/(2h)}\alpha\left(1 - \frac{1}{n_2^2}\right)^{1/2}Z.$$

Henry G. J. Moseley first measured these transitions in 1913. His results are reproduced in Figure 18.4a. They allow the simple interpretation that the atomic number $Z$ is the number of positive charges on the nucleus of the atom, since the data take the expected line in a $Z, \sqrt{\nu}$ plot. Actually, the line of the data does not follow our formula exactly. The deviation is caused by the screening – even though small for the inner atomic shells – of the nuclear Coulomb field by other inner electrons.

Another test for the viability of the shell model of the atomic hull is indicated by the formula

$$E(Z,n) = -\frac{1}{2}Mc^2\alpha^2\frac{Z^2}{n^2}$$

for the energy of the outermost electron with principal quantum number $n$ in the hull of an atom with nuclear charge $Z$. This energy is called ionization energy. The expression for $E(Z,n)$ is actually only a rough estimate of the

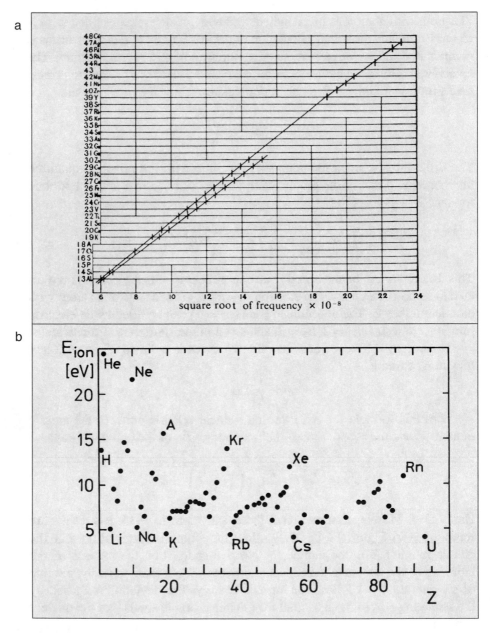

**Fig. 18.4. (a) Moseley's plot showing the square root of the X-ray frequency versus the atomic number** $Z$ **for** $K_\alpha$ **radiation,** $n_2 = 2$ **(upper line), and for** $K_\beta$ **radiation,** $n_2 = 3$ **(lower line).** From H. G. J. Moseley, *The Philosophical Magazine* **27** (1914) 703, copyright © 1914 by Taylor and Francis, Ltd., London, reprinted by permission. **(b) Ionization energies for atoms as a function of the atomic number** $Z$**. The maxima for noble gases, which have closed shells with** $Z = Z_c$ **electrons –** $Z_c = 2$ **for helium,** $Z_c = 10$ **for neon, and so on – are pronounced, and the drop from** $Z = Z_c$ **to** $Z = Z_c + 1$ **– that is, from helium to lithium, from neon to sodium, and so on – is sharp.**

ionization energy, since it does not take into account the mutual interaction of the electrons in the atomic hull. Nevertheless, for atoms with low atomic number it suffices to demonstrate how the values of the ionization energies indicate the closure of atomic shells.

In the process of "constructing" the chemical elements by filling the levels with electrons, the ionization energy $E(Z,n)$ rises with $Z$ as long as levels are filled with the same principal quantum number $n$. The highest value $E(Z_c,n)$ within each shell is reached in the element that has a closed shell with atomic number $Z_c$, that is, a noble gas. For the element with the next atomic number, a new shell with the principal quantum number $n+1$ begins to be occupied. Even though $Z$ increases in this step from $Z_c$ to $Z_c+1$, the increase from $n$ to $n+1$ means a definite decrease in ionization energy $E(Z_c+1,n+1)$ for the first element in the new shell compared to the value $E(Z_c,n)$ for the noble gas. Because there are many states belonging to each principal quantum number $n$, for each electron the principal quantum number is smaller than $Z_c$. The ratio of the two ionization energies is

$$r(Z_c) = \frac{E(Z_c+1,n+1)}{E(Z_c,n)} = \frac{(Z_c+1)^2\, n^2}{(n+1)^2\, Z_c^2}$$

$$= \frac{(1+1/Z_c)^2}{(1+1/n)^2} < 1 \quad,$$

because $Z_c$ is larger than $n$. For the jump from helium to lithium, neon to sodium, and argon to potassium, we find these values:

$$\begin{aligned}
\text{lithium/helium} &\quad r(2) &=& \quad 0.56 \quad, \\
\text{sodium/neon} &\quad r(10) &=& \quad 0.54 \quad, \\
\text{potassium/argon} &\quad r(18) &=& \quad 0.63 \quad.
\end{aligned}$$

In contrast, the ratio of the ionization energy of an element closing a shell to the energy of the preceding element in the Periodic Table is

$$r'(Z_c) = \frac{E(Z_c,n)}{E(Z_c-1,n)} = \frac{Z_c^2}{(Z_c-1)^2} = \frac{1}{(1-1/Z_c)^2} > 1 \quad.$$

For the corresponding closures of the atomic shells, we find

$$\begin{aligned}
\text{helium/hydrogen} &\quad r'(2) &=& \quad 4 \quad, \\
\text{neon/fluorine} &\quad r'(10) &=& \quad 1.23 \quad, \\
\text{argon/chlorine} &\quad r'(18) &=& \quad 1.12 \quad,
\end{aligned}$$

that is, values larger than one. The peak behavior expected by these arguments can be immediately verified by looking at the measured ionization energies plotted in Figure 18.4b, even though the experimental values for the ratios $r$ and $r'$ are different from the ones we have given.

In the classification of nuclei, the *nuclear shell model* has been successful in explaining observed regularities. For the electrons in a light element, it was reasonable to describe their motion in the Coulomb potential of the nucleus, neglecting the repulsion between electrons. For the protons and neutrons forming the nucleus, no analogous center of force exists. Nevertheless, it has proved useful in describing the motion of a single nucleon in the nuclear potential created by all remaining nucleons. Such a potential has, as does the nuclear force of a single nucleon, short range. For our simple discussion we assume that the potential is that of a harmonic oscillator. The lowest states in this potential are filled by the nucleons. Since protons and neutrons have spin $\frac{1}{2}$ according to Pauli's exclusion principle, every state characterized by $n$, $\ell$, and $m$ can be occupied by two protons and two neutrons. The lowest state in the harmonic oscillator (see Section 13.2) has quantum numbers $n = 0$, $\ell = 0$; therefore it can accommodate at most two protons and two neutrons. This is the case for the nucleus of the element helium. This nucleus, also called the $\alpha$ *particle*, is the most stable nucleus known; for its disintegration the largest amount of energy is needed. The helium nucleus has a closed proton shell and a closed neutron shell.

For the nuclei of the next heavier elements, the $n = 1$, $\ell = 1$ shell of the oscillator potential is successively filled. It offers $2 \times (2\ell + 1) = 6$ states for protons as well as six states for neutrons, so that the next closure of the proton shell, as well as of the neutron shell, is reached for $Z = 8$ and $N = 8$. Here $Z$, as before, gives the number of protons in the nucleus and $N$ gives the number of neutrons. The *nucleon number* $A = Z + N$ together with the chemical symbol which itself contains the information about $Z$ is commonly used to characterize the nucleus. The shells $Z = 8$ and $N = 8$ are those of the oxygen nucleus, $^{16}$O. As we know from Section 13.2, in the harmonic-oscillator potential the states with principal quantum number $n = 2$ are degenerate for $\ell = 0$ and $\ell = 2$. The nuclear shell with $n = 2$ contains $2 \times 1 + 2 \times 5 = 12$ states for protons and for neutrons. Thus the next closed shell is reached for $Z = N = 20$, the nucleus of the element cadmium, $^{40}$Ca. As was true of the atomic hull, this simple constructive scheme for finding the closed nuclear shells works for the lighter nuclei only.

It was the achievement of Maria Goeppert-Mayer and of Otto Haxel, Hans Jensen, and Hans Suess to discover the physical reason for the structure of the higher closed shells. They are reached at the higher *magic numbers* 28, 50, 82, 126, which cannot be obtained from the oscillator potential. In fact, these numbers are "magic" because they denote a large spin-orbit interaction, that is, a large interaction between spin $s$ and angular momentum $\ell$ of the nucleons. This coupling gives rise to an additional potential energy term in the Schrödinger equation. Evidence for nuclear shells comes from experiments in nuclear spectroscopy. We do not present them here, for their interpretation would require discussing additional details of nuclear physics.

## 18.4 Resonance Scattering off Molecules, Atoms, Nuclei, and Particles

In Chapter 15 we studied resonance phenomena in some detail. We have seen, in particular, that the total cross section for elastic scattering of a particle by a spherically symmetric potential may have pronounced maxima, as a function of the energy of the particle (see Figures 15.6 and 15.12). Such resonance phenomena are not restricted to simple potential scattering. They are observed in a variety of physical situations. In a more general situation, the collision of two particles, the total cross section is a measure of the probability that they will react. One or both particles may even be compound systems. The total cross section then is a measure of the probability for a reaction between these systems. In fact, we have seen evidence for such reactions earlier in this chapter when looking at the absorption spectrum of hydrogen (see Section 18.2, Figure 18.2b). The process is actually a collision between a photon and the hydrogen atom, which excites the electron in the atom into a higher energy level. The photograph of the spectrum shows that the absorption probability, that is, the total cross section, has pronounced maxima at particular photon energies. These energies correspond to the differences between the bound-state energies of the hydrogen atom. It turns out that in this process the higher bound states of the hydrogen atom are not absolutely stable. After excitation by absorbing a photon, a higher bound state, through photon emission, decays with a certain average lifetime into a state of lower energy and finally into the ground state. In our original calculations of the hydrogen atom (Section 13.4), only the Coulomb interaction between electron and proton was taken into account. Now we are also considering the interaction of photons and electrons. The total process of absorption and emission of a photon is nothing but the resonance scattering of a photon by the atom. We expect the process to show the qualitative features of resonance scattering discussed in Section 5.4 and Chapter 15.

Of course, similar resonance structures in total cross sections can be observed in more complicated atoms and even in molecules. Figure 18.5a shows the absorption spectra of infrared light by different paraffin molecules,

$$
\begin{aligned}
\text{n-pentane} \quad & CH_3 - CH_2 - CH_2 - CH_2 - CH_3, \\
\text{n-hexane} \quad & CH_3 - CH_2 - CH_2 - CH_2 - CH_2 - CH_3,
\end{aligned}
$$

$\vdots$

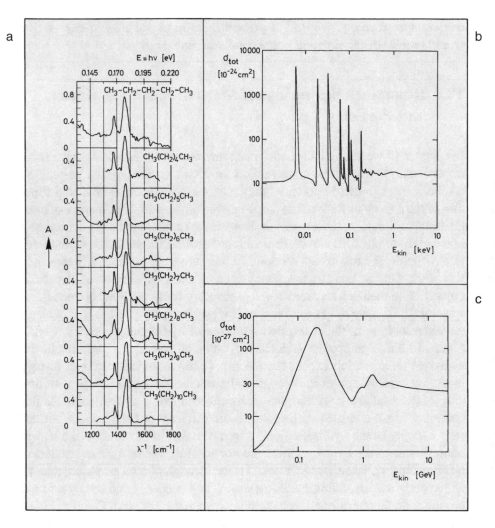

**Fig. 18.5. Total cross sections for various reactions as a function of the kinetic energy of the incident particle in the laboratory frame. (a) The absorption coefficient $A$ for infrared light passing through a layer of paraffin 0.02 mm thick. The abscissa is the wave number $\lambda^{-1} = v/c$ (bottom), which is proportional to the energy $E = h\nu$ of the light quanta (top). A high rate of absorption corresponds to a large total cross section. Thus the graphs can be interpreted as measurements of the total cross sections as a function of energy. Two characteristic resonances near $E = 0.17$ eV, which are associated with the vibrations of neighboring $CH_2$ groups, are present in all paraffins considered.** From Landolt-Börnstein, *Zahlenwerte und Funktionen,* sixth edition, Volume 1, part 2 (A. Eucken and K. H. Hellwege, editors), Figure 33, p. 365, Copyright ©1951 by Springer-Verlag, Berlin, Göttingen, Heidelberg, reprinted by permission. **(b) Total cross section for neutrons scattered off lead nuclei. There are many resonances at low energies corresponding to the formation of various metastable states of lead isotopes. (c) Total cross section for positive pions scattered on protons. The wide resonance near $E_{kin} \approx 0.2$ GeV corresponds to the excitation of the metastable state $\Delta^{++}(1232)$.**

There is a strong similarity in the absorption spectra, indicating the excitation of very similar resonances in the different molecules. They correspond to vibrations between neighboring $CH_2$ groups, which are common to all the paraffin molecules.

Figure 18.5b presents an example from nuclear physics, the total cross section of neutrons scattered off lead nuclei. The many resonances indicate that the nuclei can exist in a variety of metastable states, covering a rather wide range of energies.

We have seen that resonance scattering reveals the presence of excited states in molecules, atoms, and nuclei. Single nucleons, for example, protons, can also be investigated by scattering different projectiles on them. We choose here positive pions, also called $\pi$ mesons. These particles are lighter than protons but heavier than electrons. They play an important role in explaining nuclear forces. In Figure 18.5c the total cross section for the scattering of positive pions on protons is shown as a function of the pion energy. The pronounced resonance at the left side of the picture we interpret as a metastable state. Actually, it corresponds to a short-lived particle called the $\Delta^{++}$ baryon. The sequence of its production in a pion–proton collision and its subsequent decay into a pion and a proton is written as

$$\pi^+ p \to \Delta^{++} \to \pi^+ p \quad .$$

Electrons and positrons can be accelerated to very high energy, more than 50 GeV; they can be accumulated in a storage ring and be brought to head-on collisions. The total cross section as a function of the center-of-mass energy of the $e^+ e^-$ system has characteristic resonances. Figure 18.6 shows two series of resonances which are located near 3 and 10 GeV. They are evidence for short-lived particles called the $J/\psi$ family and the $\Upsilon$ family. The first one found is the $J/\psi$ particle with a mass of 3.1 GeV. Its production and subsequent decay into electron and positron is a resonance scattering of the form

$$e^+ e^- \to J/\psi \to e^+ e^- \quad .$$

Besides this elastic process, the inelastic one,

$$e^+ e^- \to J/\psi \to \text{hadrons} \quad ,$$

is also observed. Hadrons are particles that interact strongly in the nucleus, in particular pions, protons, and neutrons. All hadrons are assumed to be composed of only a few constituent particles called *quarks q* and *antiquarks* $\bar{q}$. The $J/\psi$ particle is composed of the very heavy *charm quark c* and its antiparticle $\bar{c}$, so that the reaction above reads

$$e^+ e^- \to (c\bar{c}) \to \text{hadrons} \quad ,$$

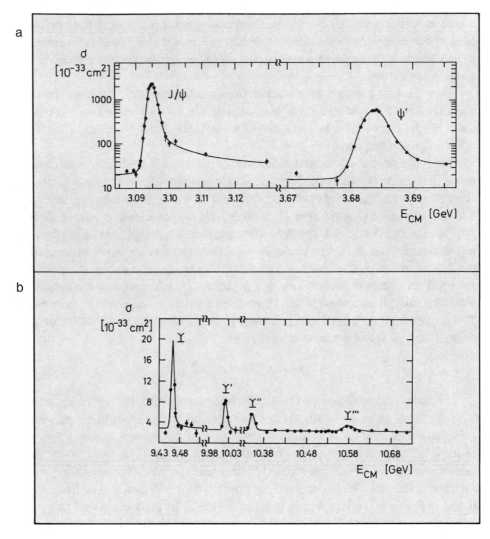

**Fig. 18.6.** Total cross section observed for the reaction in which an electron $(e^-)$ and a positron $(e^+)$ annihilate each other to form a number of strongly interacting particles, such as $\pi$ mesons. The cross section shows very sharp resonances near (a) $E_{CM} \approx 3\,\text{GeV}$ and (b) $E_{CM} \approx 10\,\text{GeV}$. Here $E_{CM}$ is the total energy in the center-of-mass system, the system in which $e^+$ and $e^-$ have equal and opposite momenta. The unexpectedly sharp resonances are interpreted as evidence that metastable states consisting of a quark–antiquark pair have formed. The $J/\psi$ family of states is composed of a "charm" quark and its antiparticle. The $\Upsilon$ family of states is a bound system of a "beauty" quark and the corresponding antiquark. *Sources*: (a) From A. M. Boyarski et al., *Physical Review Letters* **34** (1975) 1357 and from V. Lüth et al., *Physical Review Letters* **35** (1975) 1124, copyright © 1975 by American Physical Society, reprinted by permission. (b) From D. Andrews et al., *Physical Review Letters* **44** (1980) 1108 and **45** (1980) 219, copyright © 1980 by American Physical Society, reprinted by permission.

where $(c\bar{c})$ symbolizes the metastable state $J/\psi$ of $c$ and $\bar{c}$. The next resonance, $\psi'$ at 3.7 GeV, is another resonance of the $c\bar{c}$ system that can be
regarded as an exited state of the $J/\psi$ particle. In fact, these and the other
observed $(c\bar{c})$ states can be explained as bound states in a potential describing
the interaction of $c$ and $\bar{c}$. The discovery of these states has led to a much better understanding of quark bound states and to deeper insight into the structure
of matter.

A similar series of resonances in electron–positron scattering is observed
at 9.46 GeV and beyond. The family of $\Upsilon$ particles are understood to be bound
states of the even heavier beauty quark $b$ and its antiquark $\bar{b}$.

It is interesting to note that the quantum-mechanical phenomena studied in
this section span an energy range of eleven orders of magnitude, from infrared
radiation at 0.2 eV to high-energy electron storage rings at 10 GeV.

## 18.5 Phase-Shift Analysis in Nuclear and Particle Physics

In the preceding section we identified a resonance in the total cross section
as evidence for the existence of a metastable state. Figure 15.6 showed that a
maximum in the total cross section is usually an indication for a resonance in
a single partial wave. The quantum numbers of the resonant partial wave are
therefore those of the metastable state which, in elementary particle physics,
we have also called a particle. To determine the quantum numbers of such a
particle, we use the method of phase-shift analysis outlined in Section 15.3.
We decompose the measured differential cross section into partial waves, obtaining the complex partial-wave amplitudes $f_\ell$, as a function of the energy,
or equivalently, the momentum of the incident particle. From the different
partial-wave amplitudes Argand diagrams analogous to those in Figure 15.14
can be constructed.

Figure 18.7a shows the differential cross section for the elastic scattering
of positive pions on protons for various pion energies $E$. Near $E = 200$ MeV
the cross section has a simple parabolic form, indicating the dominance of the
Legendre polynomial $P_1(\cos\vartheta) = \cos\vartheta$ in the expression

$$\frac{d\sigma}{d\Omega} = |f(\vartheta)|^2 \sim |P_1(\cos\vartheta)|^2$$

for the differential scattering cross section. This parabola indicates a resonance with angular momentum $\ell = 1$. The proton has an intrinsic angular
momentum, that is, a spin of $\frac{1}{2}\hbar$. It turns out that the metastable state has
total angular momentum $\frac{3}{2}\hbar$. Figure 18.7b gives the Argand diagram for the
corresponding partial-wave amplitude. As in Figure 15.14, the partial scattering amplitude is recognized to move, as a function of energy, on the unitarity

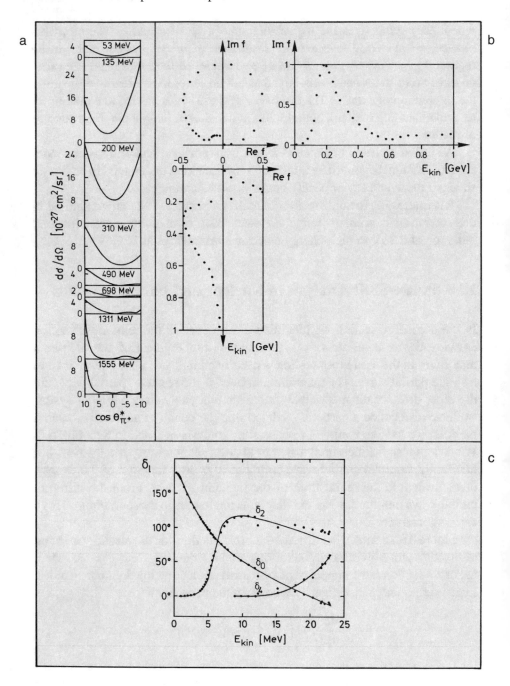

**Fig. 18.7.**

circle in the complex plane. The deviation of the experimental points from the unitarity circle designate an inelastic process. Not only elastic scattering but also the production of one or more additional pions is possible at these energies. The real and imaginary parts of the partial-wave amplitude have the characteristic features of a resonance at a center-of-mass energy of 1.232 GeV in the pion-proton system. By this phase-shift analysis the intrinsic angular momentum of the $\Delta^{++}$ hadron, which we first observed in the total cross section (Figure 18.5c), is found to be $\frac{3}{2}\hbar$.

The method of phase-shift analysis, which has proved very successful in particle physics, had already been used earlier in nuclear physics. The elastic scattering of $\alpha$ particles on helium nuclei is an interesting example. In Figure 18.7c the phase shifts $\delta_0$, $\delta_2$, and $\delta_4$ are given directly as functions of the energy of the incident particle. The phase shift $\delta_2$ shows a typical resonance, a quick rise through the value $\pi/2$. The resonance corresponds to a metastable state of angular momentum 2 of the beryllium nucleus, $^8$Be, which is formed by two $^4$He nuclei colliding.

## 18.6 Classification of Resonances on Regge Trajectories

Figure 15.12 indicated that there is a striking regularity between the energies of the lowest-lying resonances of a system and their angular momenta. In a plane spanned by energy $E$ and angular momentum $\ell$, the resonances lie on a curve in such a way that energy $E$ of the resonance increases monotonically with angular momentum $\ell$. The correlation between the energies of a family of resonances and their angular momenta in potential scattering has been derived by Tullio Regge. In elementary particle physics families of particles lying on the same *Regge trajectory* are observed. As an example, Figure 18.8

**Fig. 18.7. Phase-shift analysis. (a) The differential cross section for the elastic scattering of positive $\pi$ mesons on protons, shown for various kinetic energies of the meson, has a simple parabolic form at $E_{\text{kin}} = 200$ MeV, indicating a resonance at this energy with angular momentum $\ell = 1$. (b) The Argand diagram of the corresponding partial scattering amplitude, reconstructed from measured data. All the features of a resonance at $E_{\text{kin}} = 200$ MeV are evident. The phase shift passes swiftly through 90 degrees, while the imaginary part goes through a maximum and the real part vanishes. (c) A resonance at much lower energies. Various phase shifts for the elastic scattering of an $\alpha$ particle on a helium nucleus, that is, another $\alpha$ particle, are plotted as a function of the kinetic energy of the incoming particle. The resonance in $\delta_2$ indicates that both particles form a resonance with angular momentum $\ell = 2$.** Source: (a) From Robert C. Cence, *Pion–Nucleon Scattering*, copyright © 1969 by Princeton University Press, Figure 5.2, p. 62, reprinted by permission of Princeton University Press. (b) Adapted from G. Höhler in Landolt–Börnstein, *Numerical Data*, New Series, volume 9b2 (H. Schopper, editor), Figure 2.2.6, p. 58, copyright ©1983 by Springer-Verlag, Berlin, Heidelberg, New York, reprinted by permission. (c) From T. A. Tombrello and L. S. Senhouse, *The Physical Review* **129** (1963) 2252, copyright © 1963 by American Physical Society, reprinted by permission.

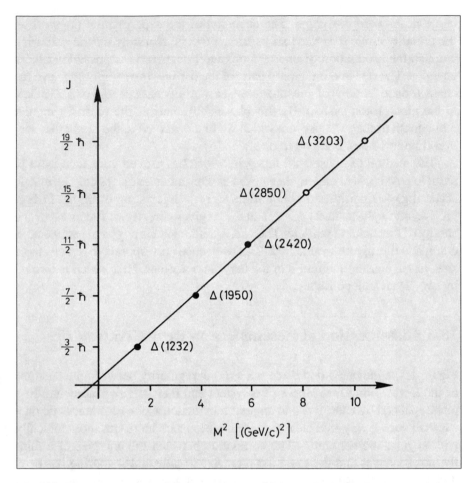

**Fig. 18.8. Regge trajectory of the** $\Delta$ **particles, which can be understood as resonances formed by a proton and a** $\pi$ **meson. The square of the resonance mass** $M^2$ **is plotted against the angular momentum** $J$ **of the resonance. For the three lowest-lying resonances (black points) both** $M$ **and** $J$ **have been experimentally determined. For the last two (open circles) only the mass has been measured so far.**

shows the Regge trajectory containing the $\Delta^{++}$ hadron, already discussed in Sections 18.4 and 18.5. We now call it more specifically $\Delta(1232)$ by indicating in brackets its mass in MeV. On the same trajectory four more resonances are shown. In this diagram, in which the square of the resonance mass is plotted on the abscissa and its spin on the ordinate, the trajectory is a straight line. From resonance to resonance, the spin is increased by two units, that is, it takes the values $\frac{3}{2}\hbar$, $\frac{7}{2}\hbar$, and so on. This complication is attributable to the half-integer spin of these resonances.

# 18.7  Radioactive Nuclei as Metastable States

The disintegration of a radioactive nucleus by the emission of an $\alpha$ particle can be considered as the decay of a metastable state. George Gamow has given a quantum-mechanical analysis of how the $\alpha$ particle behaves in the potential of the other protons and neutrons in the nucleus. The effect of the short-range nuclear forces can be approximated by a square-well potential. In addition, the $\alpha$ particle, which carries the electric charge $+2e$, experiences the repulsive long-range Coulomb force of the other protons. The total potential of both nuclear and Coulomb forces is attractive for small radii but repulsive for greater distances, as indicated in Figure 18.9a. Such a potential can contain bound states of negative energies that are stable as well as metastable states of positive energies that have a finite lifetime. In particular, metastable states with energies lower than the height of the repulsive wall are expected to have long lifetimes. An $\alpha$ particle in such a metastable state can leave the nucleus only by tunneling through the potential barrier. For an $\alpha$ particle in a metastable state, only the repulsive barrier is important. The repulsive shell studied in detail in Chapter 15 can therefore serve as a model for the potential. Figure 15.15 shows the radial wave functions of several metastable states in this potential. Figures 15.12, 15.13, and 15.14 contain the total cross sections and Argand diagrams. They indicate that resonance widths increase with resonance energy. The repulsive shell of Chapter 15 has its one-dimensional analog in the two-potential barriers of Section 5.4. The widths of the metastable states confined between two potential barriers also grow with energy, as indicated in Figure 5.12. Figures 5.9 and 5.10 examined the time dependence of the decay of metastable states and revealed that the lifetime decreases with increasing energy, that is, with increasing width. Indeed, the probability for the penetration of the barrier grows with the energy of the particle, which is tantamount to saying that the lifetime decreases with the energy of the $\alpha$ particle.

This phenomenon is observed experimentally. The energy of $\alpha$ particles is easily measured by their range in air. Figure 18.9b shows a cloud chamber photograph displaying the tracks of $\alpha$ particles emitted by radioactive polonium, $^{214}$Po. All tracks except one have very similar ranges, indicating the energy of the lowest-lying metastable state. A single track in the photograph has a considerably greater length. Its energy is that of a higher metastable state, which is already much depopulated because of its shorter lifetime. A systematic study of the relation between energy and the lifetime of $\alpha$ decays of nuclei was first carried out by Hans Geiger and John Mitchell Nuttall. Figure 18.9c shows this correlation for many radioactive elements.

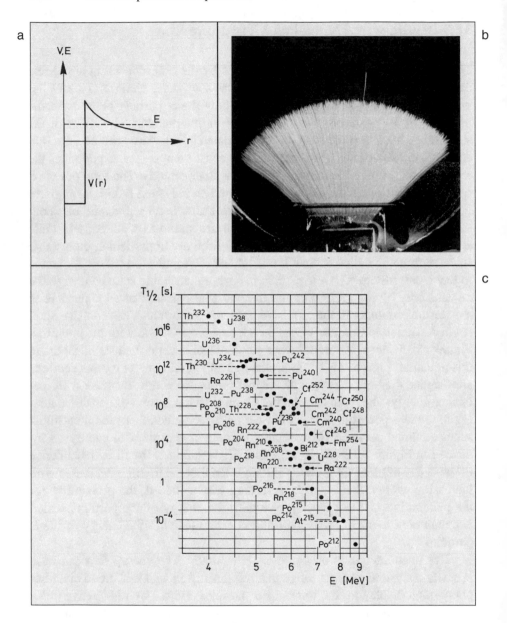

**Fig. 18.9.**

## 18.8 Magnetic-Resonance Experiments

### Units and Orders of Magnitude

The operator $\boldsymbol{\mu}$ of the magnetic moment of an electron and its spin operator $\mathbf{S}$ are simply proportional to each other,

$$\boldsymbol{\mu} = -\gamma \mathbf{S} \quad.$$

The quantity $\gamma$, the *gyromagnetic ratio of the electron*, is given as

$$\gamma = g_0 \mu_B / \hbar \quad.$$

The constant

$$\mu_B = \frac{e}{2M} \hbar = 9.274078 \times 10^{-24} \, \mathrm{A\,m^2}$$

is called the *Bohr magneton*. Here $M$ denotes the electron mass. The *gyromagnetic factor of the free electron* $g_0$ can be computed in the framework of *quantum electrodynamics* (QED). Precision measurements of $g_0$ are therefore important tests of the validity of QED. Current experimental and theoretical values are

$$g_0^{(\mathrm{exp})} = 2.002319304386 \pm 20 \times 10^{-12} \quad,$$
$$g_0^{(\mathrm{th})} = 2.002319304822 \pm 332 \times 10^{-12} \quad.$$

The astonishing accuracy of the experimental value is due to magnetic resonance experiments performed with a single electron by Hans Dehmelt and his group.

Because of the negative charge of the electron the vectors $\boldsymbol{\mu}$ and $\mathbf{S}$ are antiparallel. For particles with positive charge, in particular for atomic nuclei, they are parallel. Since one wants $\gamma$ to be positive also for nuclei one has to write

$$\boldsymbol{\mu} = \gamma \mathbf{S} \quad.$$

We shall use this relation for both electrons and nuclei. Our formulae will correspond to those given in the literature specific to the field of *electron spin resonance* (ESR) if $\gamma$ is replaced by $-\gamma$.

In order to get $g$ factors of the order of one also for atomic nuclei with a magnetic moment one writes for nuclei:

$$\gamma = g_N \mu_p / \hbar \quad.$$

**Fig. 18.9.** $\alpha$ **decay. (a) Potential energy** $V(r)$ **of an** $\alpha$ **particle in a nucleus. Although the total energy** $E$ **(dashed line) of an** $\alpha$ **particle may be positive, the particle can leave the nucleus only by tunneling through the potential barrier created by the Coulomb attraction between nucleus and** $\alpha$ **particle. Therefore metastable states of positive energy can exist. (b) Cloud chamber photograph of tracks of** $\alpha$ **particles from the decay of the polonium nucleus,** $^{214}$**Po. All particles except one have approximately the same range in the chamber gas, indicating that they possess equal energies. The single, long-range track was caused by the decay of an exited state of** $^{214}$**Po possessing a higher energy.** From K. Phillip, *Naturwissenschaften* **14** (1926) 1203, copyright © 1926 Verlag von Julius Springer, Berlin, reprinted by permission. **(c) Geiger–Nuttall diagram showing the relation between the half-life** $T_{1/2}$ **and the energy of the emitted** $\alpha$ **particles for the lowest-lying states of radioactive nuclei. The diagram indicates that the lifetime decreases very rapidly with energy.**

Here

$$\mu_p = \frac{e}{2M_p}\hbar = \frac{1}{1836}\mu_B = 5.051 \times 10^{-27}\,\text{A m}^2$$

is the *nuclear magneton* and $M_p = 1836M$ the proton mass. Measurements of the gyromagnetic factor $g_{\text{proton}}$ of the proton yield

$$g_{\text{proton}} = 5.58 \quad .$$

This value cannot be computed in the framework of QED. It is assumed that the magnetic moment of the proton results from the intrinsic magnetic moments of its constituents, the quarks, and from magnetic moments of orbital motion of the quarks within the proton. The determination of the magnetic moment of the proton, the neutron, and in general of atomic nuclei is an important field of nuclear physics.

In Section 17.4 we have discussed the phenomenon of magnetic resonance. In a homogeneous magnetic induction field $\mathbf{B}_0 = B_0\mathbf{e}_z$ the expectation value $\langle\boldsymbol{\mu}\rangle$ of a magnetic moment precesses around the field direction with the Larmor frequency $\Omega_0 = \gamma B_0$. The polar angle $\vartheta$ of $\langle\boldsymbol{\mu}\rangle$ with respect to $\mathbf{B}_0$ stays constant. If in addition to the constant field $\mathbf{B}_0$ there is a field $\mathbf{B}_1(t)$ perpendicular to $\mathbf{B}_0$ and itself rotating with a frequency $\omega$ equal to the Larmor frequency $\Omega_0$ then the angle $\vartheta$ changes by $\pi$ within the time $T/2$ where $T = \pi/\Omega_1$, $\Omega_1 = \gamma B_1$. This way the direction of $\langle\boldsymbol{\mu}\rangle$ which may originally have been parallel to $\mathbf{B}_0$ is changed to be antiparallel to $\mathbf{B}_0$.

The difference in potential energy between the two spin states in which the spin (and therefore the magnetic moment) is oriented parallel or antiparallel to the time-independent magnetic-induction field $\mathbf{B}_0$ is

$$\Delta E = |\mu B_0| \quad .$$

At resonance the frequency $\omega$ of the rotating field $\mathbf{B}_1$ is equal to the Larmor frequency $\Omega_0 = \mu B_0$, so that

$$\Delta E = \hbar\Omega_0 = \hbar\omega = |\mu B_0| \quad .$$

The transition from the state of lower energy to the state of higher energy is made possible through the *absorption* of a quantum $\hbar\omega$ of energy from the rotating field. The transition from the higher to the energetically lower level is accompanied by the emission of a quantum of electromagnetic energy $\hbar\omega$. In the presence of the rotating external field the transition is accelerated due to *stimulated emission*.

For a typical field $B_0$ of $1\,\text{T} = 1\,\text{V s m}^{-2}$ and a magnetic moment of $1\mu_p$ (1 nuclear magneton) the frequency of the oscillating field has to be

$$\nu = \omega/2\pi = \mu_p B_0/h = 0.762 \times 10^7\,\text{s}^{-1} \quad .$$

Such a frequency is easily produced with *radio frequency* (RF) technology, which is therefore used in *nuclear magnetic resonance* (NRM) experiments.

For $B_0 = 1$ T and a magnetic moment of $1\mu_B$ (1 Bohr magneton) the frequency is

$$\nu = \omega/2\pi = \mu_B B_0/h = 1.4 \times 10^{10}\, \text{s}^{-1} \quad .$$

*Microwave* techniques are required to generate fields which oscillate in this frequency range. Consequently, experiments measuring the *electron spin resonance* (ESR), sometimes also called *electron paramagnetic resonance* (EPR), use the microwave technology.

**Experiments with Atomic and Molecular Beams**

We can now discuss the magnetic resonance method with atomic and molecular beams pioneered by Isidor Rabi and collaborators. A beam of neutral atoms or molecules passes through three consecutive magnetic fields denoted A, C, and B in Figure 18.10. The fields A and B are inhomogeneous fields of the Stern–Gerlach type, cf. Section 1.4. They are identical except for the fact that the field gradient of A is directed downwards in the plane of Figure 18.10 and that of B is directed upwards. In region C there is a constant homogeneous field $\mathbf{B}_0$ directed upwards and an oscillating field $\mathbf{B}_1$. The latter is produced by the current from a radio-frequency generator which is run through a wire in the plane of Figure 18.10 parallel to the beam somewhat above the beam and returns through a parallel wire below the beam. Due to the force exerted by the inhomogeneous fields on the magnetic moment the trajectories in regions A and B are parabolae. Particles with magnetic moments pointing upwards in region A and having a certain limited range in initial direction and momentum will pass a slit in front of region C. If the orientation of the magnetic moment is not changed in C the particles pass through region B on a trajectory symmetrical to the one in region A and are registered by a detector beyond region B. If, however, the orientation of the magnetic moment is turned downwards in region C by magnetic resonance, i.e., if the oscillating frequency $\omega$ of the field $B_1$ is equal to the Larmor frequency $\Omega_0 = \mu B_0/\hbar$ of the magnetic moment in the field $\mathbf{B}_0$ and if the time the particle needs to traverse region C is about $T/2 = h/(\mu B_1)$, then the particle will be deflected downwards in region B rather than upwards and the intensity registered in the detector will decrease drastically. Figure 18.11 shows the first resonance curve reported by Rabi et al. for a beam of LiCl molecules which is due to the magnetic moment of the $^7$Li nucleus. The strength of the field $B_0$ rather than the RF frequency $\omega$ was varied in their experiment.

The method of atomic and molecular beams was and is very successfully used to measure magnetic moments of nuclei and to study the interaction between the nuclear magnetic moments and the electrons in the atomic shell

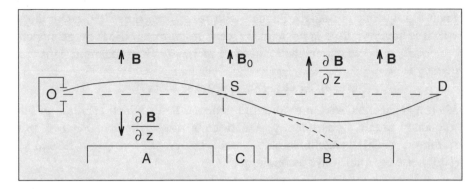

**Fig. 18.10. Rabi apparatus. The magnetic-induction field B points upwards (in the $z$ direction) in the three magnets A, B, and C. In A and B the field is inhomogeneous, the field gradient $dB_z/dz$ being negative in A and positive in B. The field in C is homogeneous. For molecules with magnetic moment in $z$ direction and with momentum within a certain range the trajectory from the source O through the slit S to the detector D is drawn as a solid line. If the direction of the magnetic-moment expectation value is changed to the $-z$ direction due to magnetic resonance in the additional oscillating field in C the trajectory changes to the broken line and the molecules no longer reach the detector.**

which gives rise to the *hyperfine structure* of atomic spectra. The distance between atoms or molecules within the beam is very large. Therefore the experiments are essentially performed with free atoms or molecules.

### Magnetic Resonance in Bulk Matter

**Apparatus.** In bulk matter (solid, liquid, or gaseous) a large number of particles are present per unit volume and the collective effect of their magnetic moments can be recorded by magnetic-resonance methods. *Nuclear magnetic-resonance* (NMR) experiments of this type were first developed by the research groups of Edward Purcell and of Felix Bloch in 1945. *Electron magnetic* (or *spin*) *resonance* (ESR) was discovered also in 1945 by E. Zavoisky.

The principal components of an apparatus for NMR experiments are shown in Figure 18.12. A large homogeneous field $\mathbf{B}_0 = B_0\mathbf{e}_z$ is provided by an electromagnet. A field $\mathbf{B}_1 = 2B_1\cos(\omega t)\mathbf{e}_x$ oscillating in $x$ direction is generated by a coil oriented in $x$ direction and connected to a radio-frequency generator. The sample of bulk matter is placed inside the coil. (In ESR experiments instead of the coil one has a microwave resonator excited by a microwave transmitter.) The complex resistance of the coil is measured with high precision with a radio-frequency version of a Wheatstone bridge. We will show further down that near the resonance frequency of the magnetic moments of the sample this resistance changes dramatically. At resonance en-

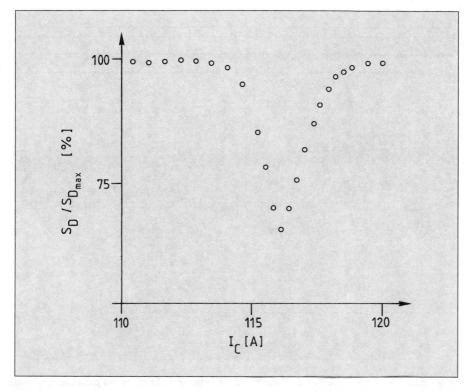

**Fig. 18.11. Magnetic-resonance curve obtained for LiCl molecules. The signal $S_D$ in the detector (in percent of the maximum signal) is shown as a function of the current $I_C$ in the exciting coil of the C magnet and thus the field $B_0$ while keeping the frequency $\omega$ of the oscillating field constant.** From I. I. Rabi, J. R. Zacharias, S. Millman, and P. Kusch, *Physical Review* **53** (1938) 318, © 1938 by American Physical Society, reprinted by permission.

ergy is absorbed from the RF field. Thus the resonance manifests itself as a maximum in the real part of the complex resistance of the coil. But at resonance energy is also emitted by the sample which is excited by the RF field. The emission of energy can be detected as an induced current in a pick-up coil the axis of which is oriented along the $y$ direction in Figure 18.12. In the following we discuss magnetic resonance in bulk matter in a little more detail.

**Magnetization.** In a homogeneous induction field $\mathbf{B}_0$ the expectation value $\langle\boldsymbol{\mu}\rangle$ of the magnetic moment of an isolated particle precesses around the field direction. The energy expectation value $-\langle\boldsymbol{\mu}\rangle\cdot\mathbf{B}_0$ stays constant. In the presence of many other particles, i.e., in bulk matter, energy exchange with other particles occurs and a statistical distribution of the potential energies $-\langle\boldsymbol{\mu}\rangle\cdot\mathbf{B}_0$ is established which depends on the temperature $T$ of the sample. To give an idea for the order of magnitude of this effect we note that the ratio of the num-

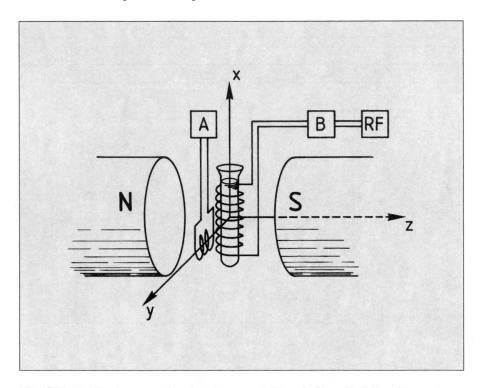

**Fig. 18.12. NMR apparatus. An electromagnet (with pole faces N, S) provides a strong homogeneous field $B_0$ in the $z$ direction. A coil oriented along the $x$ direction contains the sample. It is excited by a radio-frequency generator RF of frequency $\omega$. The complex resistance of the coil is measured in a bridge circuit B and registered while either $B_0$ or $\omega$ are varied. In addition there may be a coil oriented along the $y$ direction which can pick up magnetic-induction signals from the sample. The signals are amplified by the amplifier A and also registered as a function of $B_0$ or $\omega$.**

ber $N_+$ of particles with the highest potential energy ($\langle\boldsymbol{\mu}\rangle$ antiparallel to $\mathbf{B}_0$) to the number $N_-$ with the lowest energy ($\langle\boldsymbol{\mu}\rangle$ parallel to $\mathbf{B}_0$) is

$$\frac{N_+}{N_-} = \exp\left\{-\frac{|\mu B_0|}{kT}\right\} \approx 1 - \frac{|\mu B_0|}{kT} \quad ,$$

where $k = 1.381 \times 10^{-23}\,\mathrm{J\,K^{-1}}$ is *Boltzmann's constant*. At room temperature ($T = 300\,\mathrm{K}$) one has $kT = 4.14 \times 10^{-21}\,\mathrm{J}$ which is very large compared to $\mu B_0 = 5.05 \times 10^{-27}\,\mathrm{J}$ where we have set $\mu = \mu_\mathrm{p}$ and $B_0 = 1\,\mathrm{T}$. Therefore, $N_+/N_-$ is of the order of $1 - 10^{-6}$. If we form the statistical average over the expectation values $\langle\boldsymbol{\mu}\rangle$ of the magnetic moments of all particles in the sample in thermal equilibrium,

$$\overline{\boldsymbol{\mu}} = \overline{\langle\boldsymbol{\mu}\rangle} \quad ,$$

we find a vector parallel to $\mathbf{B}_0$. However, the magnitude of this vector is by a factor of the order of $10^{-6}$ smaller than the magnitude of $\langle\boldsymbol{\mu}\rangle$.

The *magnetization* of a sample is the magnetic moment per unit volume,

$$\mathbf{M} = n\overline{\mu} \quad ,$$

with $n$ being the number of particles per unit volume which carry the magnetic moment of interest.

**The Bloch Equations.** Since the magnetization $\mathbf{M}$ is a sum over magnetic-moment expectation values $\langle\mu\rangle$ which in turn are proportional to spin-vector expectation values $\langle\mathbf{S}\rangle$, the equation of motion for $\mathbf{M}$ is identical to that for $\langle\mathbf{S}\rangle$ discussed at the end of Section 17.4 and the beginning of Section 17.5,

$$\left(\frac{d\mathbf{M}}{dt}\right)_{\mathrm{L}} = \gamma \mathbf{M} \times \mathbf{B} \quad .$$

The index L indicates that this equation describes a Larmor precession of $\mathbf{M}$ about the direction of $\mathbf{B}$. The equation holds as long as the magnetization is influenced only by the field $\mathbf{B}$. We have to extend it in order to take into account in a global way the *relaxation effects* that take place in the sample. The magnetization is the sum over all magnetic moments $\langle\mu_i\rangle$ in a unit volume. The $\langle\mu_i\rangle$ are not only influenced by the external field but also by the fields originating from the components of the sample. As a result of such interaction within the sample the magnetization tends in an irreversible way towards an equilibrium magnetization.

As before we consider a field of the form

$$\mathbf{B}(t) = \mathbf{B}_0 + \mathbf{B}_1(t)$$

with

$$\mathbf{B}_0 = B_0 \mathbf{e}_z$$

and

$$\mathbf{B}_1(t) = B_1 \cos(\omega t)\mathbf{e}_x + B_1 \sin(\omega t)\mathbf{e}_y \quad .$$

If initially (at time $t = 0$) there is a nonvanishing *transverse magnetization*

$$\mathbf{M}_\perp = M_x \mathbf{e}_x + M_y \mathbf{e}_y \quad ,$$

then it will decrease exponentially,

$$\left(\frac{d\mathbf{M}_\perp}{dt}\right)_{T_2} = -\frac{\mathbf{M}_\perp}{T_2} \quad .$$

This effect is called *spin–spin relaxation*. It is characterized by the *spin–spin relaxation time $T_2$*.

Spin–spin relaxation does not change the $z$ component $\langle \mu_i \rangle_z$ of the individual moments and therefore it does not change the *magnetic energy density*

$$w_m = -\mathbf{M} \cdot \mathbf{B}_0 = -M_z B_0 \quad .$$

If the sample was for a long time in the constant field $\mathbf{B}_0$ then $\mathbf{M}$ will be the vector of *equilibrium magnetization*

$$\mathbf{M}_0 = M_0 \mathbf{e}_z \quad .$$

A *longitudinal magnetization*

$$\mathbf{M}_\| = M_z \mathbf{e}_z$$

different from the equilibrium magnetization $\mathbf{M}_0$ will develop towards the equilibrium in an exponential way,

$$\left( \frac{d\mathbf{M}_\|}{dt} \right)_{T_1} = \frac{M_0 - M_z}{T_1} \mathbf{e}_z \quad .$$

In this process energy is transferred between the magnetic moments and their surrounding atoms. Since these atoms in many cases form a regular lattice the process is called *spin-lattice relaxation*. Because in contrast to spin–spin relaxation energy transfer is involved the *spin-lattice relaxation time* $T_1$ is usually much longer than the spin–spin relaxation time $T_2$,

$$T_1 \gg T_2 \quad .$$

The model sketched above was developed by Felix Bloch in 1946. It has proved to be very successful for the understanding of magnetic resonance in bulk matter although it needs to be refined in particular situations. In summary, it yields

$$\frac{d\mathbf{M}}{dt} = \gamma \mathbf{M} \times \mathbf{B} - \frac{\mathbf{M}_\perp}{T_2} + \frac{M_0 - M_z}{T_1} \mathbf{e}_z$$

as equation of motion for the vector of magnetization. Writing this equation in components we obtain the *Bloch equations*,

$$
\begin{aligned}
\frac{dM_x}{dt} &= \gamma (\mathbf{M} \times \mathbf{B})_x - \frac{M_x}{T_2} \quad , \\
\frac{dM_y}{dt} &= \gamma (\mathbf{M} \times \mathbf{B})_y - \frac{M_y}{T_2} \quad , \\
\frac{dM_z}{dt} &= \gamma (\mathbf{M} \times \mathbf{B})_z + \frac{M_0 - M_z}{T_1} \quad .
\end{aligned}
$$

**Complex Susceptibility.** In the following we recall some results from classical electrodynamics. The relation between the magnetic field $\mathbf{H}$, the induction field $\mathbf{B}$, and the magnetization $\mathbf{M}$ is

$$\mathbf{H} = \frac{1}{\mu_0}\mathbf{B} - \mathbf{M}$$

with $\mu_0 = 4\pi \times 10^{-7}\,\mathrm{V\,s\,A^{-1}\,m^{-1}}$ being the permeability of free space. For a sample with relative permeability[1] $\mu$ the field strength is

$$\mathbf{H} = \mathbf{B}/(\mu\mu_0)\quad,\qquad \text{i.e.,}\qquad \mu\mu_0\mathbf{H} = \mathbf{B} = \mu\mathbf{B} - \mu\mu_0\mathbf{M}\quad.$$

For all types of samples of interest in magnetic-resonance experiments one has $\mu \approx 1$ and, therefore,

$$\mathbf{M} = \frac{1}{\mu_0}(\mu - 1)\mathbf{B} = \frac{1}{\mu_0}\chi\mathbf{B}\quad,$$

where $\chi = \mu - 1$ is the *magnetic susceptibility* of the sample.

We now consider the time dependence of the vector $\mathbf{B}_\perp$ in the $x, y$ plane,

$$
\begin{aligned}
\mathbf{B}_\perp &= B_\perp\cos(\omega t)\mathbf{e}_x + B_\perp\sin(\omega t)\mathbf{e}_y \\
&= \mathrm{Re}\left\{B_\perp e^{i\omega t}\right\}\mathrm{Re}\left\{\mathbf{e}_c\right\} - \mathrm{Im}\left\{B_\perp e^{i\omega t}\right\}\mathrm{Im}\left\{\mathbf{e}_c\right\} \\
&= \mathrm{Re}\left\{B_\perp e^{i\omega t}\mathbf{e}_c\right\} = \mathrm{Re}\left\{B_{c\perp}\mathbf{e}_c\right\}
\end{aligned}
$$

with

$$\mathbf{e}_c = \mathbf{e}_x - i\mathbf{e}_y\quad.$$

We call

$$B_{c\perp} = B_\perp e^{i\omega t} = B_\perp\cos\omega t + iB_\perp\sin\omega t$$

the complex magnetic induction in the transverse plane.

Correspondingly, we have

$$\mathbf{M}_\perp = \mathrm{Re}\left\{M_{c\perp}\mathbf{e}_c\right\}\quad,$$

however, with the representation

$$M_{c\perp} = M_\perp e^{i(\omega t - \delta)}\quad.$$

The angle $\delta$ allows for a phase shift between $M_{c\perp}$ and $B_{c\perp}$. It describes the fact that while rotating the vectors $\mathbf{B}_\perp$ and $\mathbf{M}_\perp$ need not have the same direction at a given time $t$. This is consequence of a complex susceptibility

---

[1] Of course, the symbols $\mu_0$ and $\mu$ used in this context must not be confused with magnetic moments.

$$\chi = |\chi| e^{-i\delta} = \chi' - i\chi''$$

with real and imaginary part being equal to

$$\chi' = |\chi| \cos\delta \quad , \qquad \chi'' = |\chi| \sin\delta \quad .$$

It relates the complex magnetization $M_{c\perp}$ and the complex magnetic induction $B_{c\perp}$ by

$$M_{c\perp} = \frac{1}{\mu_0} \chi B_{c\perp} \quad .$$

After cancellation of the factor $\exp\{i\omega t\}$ on both sides of this equation we find

$$M_\perp e^{-i\delta} = M_\perp \cos\delta - iM_\perp \sin\delta = \frac{1}{\mu_0}\chi' B_\perp - i\frac{1}{\mu_0}\chi'' B_\perp$$

yielding

$$\chi' = \mu_0 \frac{M_\perp \cos\delta}{B_\perp} \quad , \qquad \chi'' = \mu_0 \frac{M_\perp \sin\delta}{B_\perp} \quad .$$

The quantities $\chi'$ and $\chi''$ are called *dispersive part* and *absorptive part* of the susceptibility, respectively. In the rotating frame of reference of Section 17.5 one has $B_\perp = B_1 = B_{x'}$ and $M_\perp \cos\delta = M_{x'}$, $M_\perp \sin\delta = M_{y'}$ so that

$$\chi' = \mu_0 M_{x'}/B_1 \quad , \qquad \chi'' = \mu_0 M_{y'}/B_1 \quad .$$

**NMR Spectra Obtained by Slow Passage through Resonance Conditions.** In order to detect a resonance the external field conditions have to be changed with time so that a *passage* through the resonance region at

$$\omega = \Omega_0 = \gamma B_0$$

is achieved. The reaction of the sample depends on the speed of passage. We discuss here only the case of *slow passage*.

In the rotating reference frame the effective field vector $\mathbf{B}_{\text{eff}}$, cf. Section 17.5, moves in the $x',z'$ plane. This motion can be made so slow that the magnetization is at any moment in equilibrium, so that the time derivative of $\mathbf{M}$ can be neglected. In the rotating frame Bloch's equations thus read in components

$$0 = \frac{dM_{x'}}{dt} = \gamma(\mathbf{B}_{\text{eff}} \times \mathbf{M})_{x'} - \frac{M_{x'}}{T_2} = -(\gamma B_0 - \omega)M_{y'} - \frac{M_{x'}}{T_2} \quad ,$$

$$0 = \frac{dM_{y'}}{dt} = \gamma(\mathbf{B}_{\text{eff}} \times \mathbf{M})_{y'} - \frac{M_{y'}}{T_2} = (\gamma B_0 - \omega)M_{x'} - \gamma B_{x'}M_{z'} - \frac{M_{y'}}{T_2} \quad ,$$

$$0 = \frac{dM_{z'}}{dt} = \gamma(\mathbf{B}_{\text{eff}} \times \mathbf{M})_{z'} - \frac{M_0 - M_{z'}}{T_1} = \gamma B_{x'}M_{y'} - \frac{M_0 - M_{z'}}{T_1} \quad .$$

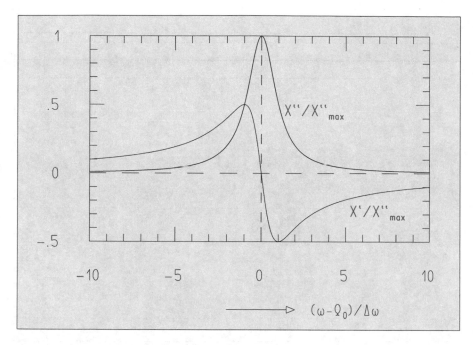

**Fig. 18.13. Frequency dependence of the real part** $\chi'$ **(thin line) and the imaginary part** $\chi''$ **(thick line) of the complex susceptibility.**

Solution of this set of equations for the components of $\mathbf{M}$ and multiplication by $\mu_0/B_{x'} = \mu_0/B_1$ yields

$$\chi' = \mu_0\frac{M_{x'}}{B_1} = -\mu_0\frac{\gamma(\gamma B_0 - \omega)T_2^2 M_0}{1 + (\gamma B_0 - \omega)^2 T_2^2 + \gamma^2 B_1^2 T_1 T_2} \quad ,$$

$$\chi'' = -\mu_0\frac{M_{y'}}{B_1} = \mu_0\frac{\gamma T_2 M_0}{1 + (\gamma B_0 - \omega)^2 T_2^2 + \gamma^2 B_1^2 T_1 T_2} \quad .$$

For small values of $B_1$, i.e., $\gamma^2 B_1^2 T_1 T_2 \ll 1$, one gets for the frequency dependence of $\chi'$ and $\chi''$ the relations

$$\chi'' = \mu_0\frac{\gamma T_2}{1 + (\gamma B_0 - \omega)^2 T_2^2} M_0 \quad ,$$

$$\chi' = -(\gamma B_0 - \omega)T_2 \chi'' \quad .$$

They are shown graphically in Figure 18.13.

The imaginary part $\chi''$ of the complex susceptibility gives rise to a real part in the complex resistance

$$Z = i\omega L = i\omega\mu_0(1 + \chi' - i\chi'')N^2 a/\ell$$

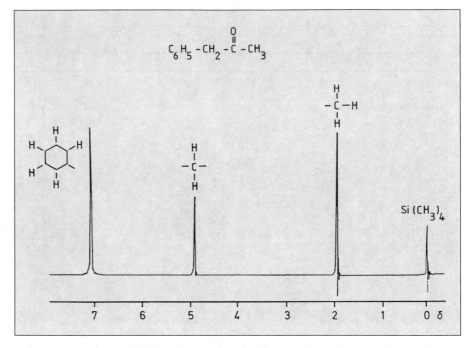

**Fig. 18.14.** NMR spectrum due to the protons in benzylacetate (first three peaks from the left) in the presence of tetramethylsilane as a reference substance (peak on the far right). On the scale at the bottom of the figure the relative difference of the resonance frequency to the reference peak is given in parts per million (ppm). Adapted from H. Günter, *NMR-Spektroskopie*, 1993 © by Georg Thieme Verlag, Stuttgart, reprinted by permission.

of the coil ($N$ windings, length $\ell$, cross section $a$, inductivity $L = \mu_0 \mu N^2 a / \ell$) which generates the time-dependent field and thus to an energy absorption by the sample. The function $\chi''(\omega)$ has its maximum for $\omega = \gamma B_0$. It drops to half the maximum value at the frequencies

$$\omega = \omega_\pm = \gamma B_0 \pm \Delta\omega \quad , \qquad \Delta\omega = \frac{1}{T_2} \quad .$$

The quantity $\Delta\omega$ is a measure for the width of the peak in $\chi''(\omega)$. Therefore two peaks at positions $\omega_1$ and $\omega_2$ can be separated if $|\omega_2 - \omega_1| > \Delta\omega$.

In Figure 18.14 we present a spectrum obtained for a sample of benzylacetate $C_8H_{10}O_2$ for $B_0 = 1.4\,T$ and frequencies $\nu = \omega/2\pi$ around 60 MHz. These are resonance conditions for the proton. Rather than to observe one absorption line at the exact resonance frequency $\omega_0$ of the free proton we see three lines which are shifted to slightly different frequencies $\omega_1$, $\omega_2$, $\omega_3$. It is customary to express these shifts in dimensionless units, i.e., to define the ratios

$$\delta_i = \frac{\omega_i - \omega_0}{\omega_0} \quad .$$

In practice not the free proton resonance frequency is used as reference frequency $\omega_0$ but the frequency of a sharp absorption line produced by a reference substance (e.g., tetramethylsilane $Si(CH_3)_4$) which is simply added to the sample substance. The relative shifts $\delta_i$ are of the order of a few times $10^{-6}$ (or a few parts per million (ppm)).

The reason for the *chemical shift* in resonance frequency is due to the presence of electrons in the molecule. Let us suppose for the moment that the sample was simply atomic hydrogen. Then the time-dependent external field would induce an orbital magnetic moment in the ground state of the electron. This moment would give rise to an additional magnetic field at the position of the proton and thus shift the resonance frequency. In complex molecules these frequency shifts are different for protons in different positions within the molecule. In the example of Figure 18.14 the three lines are characteristic for the environment experienced by a proton within a benzol ring (left), in a $CH_2$ group (middle), and in a $CH_3$ group (right). This interpretation is verified by the fact that the integrals over the three peaks are in the ratio 5:2:3 just as the numbers of protons in the three groups are. It is obvious from this very simple example that NMR measurements are an important tool used to determine the structure of organic molecules.

**Spin Echo. Measurement of Relaxation Times.** Information about the surroundings of a nucleus with a magnetic moment is not only contained in the exact resonance frequency $\omega$ (for a given external field $B_0$) but also in the spin–spin relaxation time $T_2$ and in the spin-lattice relaxation time $T_1$. Measurements of $T_1$ and $T_2$ require good time resolution and cannot, of course, be done using the method of slow passage described before.

We begin by discussing the sudden application of resonance conditions to the sample and the observation of the time dependence of the change of magnetization brought about by it. If a sample with equilibrium magnetization $\mathbf{M} = M_0 \mathbf{e}_z$ is exposed suddenly to resonance conditions, e.g., by switching on the field $\mathbf{B}_1$ at resonance frequency for a time which is short compared to the relaxation times $T_1$ and $T_2$ and if the system is studied for a time which is also short compared to $T_1$ and $T_2$, then the terms containing $T_1$ and $T_2$ can be neglected in Bloch's equations. The magnetization $\mathbf{M}$, originally parallel to the $z$ axis, precesses around the $x'$ axis in the rotating frame. If the resonance condition is applied exactly for the time $T/4 = \pi/(2\gamma B_1)$ then $\mathbf{M}$ is rotated by exactly 90° (we speak of the application of a *90° pulse*) and falls onto the negative $y'$ axis. In the laboratory system it then rotates in the $x, y$ plane with the resonance frequency. After the time $T/4$ the field $\mathbf{B}_1$ is switched off, the vector $\mathbf{M}$ keeps rotating in the $x, y$ plane and radiates off electromagnetic waves of frequency $\omega$. These can be detected for instance by the signal they induce (the *free induction signal*) in the additional coil shown in Figure 18.12

or also in the exciting coil since the exciting radio frequency is now switched off.

The signal detected indeed falls rapidly as expected from the decay of the transverse magnetization in Bloch's equations. The time constant $\widetilde{T}_2$ of this decay is, however, considerably shorter than the spin–spin relaxation time $T_2$. The reason for this effect is that, in addition to the *irreversible* decrease of the transverse magnetization (described by $T_2$ in Bloch's equations) there is a *reversible* decrease. The latter can for instance be due to a small inhomogeneity of the static field $\mathbf{B}_0$. The vectors $\mathbf{M}_i$ of local magnetization at different locations $i$ within the sample which are in phase directly after the 90° pulse then rotate in the $x, y$ plane with slightly different angular velocities. With time they develop larger and larger relative phase differences so that the magnetization being the average over the $\mathbf{M}_i$ goes to zero. Since the time constant $\widetilde{T}_2$ of this process is smaller than $T_2$ the latter cannot be measured from the decay of the free induction signal.

This difficulty is overcome by the *spin-echo* technique invented by Erwin Hahn. Of the many clever schemes in use now we mention but two:

*Measurement of $T_2$ with a pulse sequence 90°–180°.* Figure 18.15 displays several vectors $\mathbf{M}_i$ of local magnetization at various times during the experiment. Initially all $\mathbf{M}_i$ are along the $z'$ direction. By the 90° pulse (of length $T/4$) they are rotated and fall onto the negative $y'$ axis. In the $x', y'$ plane the vectors spread out because they rotate with slightly different angular velocities in the laboratory system, i.e., they are not all exactly stationary in the rotating frame. At the same time the magnitude of the $\mathbf{M}_i$ decreases with the time constant $T_2$ according to Bloch's equations. At the time $T/4 + \tau, \tau \gg T$, a 180° pulse (of length $T/2$) is applied, i.e., all $\mathbf{M}_i$ are rotated by 180° about the $x'$ axis. During the time from $t = 3T/4 + \tau$ to $t = 3T/4 + 2\tau$ they get back into phase again along the $y'$ direction, since the $\mathbf{M}_i$ with the highest relative angular velocity which have got furthest away from the $-y'$ direction in the spread-out period have the largest angle to the $y'$ direction at the beginning of the rephasing period. The result is that at $t = 3T/4 + 2\tau \approx 2\tau$ another induction signal, the *spin echo*, is detected. Its amplitude, however, has decreased by the factor $\exp\{-2\tau/T_2\}$. By repeating the measurement for various values of $\tau$ signals similar to the curves shown in Figure 18.16 are obtained and the spin–spin relaxation time $T_2$ can be easily extracted.

*Measurement of $T_1$ with a pulse sequence 180°–90°.* If at $t = 0$ a 180° pulse is applied the magnetization is turned from equilibrium $\mathbf{M} = M_0 \mathbf{e}_z$ to $\mathbf{M} = -M_0 \mathbf{e}_z$. According to Bloch's equations it develops back towards equilibrium, the $z$ component being

$$M_z(t) = -M_0(2\exp\{-t/T_1\} - 1) \quad .$$

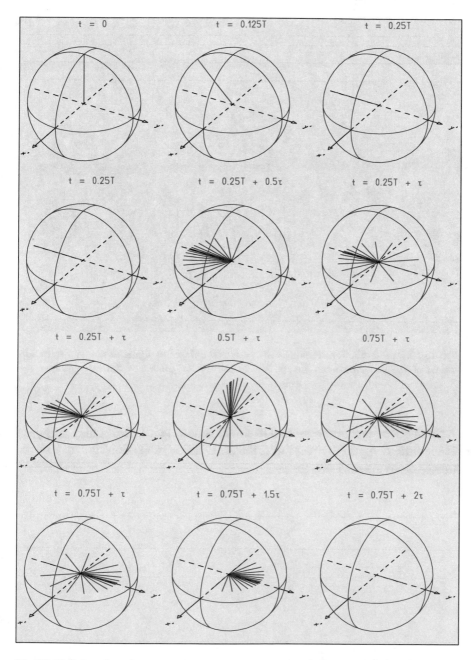

**Fig. 18.15. Spin echo with** $90°$–$180°$ **pulse sequence shown in the rotating frame of refer-ence. Top row: Application of** $90°$ **pulse; local magnetization vectors** $\mathbf{M}_i$**, initially parallel to each other and to the** $z'$ **direction, are rotated onto the** $-y'$ **direction. Second row: The** $\mathbf{M}_i$ **get out of phase and spread out in the** $x', y'$ **plane. Third row: Application of** $180°$ **pulse; each** $\mathbf{M}_i$ **is rotated by** $180°$ **about the** $x'$ **axis. Bottom row: The** $\mathbf{M}_i$ **back into phase. Because of spin–spin relaxation the magnitudes** $M_i$ **of all local magnetization vectors decrease with time.**

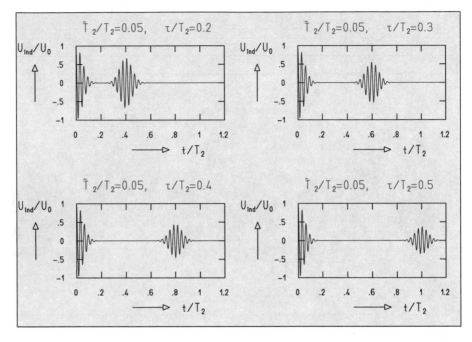

**Fig. 18.16. Free induction signal (near $t = 0$) and spin-echo signal at $t = 2\tau$ for various values of the time $\tau$ between the applications of the 90° pulse and the 180° pulse.**

An induction signal proportional to $M_z(t)$ is detected if a 90° pulse is applied at the time $t$. Again, by varying $t$, the spin-lattice relaxation time can be extracted from the measurements.

# A. Simple Aspects of the Structure of Quantum Mechanics

In Chapters 2 to 16 we have used the formulation of quantum mechanics in terms of wave functions and differential operators. This is but one of many equivalent representations of quantum mechanics. In this appendix we shall briefly review that representation and develop an alternative representation in which state vectors correspond to the wave functions and matrices to the operators. To keep things simple we shall restrict ourselves to systems with discrete energy spectra exemplified on the one-dimensional harmonic oscillator.

## A.1 Wave Mechanics

In Section 6.3 the *stationary Schrödinger equation*

$$\left(-\frac{\hbar}{2m}\frac{d^2}{dx^2} + \frac{m}{2}\omega^2 x^2\right)\varphi_n = E_n\varphi_n(x)$$

of the harmonic oscillator has been solved. The *eigenvalues* $E_n$ were found to be

$$E_n = (n + \tfrac{1}{2})\hbar\omega$$

together with the corresponding *eigenfunctions*

$$\varphi_n(x) = \frac{1}{(\sqrt{\pi}\sigma_0 2^n n!)^{1/2}} H_n\left(\frac{x}{\sigma_0}\right)\exp\left\{-\frac{x^2}{2\sigma_0^2}\right\} \quad , \qquad \sigma_0 = \sqrt{\frac{\hbar}{m\omega}} \quad .$$

Quite generally, we can write the stationary Schrödinger equation as an *eigenvalue equation*

$$H\varphi_n = E_n\varphi_n \quad ,$$

where the *Hamiltonian $H$* – as in classical mechanics – is the sum

$$H = T + V$$

of the kinetic energy

S. Brandt and H.D. Dahmen, *The Picture Book of Quantum Mechanics*,
DOI 10.1007/978-1-4614-3951-6, © Springer Science+Business Media New York 2012

$$T = \frac{\hat{p}^2}{2m}$$

and the potential energy

$$V = \frac{m}{2}\omega^2 x^2 \quad .$$

The difference to classical mechanics consists in the momentum being given in one-dimensional quantum mechanics by the *differential operator*

$$\hat{p} = \frac{\hbar}{i}\frac{d}{dx}$$

so that the kinetic energy takes the form

$$T = -\frac{\hbar^2}{2m}\frac{d^2}{dx^2} \quad .$$

Two eigenfunctions $\varphi_n(x)$, $\varphi_m(x)$, $m \neq n$ belonging to different eigenvalues $E_m \neq E_n$ are *orthogonal*, i.e.,[1]

$$\int_{-\infty}^{+\infty} \varphi_m^*(x)\varphi_n(x)\,dx = 0 \quad .$$

Conventionally, for $m = n$ the eigenfunctions are *normalized* to one, i.e.,

$$\int_{-\infty}^{+\infty} \varphi_n^*(x)\varphi_n(x)\,dx = 1 \quad ,$$

so that we may summarize

$$\int_{-\infty}^{+\infty} \varphi_m^*(x)\varphi_n(x)\,dx = \delta_{mn} \quad ,$$

where we have used the *Kronecker symbol*

$$\delta_{mn} = \begin{cases} 1 & , \quad m = n \\ 0 & , \quad m \neq n \end{cases} \quad .$$

The infinite set of mutually orthogonal and normalized eigenfunctions $\varphi_n(x)$, $n = 0, 1, 2, \ldots$, forms a *complete orthonormal basis* of all complex-valued functions $f(x)$ which are square integrable, i.e.,

$$\int_{-\infty}^{+\infty} f^*(x)f(x)\,dx = N^2 \quad , \qquad N < \infty \quad .$$

$N$ is called the *norm* of the function $f(x)$. Functions with norm $N$ can be normalized to one,

---

[1] The functions $\varphi_n(x)$ are real functions. We add an asterisk (indicating the complex conjugate) to the function $\varphi_m(x)$ in the integral, since in other cases one often has to deal with complex functions.

$$\int_{-\infty}^{+\infty} \varphi^*(x)\varphi(x)\,dx = 1 \quad ,$$

by dividing them by the normalization factor $N$,

$$\varphi(x) = \frac{1}{N}f(x) \quad .$$

The completeness of the set $\varphi_n(x)$, $n = 0, 1, 2, \ldots$, allows the expansion

$$f(x) = \sum_{n=0}^{\infty} f_n \varphi_n(x) \quad .$$

Because of the orthonormality of the eigenfunctions the complex coefficients $f_n$ are simply

$$f_n = \int_{-\infty}^{+\infty} \varphi_n^*(x) f(x)\,dx \quad .$$

We also get

$$N^2 = \int_{-\infty}^{+\infty} f^*(x) f(x)\,dx = \sum_{n=0}^{\infty} |f_n|^2 \quad .$$

The *superposition* of two normalizable functions

$$f(x) = \sum_{n=0}^{\infty} f_n \varphi_n(x) \quad , \qquad g(x) = \sum_{n=0}^{\infty} g_n \varphi_n(x)$$

with complex coefficients $\alpha$, $\beta$ may be expressed by

$$\alpha f(x) + \beta g(x) = \sum_{n=0}^{\infty} (\alpha f_n + \beta g_n)\varphi_n(x) \quad .$$

Their *scalar product* is defined as

$$\int_{-\infty}^{+\infty} g^*(x) f(x)\,dx = \sum_{n=0}^{\infty} g_n^* f_n \quad .$$

## A.2  Matrix Mechanics in an Infinite Vector Space

The normalizable functions $f(x)$ form a *linear vector space of infinite dimensionality*, i.e., each function $f(x)$ can be represented by a vector $\mathbf{f}$ in that space,

$$f(x) \rightarrow \mathbf{f} \quad .$$

With the *base vectors*

$$\boldsymbol{\varphi}_0 = \begin{pmatrix} 1 \\ 0 \\ 0 \\ \vdots \end{pmatrix}, \quad \boldsymbol{\varphi}_1 = \begin{pmatrix} 0 \\ 1 \\ 0 \\ \vdots \end{pmatrix}, \quad \boldsymbol{\varphi}_2 = \begin{pmatrix} 0 \\ 0 \\ 1 \\ \vdots \end{pmatrix}, \quad \cdots$$

a general vector **f** takes the form

$$\mathbf{f} = \sum_{n=0}^{\infty} f_n \boldsymbol{\varphi}_n = \begin{pmatrix} f_0 \\ f_1 \\ f_2 \\ \vdots \end{pmatrix}.$$

The axioms of the infinite space of complex column vectors are the natural extension of the ones for finite complex vectors:

(i) *Linear superposition* ($\alpha$, $\beta$ complex numbers):

$$\alpha \mathbf{f} + \beta \mathbf{g} = \begin{pmatrix} \alpha f_0 + \beta g_0 \\ \alpha f_1 + \beta g_1 \\ \alpha f_2 + \beta g_2 \\ \vdots \end{pmatrix}.$$

(ii) *Scalar product*:

$$\mathbf{g}^+ \cdot \mathbf{f} = (g_0^*, g_1^*, g_2^*, \ldots) \begin{pmatrix} f_0 \\ f_1 \\ f_2 \\ \vdots \end{pmatrix} = \sum_{n=0}^{\infty} g_n^* f_n.$$

Here the adjoint $\mathbf{g}^+$ of the vector $\mathbf{g}$ has been introduced, $\mathbf{g}^+ = (g_0^*, g_1^*, g_2^*, \ldots)$, as the line vector of the complex conjugates $g_0^*, g_1^*, g_2^*, \ldots$ of the components of the column vector

$$\mathbf{g} = \begin{pmatrix} g_0 \\ g_1 \\ g_2 \\ \vdots \end{pmatrix}.$$

Because of the infinity of the set of natural numbers an additional axiom has to be added:

(iii)  The *norm* $|\mathbf{f}|$ of the vectors $\mathbf{f}$ is finite,

$$|\mathbf{f}|^2 = \mathbf{f}^+ \cdot \mathbf{f} = \sum_{n=0}^{\infty} f_n^* f_n = N^2 \quad , \qquad N < \infty \quad ,$$

i.e., the infinite sum has to converge. Because of *Schwartz's inequality*

$$|\mathbf{g}^+ \cdot \mathbf{f}| \le |\mathbf{g}||\mathbf{f}|$$

all scalar products of vectors $\mathbf{f}, \mathbf{g}$ of the space are finite.

As in the ordinary finite-dimensional vector spaces we call a *linear trans-formation* $A$ of a function $f(x)$ into a function $g(x)$,

$$g(x) = Af(x) \quad ,$$

the *linear operator* $A$. Examples of linear transformations are

- the *momentum operator* $\hat{p} = -\mathrm{i}\hbar \, \mathrm{d}/\mathrm{d}x$,

$$\hat{p} f = \frac{\hbar}{\mathrm{i}} \frac{\mathrm{d}f}{\mathrm{d}x}(x) \quad ,$$

- the *Hamiltonian* $H = -(\hbar^2/2m)\mathrm{d}^2/\mathrm{d}x^2 + V(x)$,

$$Hf = -\frac{\hbar^2}{2m} \frac{\mathrm{d}^2 f(x)}{\mathrm{d}x^2} + V(x)f(x) \quad ,$$

- the *position operator* $\hat{x} = x$,

$$\hat{x} f = x f(x) \quad .$$

Linear operators can be represented by *matrices*. We show this by the following argument. The function $g$ is represented by the coefficients $g_m$,

$$g(x) = \sum_{m=0}^{\infty} g_m \varphi_m(x) \quad , \qquad g_m = \int_{-\infty}^{+\infty} \varphi_m^*(x) g(x) \, \mathrm{d}x \quad .$$

The *image function* $g$ of $f$ is given by

$$g(x) = Af(x) = A\left(\sum_{n=0}^{\infty} f_n \varphi_n(x)\right) = \sum_{n=0}^{\infty} A\varphi_n(x) f_n \quad ,$$

i.e., by a linear combination of the images $A\varphi_n$ of the elements $\varphi_n$ of the orthonormal basis. The $A\varphi_n$ themselves can be represented by a linear combination of the basis vectors

$$A\varphi_n = \sum_{m=0}^{\infty} \varphi_m(x) A_{mn} \quad , \qquad n = 0, 1, 2, \ldots \quad ,$$

with the coefficients

$$A_{mn} = \int_{-\infty}^{+\infty} \varphi_m^*(x) A \varphi_n(x) \, \mathrm{d}x \quad .$$

Inserting this into the expression for $g(x)$ we obtain

$$g(x) = \sum_{m=0}^{\infty} \sum_{n=0}^{\infty} \varphi_m(x) A_{mn} f_n \quad .$$

Comparing with the representation for $g(x)$ we find for the coefficients $g_m$ the expression

$$g_m = \sum_{n=0}^{\infty} A_{mn} f_n \quad .$$

We arrange the coefficients $A_{mn}$ like matrix elements in an infinite matrix scheme

$$A = \begin{pmatrix} A_{00} & A_{01} & A_{02} & \cdots \\ A_{10} & A_{11} & A_{12} & \cdots \\ A_{20} & A_{21} & A_{22} & \cdots \\ \vdots & \vdots & \vdots & \ddots \end{pmatrix}$$

and recover an infinite dimensional extension of the *matrix multiplication*

$$\mathbf{g} = A\mathbf{f}$$

in the form

$$\begin{pmatrix} g_0 \\ g_1 \\ g_2 \\ \vdots \end{pmatrix} = \begin{pmatrix} A_{00} & A_{01} & A_{02} & \cdots \\ A_{10} & A_{11} & A_{12} & \cdots \\ A_{20} & A_{21} & A_{22} & \cdots \\ \vdots & \vdots & \vdots & \ddots \end{pmatrix} \begin{pmatrix} f_0 \\ f_1 \\ f_2 \\ \vdots \end{pmatrix} = \begin{pmatrix} \sum_{n=0}^{\infty} A_{0n} f_n \\ \sum_{n=0}^{\infty} A_{1n} f_n \\ \sum_{n=0}^{\infty} A_{2n} f_n \\ \vdots \end{pmatrix} \quad .$$

It should be noted that the two descriptions by wave functions and operators or by vectors and matrices are equivalent. The correspondence relations

$$\varphi(x) = \sum_{n=0}^{\infty} a_n \varphi_n(x) \leftrightarrow \boldsymbol{\varphi} = \begin{pmatrix} a_0 \\ a_1 \\ a_2 \\ \vdots \end{pmatrix}$$

with

$$a_n = \int_{-\infty}^{+\infty} \varphi_n^*(x) \varphi(x) \, \mathrm{d}x$$

for wave functions and vectors work in both directions. For a given wave function $\varphi(x)$ we can uniquely determine the vector $\boldsymbol{\varphi}$ relative to the basis $\varphi_n(x)$, $n = 0, 1, 2, \ldots$ . Conversely, for a given vector $\boldsymbol{\varphi}$ relative to the basis $\varphi_n(x)$ we can reconstruct the wave function $\varphi(x)$ as the above superposition of the $\varphi_n(x)$. The descriptions in terms of $\varphi(x)$ and $\boldsymbol{\varphi}$ contain the same information about the state the system is in.

Thus, generally one does not distinguish the two descriptions and says the system is in the state $\varphi$, often denoted by the *ket* $|\varphi\rangle$ as introduced by Paul A. M. Dirac.

The wave function $\varphi(x)$ or the vector $\boldsymbol{\varphi}$ are considered merely as two *representations* out of which many can be invented. The same statements hold true for the representation of operators in terms of differential operators or matrices. Also these are only representations of one and the same linear transformation called linear operator. The states $\varphi$ like the wave functions or vectors $\boldsymbol{\varphi}$ form a linear vector space with scalar product. This general space is called *Hilbert space*. The linear operators transform a state of the Hilbert space into another state.

## A.3 Matrix Representation of the Harmonic Oscillator

Since the $\varphi_n(x)$ are normalized eigenfunctions of the Hamiltonian, we find for its matrix elements

$$
\begin{aligned}
H_{mn} &= \int_{-\infty}^{+\infty} \varphi_m^*(x) H \varphi_n(x)\, dx \\
&= (n + \tfrac{1}{2})\hbar\omega \int_{-\infty}^{+\infty} \varphi_m^*(x) \varphi_n(x)\, dx = (n + \tfrac{1}{2})\hbar\omega \delta_{mn} \quad .
\end{aligned}
$$

Thus, the matrix representation of the Hamiltonian of the harmonic oscillator in its eigenfunction basis is diagonal:

$$
H = \hbar\omega \begin{pmatrix} \frac{1}{2} & 0 & 0 & \cdots \\ 0 & \frac{3}{2} & 0 & \cdots \\ 0 & 0 & \frac{5}{2} & \cdots \\ \vdots & \vdots & \vdots & \ddots \end{pmatrix} \quad .
$$

The representation of the eigenfunctions $\varphi_n(x)$ in their own basis are given by the standard columns

$$
\boldsymbol{\varphi}_0 = \begin{pmatrix} 1 \\ 0 \\ 0 \\ \vdots \end{pmatrix} \quad , \qquad \boldsymbol{\varphi}_1 = \begin{pmatrix} 0 \\ 1 \\ 0 \\ \vdots \end{pmatrix} \quad , \qquad \boldsymbol{\varphi}_2 = \begin{pmatrix} 0 \\ 0 \\ 1 \\ \vdots \end{pmatrix} \quad , \qquad \cdots \quad .
$$

Of course, the eigenvector equation is recovered also in matrix representation,

$$H\varphi_n = (n + \tfrac{1}{2})\hbar\omega\varphi_n \quad .$$

With the help of the *recurrence relations for Hermite polynomials*,

$$\frac{dH_n(x)}{dx} = 2n H_{n-1}(x) \quad ,$$
$$H_{n+1}(x) = 2x H_n(x) - 2n H_{n-1}(x) \quad ,$$

we find the matrix representations for the position operator $\hat{x}$ and the momentum operator $\hat{p}$ in harmonic-oscillator representation,

$$\hat{x}\varphi_n = x\varphi_n(x) = \frac{\sigma_0}{(\sqrt{\pi}\sigma_0 2^n n!)^{1/2}} \left(n H_{n-1}(x) + \tfrac{1}{2} H_{n+1}(x)\right) \exp\left\{-\frac{x^2}{2\sigma_0^2}\right\}$$
$$= \frac{\sigma_0}{\sqrt{2}} \left(\sqrt{n}\,\varphi_{n-1}(x) + \sqrt{n+1}\,\varphi_{n+1}(x)\right) \quad .$$

The coefficients $x_{mn}$ are given by

$$x_{mn} = \int_{-\infty}^{+\infty} \varphi_m^*(x) x \varphi_n(x)\,dx = \frac{\sigma_0}{\sqrt{2}} \left(\sqrt{n}\,\delta_{m(n-1)} + \sqrt{n+1}\,\delta_{m(n+1)}\right) \quad ,$$

and the matrix representation of the position operator is

$$x = \frac{\sigma_0}{\sqrt{2}} \begin{pmatrix} 0 & 1 & 0 & 0 & \cdots \\ 1 & 0 & \sqrt{2} & 0 & \cdots \\ 0 & \sqrt{2} & 0 & \sqrt{3} & \cdots \\ 0 & 0 & \sqrt{3} & 0 & \cdots \\ \vdots & \vdots & \vdots & \vdots & \ddots \end{pmatrix} \quad .$$

For the momentum operator we obtain in a similar way the matrix representation

$$p = \frac{\hbar}{\sqrt{2}\sigma_0} \begin{pmatrix} 0 & i & 0 & 0 & \cdots \\ -i & 0 & i\sqrt{2} & 0 & \cdots \\ 0 & -i\sqrt{2} & 0 & i\sqrt{3} & \cdots \\ 0 & 0 & -i\sqrt{3} & 0 & \cdots \\ \vdots & \vdots & \vdots & \vdots & \ddots \end{pmatrix} \quad .$$

One easily verifies the *commutation relation*

$$[p, x] = px - xp = \frac{\hbar}{i}$$

also for the infinite matrix representations of $x$ and $\hat{p}$. Both matrices for $x$ and $\hat{p}$ are *Hermitean*, i.e.,

$$x_{nm}^* = x_{mn} \quad , \qquad p_{nm}^* = p_{mn} \quad .$$

## A.4 Time-Dependent Schrödinger Equation

The time dependence of wave functions is determined by the *time-dependent Schrödinger equation*

$$i\hbar \frac{\partial}{\partial t} \psi(x,t) = H\psi(x,t) \quad .$$

The eigenstates $\varphi_n(x)$ of the Hamiltonian are the space-dependent factors in an *ansatz*

$$\psi_n(x,t) = \exp\left\{-\frac{i}{\hbar}E_n t\right\}\varphi_n(x) \quad ,$$

and the eigenvalues $E_n$ determine the time dependence of the phase factor.

In the matrix representation of the harmonic oscillator the time-dependent Schrödinger equation simply reads

$$i\hbar \frac{d}{dt}\boldsymbol{\psi}(t) = H\boldsymbol{\psi}(t) \quad ,$$

where $\boldsymbol{\psi}(t)$ is a vector in Hilbert space,

$$\boldsymbol{\psi}(t) = \begin{pmatrix} \psi_0(t) \\ \psi_1(t) \\ \psi_2(t) \\ \vdots \end{pmatrix} \quad .$$

Because of the linearity of the Schrödinger equation any linear combination

$$\psi(x,t) = \sum_{n=0}^{\infty} a_n \psi_n(x,t) = \sum_{n=0}^{\infty} a_n \exp\left\{-\frac{i}{\hbar}E_n t\right\}\varphi_n(x)$$

also solves the Schrödinger equation. In vectorial representation we have

$$\boldsymbol{\psi}(t) = \sum_{n=0}^{\infty} a_n \exp\left\{-\frac{i}{\hbar}E_n t\right\}\boldsymbol{\varphi}_n \quad , \qquad E_n = (n + \tfrac{1}{2})\hbar\omega \quad .$$

The initial condition at $t = 0$ for the time-dependent Schrödinger equation is the initial wave function

$$\psi(x,0) = \psi_i(x) = \sum_{n=0}^{\infty} \psi_{in}\varphi_n(x) \quad .$$

In vector notation this is an initial state vector $\boldsymbol{\psi}_i$. Its decomposition into eigenvectors $\boldsymbol{\varphi}_n$,

$$
\begin{pmatrix} \psi_{i0} \\ \psi_{i1} \\ \psi_{i2} \\ \vdots \end{pmatrix} = \boldsymbol{\psi}_i = \sum_{n=0}^{\infty} a_n \boldsymbol{\varphi}_n = \begin{pmatrix} a_0 \\ a_1 \\ a_2 \\ \vdots \end{pmatrix} \quad ,
$$

directly provides the identification of the expansion coefficients $a_n$ with the components $\psi_{in}$ of the initial vector $\boldsymbol{\psi}_i$,

$$
a_n = \psi_{in} \quad .
$$

This way the time-dependent Schrödinger equation is solved by the expression

$$
\boldsymbol{\psi}(t) = \sum_{n=0}^{\infty} \psi_{in} \exp\left\{ -\frac{i}{\hbar} E_n t \right\} \boldsymbol{\varphi}_n
$$

for the initial condition

$$
\boldsymbol{\psi}(0) = \boldsymbol{\psi}_i \quad .
$$

The time-dependent vector $\boldsymbol{\psi}_n(t)$ corresponding to $\psi_n(x,t)$ is

$$
\boldsymbol{\psi}_n(t) = \exp\left\{ -\frac{i}{\hbar} E_n t \right\} \boldsymbol{\varphi}_n = \exp\left\{ -\frac{i}{\hbar} H t \right\} \boldsymbol{\varphi}_n \quad ,
$$

where $E_n$ is the energy eigenvalue corresponding to the eigenvector $\boldsymbol{\varphi}_n$, i.e., $E_n = (n + \frac{1}{2})\hbar\omega$ for the harmonic oscillator. The last equality is meaningful if we define the *exponential of a matrix* by its Taylor series

$$
\exp\left\{ -\frac{i}{\hbar} H t \right\} = \sum_{n=0}^{\infty} \frac{1}{n!} \left( -\frac{i}{\hbar} H t \right)^n \quad .
$$

For the case of the diagonal matrix $H$ the $n$th power is trivial,

$$
H^n = \begin{pmatrix} E_0^n & 0 & 0 & \cdots \\ 0 & E_1^n & 0 & \cdots \\ 0 & 0 & E_2^n & \cdots \\ \vdots & \vdots & \vdots & \ddots \end{pmatrix} \quad ,
$$

and the explicit matrix form is

$$
\exp\left\{ -\frac{i}{\hbar} H t \right\} = \begin{pmatrix} \exp\left\{ -\frac{i}{\hbar} E_0 t \right\} & 0 & 0 & \cdots \\ 0 & \exp\left\{ -\frac{i}{\hbar} E_1 t \right\} & 0 & \cdots \\ 0 & 0 & \exp\left\{ -\frac{i}{\hbar} E_2 t \right\} & \cdots \\ \vdots & \vdots & \vdots & \ddots \end{pmatrix} \quad .
$$

Using the operator representation of

$$\psi_n(x,t) = \exp\left\{-\frac{i}{\hbar}E_n t\right\}\varphi_n = \exp\left\{-\frac{i}{\hbar}Ht\right\}\varphi_n$$

as derived above we may rewrite $\psi(t)$ into the form

$$
\begin{aligned}
\psi(t) &= \sum_{n=0}^{\infty}\psi_{in}\exp\left\{-\frac{i}{\hbar}E_n\right\}\varphi_n \\
&= \exp\left\{-\frac{i}{\hbar}Ht\right\}\sum_{n=0}^{\infty}\psi_{in}\varphi_n \\
&= \exp\left\{-\frac{i}{\hbar}Ht\right\}\psi(0) \\
&= U_H(t)\psi(0) \quad .
\end{aligned}
$$

The operator

$$U_H(t) = \exp\left\{-\frac{i}{\hbar}Ht\right\}$$

is called the *temporal-evolution operator*.

## A.5 Probability Interpretation

The eigenfunctions $\varphi_n(x)$, equivalently the eigenvectors $\varphi_n$, describe a state of the physical system with the energy eigenvalue $E_n$. Thus, a precise measurement of the energy of this system in the state $\varphi_n$ should be devised to produce as a result the value $E_n$. In order to preserve the reproducibility of the measurement it should not change the eigenstate $\varphi_n$ of the system during the measurement, i.e., immediately after the energy measurement the state of the system should still be $\varphi_n$.

The question arises what result will be found in the same energy measurement carried out at a system in a state $\varphi$ described by the wave function $\varphi(x)$, or equivalently, by the vector $\varphi$ being a superposition of eigenfunctions $\varphi_n(x)$ or eigenvectors $\varphi_n$,

$$\varphi = \sum_{n=0}^{\infty}a_n\varphi_n \quad ,$$

with norm one, i.e.,

$$\sum_{n=0}^{\infty}|a_n|^2 = 1 \quad .$$

The single measurement of the energy will result in one of the energy eigen-values which we call $E_m$. Reproducibility of the measurement then requires that the system is in the state $\varphi_m$ after the measurement.

The absolute square $|a_m|^2$ of the coefficients $a_m$ in the superposition of the $\varphi_m$ defining $\varphi$ is the *probability* with which the energy eigenvalue $E_m$ will be determined in the single measurement.

Let us assume that we prepare a large assembly of $N$ identical systems, all in the same state $\varphi$. If we carry out single measurements on these vari-ous identical systems we shall measure the energy eigenvalue $E_m$ with the abundance $|a_m|^2 N$.

Performing a weighted average over the results of all measurements yields the *expectation value* of the energy

$$\langle E \rangle = \frac{1}{N} \sum_{n=0}^{\infty} |a_n|^2 N E_n = \sum_{n=0}^{\infty} |a_n|^2 E_n \quad .$$

Using the state-vector representation of $\varphi$,

$$\varphi = \begin{pmatrix} a_0 \\ a_1 \\ a_2 \\ \vdots \end{pmatrix} ,$$

we find that the energy expectation value is simply

$$\varphi^+ H \varphi = (a_0^*, a_1^*, a_2^*, \ldots) \begin{pmatrix} E_0 & 0 & 0 & \cdots \\ 0 & E_1 & 0 & \cdots \\ 0 & 0 & E_2 & \cdots \\ \vdots & \vdots & \vdots & \ddots \end{pmatrix} \begin{pmatrix} a_0 \\ a_1 \\ a_2 \\ \vdots \end{pmatrix}$$

$$= \sum_{n=0}^{\infty} a_n^* E_n a_n = \langle E \rangle \quad .$$

Equivalently, in wave-function formulation, we have

$$\int_{-\infty}^{+\infty} \varphi^*(x) H \varphi(x) \, dx = \int_{-\infty}^{+\infty} \varphi^*(x) \sum_{n=0}^{\infty} E_n a_n \varphi_n(x)$$

$$= \sum_{n=0}^{\infty} E_n a_n \int_{-\infty}^{+\infty} \varphi^*(x) \varphi_n(x) \, dx$$

$$= \sum_{n=0}^{\infty} E_n a_n a_n^* = \langle E \rangle \quad .$$

# B. Two-Level System

In Appendix A the equivalence of wave-function and matrix representation of quantum mechanics was shown. The simplest matrix structure is the one in two dimensions, i.e., in a space with two *base states*:

$$\eta_1 = \begin{pmatrix} 1 \\ 0 \end{pmatrix} \quad , \qquad \eta_{-1} = \begin{pmatrix} 0 \\ 1 \end{pmatrix} \quad .$$

The linear space consists of all linear combinations

$$\chi = \chi_1 \eta_1 + \chi_{-1} \eta_{-1} = \begin{pmatrix} \chi_1 \\ \chi_{-1} \end{pmatrix}$$

of the base states with complex coefficients $\chi_1$ and $\chi_{-1}$. The two states $\eta_1$ and $\eta_{-1}$ form an *orthonormal basis* of this space, i.e.,

$$\eta_1^+ \cdot \eta_1 = 1 \quad , \qquad \eta_{-1}^+ \cdot \eta_{-1} = 1 \quad , \qquad \eta_1^+ \cdot \eta_{-1} = \eta_{-1}^+ \cdot \eta_1 = 0 \quad .$$

For the linear combination $\chi$ to be normalized to one we have

$$\chi^+ \cdot \chi = \chi_1^* \chi_1 + \chi_{-1}^* \chi_{-1} = |\chi_1|^2 + |\chi_{-1}|^2 = 1 \quad .$$

This suggests a representation of the absolute values $|\chi_r|$, $r = 1, -1$, of the complex coefficients by trigonometric functions:

$$|\chi_1| = \cos\frac{\Theta}{2} \quad , \qquad |\chi_{-1}| = \sin\frac{\Theta}{2} \quad .$$

The use of the half-angle $\Theta/2$ is a convention, the usefulness of which will become obvious in the sequel. The complex coefficients themselves are obtained by multiplication of the moduli $|\chi_r|$ with arbitrary phase factors:

$$\chi_1 = e^{-i\Phi_1/2} \cos\frac{\Theta}{2} \quad , \qquad \chi_{-1} = e^{-i\Phi_{-1}/2} \sin\frac{\Theta}{2} \quad .$$

Since a common phase factor is irrelevant the general form can be restricted to

S. Brandt and H.D. Dahmen, *The Picture Book of Quantum Mechanics*,
DOI 10.1007/978-1-4614-3951-6, © Springer Science+Business Media New York 2012

$$\chi_1 = e^{-i\Phi/2} \cos\frac{\Theta}{2} \quad , \qquad \chi_{-1} = e^{i\Phi/2} \sin\frac{\Theta}{2}$$

with

$$\Phi = (\Phi_1 - \Phi_{-1})/2 \quad .$$

The general linear combination is therefore

$$\chi(\Theta,\Phi) = e^{-i\Phi/2}\cos\frac{\Theta}{2}\eta_1 + e^{i\Phi/2}\sin\frac{\Theta}{2}\eta_{-1} = \begin{pmatrix} e^{-i\Phi/2}\cos\frac{\Theta}{2} \\ e^{i\Phi/2}\sin\frac{\Theta}{2} \end{pmatrix} \quad .$$

The operators corresponding to physical quantities are *Hermitean matrices*

$$A = \begin{pmatrix} A_{1,1} & A_{1,-1} \\ A_{-1,1} & A_{-1,-1} \end{pmatrix} \quad .$$

The *Hermitean conjugate* of $A$ is defined as

$$A^+ = \begin{pmatrix} A_{1,1}^* & A_{-1,1}^* \\ A_{1,-1}^* & A_{-1,-1}^* \end{pmatrix} \quad , \qquad \text{i.e.,} \qquad A_{rs}^+ = A_{sr}^* \quad .$$

The condition of Hermiticity,

$$A^+ = A \quad , \qquad \text{i.e.,} \qquad A_{sr}^* = A_{rs} \quad ,$$

requires

$$A_{1,1}^* = A_{1,1} \quad , \qquad A_{1,-1}^* = A_{-1,1} \quad ,$$
$$A_{-1,1}^* = A_{1,-1} \quad , \qquad A_{-1,-1}^* = A_{-1,-1} \quad .$$

Thus, the diagonal elements $A_{1,1}$, $A_{-1,-1}$ are real quantities, the off-diagonal elements $A_{1,-1}$, $A_{-1,1}$ are complex conjugates of each other. Hermiticity of the operator $A$ ensures that the *expectation value* of $A$ for a given general state is real,

$$
\begin{aligned}
\chi^+ A \chi &= \sum_{i,j=1,-1} \chi_i^* A_{ij} \chi_j \\
&= \chi_1^* A_{1,1} \chi_1 + \chi_1^* A_{1,-1} \chi_{-1} + \chi_{-1}^* A_{-1,1}^* \chi_1 + \chi_{-1}^* A_{-1,-1} \chi_{-1} \quad .
\end{aligned}
$$

All Hermitean matrices can be linearly combined as the superpositions

$$A = a_0 \sigma_0 + a_1 \sigma_1 + a_2 \sigma_2 + a_3 \sigma_3$$

(with real coefficients $a_0$, ..., $a_3$) of the unit matrix

$$\sigma_0 = \begin{pmatrix} 1 & 0 \\ 0 & 1 \end{pmatrix}$$

and the *Pauli matrices*

$$\sigma_1 = \begin{pmatrix} 0 & 1 \\ 1 & 0 \end{pmatrix} \quad , \quad \sigma_2 = \begin{pmatrix} 0 & -i \\ i & 0 \end{pmatrix} \quad , \quad \sigma_3 = \begin{pmatrix} 1 & 0 \\ 0 & -1 \end{pmatrix} \quad ,$$

since the four matrices $\sigma_0, \ldots, \sigma_3$ are Hermitean. One directly verifies the relations

$$\sigma_i^2 = \sigma_0 \quad , \qquad i = 0, 1, 2, 3 \quad ,$$

and

$$\sigma_1 \sigma_2 = i\sigma_3 \quad , \qquad \sigma_2 \sigma_3 = i\sigma_1 \quad , \qquad \sigma_3 \sigma_1 = i\sigma_2 \quad .$$

These yield the *commutation relations*

$$[\sigma_1, \sigma_2] = \sigma_1 \sigma_2 - \sigma_2 \sigma_1 = 2i\sigma_3$$

and cyclic permutations.

The three Pauli matrices can be grouped into a *vector* in three dimensions,

$$\boldsymbol{\sigma} = (\sigma_1, \sigma_2, \sigma_3) \quad ,$$

with the square

$$\boldsymbol{\sigma}^2 = \sigma_1^2 + \sigma_2^2 + \sigma_3^2 = 3\sigma_0^2 \quad .$$

The base states $\boldsymbol{\eta}_1, \boldsymbol{\eta}_{-1}$ are eigenstates of the Pauli matrix $\sigma_3$ and of the sum of their squares $\boldsymbol{\sigma}^2$,

$$\sigma_3 \boldsymbol{\eta}_r = r\boldsymbol{\eta}_r \quad , \qquad \boldsymbol{\sigma}^2 \boldsymbol{\eta}_r = 3\sigma_0 \boldsymbol{\eta}_r = 3\boldsymbol{\eta}_r \quad , \qquad r = 1, -1 \quad ,$$

since $\sigma_3$ and $\sigma_0$ are diagonal matrices.

According to Section A the *time-dependent Schrödinger equation* reads

$$i\hbar \frac{d}{dt} \boldsymbol{\xi}(t) = H\boldsymbol{\xi}(t) \quad ,$$

where the Hamiltonian is a Hermitean $2 \times 2$ matrix,

$$H = \begin{pmatrix} H_{1,1} & H_{1,-1} \\ H_{-1,1} & H_{-1,-1} \end{pmatrix} \quad ,$$

with real diagonal matrix elements $H_{1,1}$, $H_{-1,-1}$ and with off-diagonal elements $H_{-1,1} = H_{1,-1}^*$. It can be represented by a superposition of the $\sigma$ matrices,

$$H = h_0 \sigma_0 + h_3 \sigma_3 + h_1 \sigma_1 + h_2 \sigma_2 \quad ,$$

where

$$h_0 = \frac{1}{2}(H_{1,1} + H_{-1,-1}) \quad , \qquad h_3 = \frac{1}{2}(H_{1,1} - H_{-1,-1}) \quad ,$$

and

$$h_1 = \frac{1}{2}(H_{1,-1} + H_{-1,1}) = \operatorname{Re} H_{1,-1} \quad,$$

$$h_2 = \frac{i}{2}(H_{1,-1} - H_{-1,1}) = -\operatorname{Im} H_{1,-1} \quad.$$

Introducing the $h_i$ ($i = 0, 1, 2, 3$) into the matrix $H$ we obtain

$$H = \begin{pmatrix} h_0 + h_3 & h_1 - ih_2 \\ h_1 + ih_2 & h_0 - h_3 \end{pmatrix} \quad.$$

Introducing the factorization

$$\xi_r(t) = \exp\left\{-\frac{i}{\hbar} E_r t\right\} \chi_r \quad, \qquad r = 1, -1 \quad,$$

into the time-dependent phase factor and the stationary state $\chi_r$ we obtain the *stationary Schrödinger equation*

$$H \chi_r = E_r \chi_r \quad, \qquad r = 1, -1 \quad,$$

for the *eigenstate* $\chi_r$ belonging to the *energy eigenvalue* $E_r$. For the eigenvalues we find

$$E_{\pm 1} = h_0 \pm |\mathbf{h}| \quad, \qquad |\mathbf{h}| = \sqrt{h_1^2 + h_2^2 + h_3^2} \quad.$$

Since there are only two eigenvalues our system is called a *two-level system*.

The eigenstates are

$$\chi_1 = \frac{1}{\sqrt{2|\mathbf{h}|}} \begin{pmatrix} \sqrt{|\mathbf{h}| + h_3} \; e^{-i\Phi/2} \\ \sqrt{|\mathbf{h}| - h_3} \; e^{i\Phi/2} \end{pmatrix} \quad,$$

$$\chi_{-1} = \frac{1}{\sqrt{2|\mathbf{h}|}} \begin{pmatrix} -\sqrt{|\mathbf{h}| - h_3} \; e^{-i\Phi/2} \\ \sqrt{|\mathbf{h}| + h_3} \; e^{i\Phi/2} \end{pmatrix} \quad,$$

with the phase factor determined by

$$e^{i2\Phi} = \frac{h_1 + ih_2}{h_1 - ih_2} \quad.$$

Introducing the angle $\Theta$ by

$$\cos\frac{\Theta}{2} = \sqrt{\frac{|\mathbf{h}| + h_3}{2|\mathbf{h}|}} \quad, \qquad \sin\frac{\Theta}{2} = \sqrt{\frac{|\mathbf{h}| - h_3}{2|\mathbf{h}|}}$$

we may write the eigenstates in the form

$$\chi_1 = e^{-i\Phi/2}\cos\frac{\Theta}{2}\eta_1 + e^{i\Phi/2}\sin\frac{\Theta}{2}\eta_{-1} \quad,$$

$$\chi_{-1} = -e^{-i\Phi/2}\sin\frac{\Theta}{2}\eta_1 + e^{i\Phi/2}\cos\frac{\Theta}{2}\eta_{-1} \quad.$$

They are normalized and orthogonal to each other.

The eigenstates $\chi_1$, $\chi_{-1}$ of the two-level system exhibit a time dependence which is given by a phase factor only,

$$\xi_r(t) = \exp\left\{-\frac{i}{\hbar}E_r t\right\}\chi_r \quad, \qquad r = 1, -1 \quad.$$

If initially the system is not in an eigenstate the state oscillates. We assume that the initial state is

$$\varphi(0) = \eta_{-1} \quad.$$

Decomposition into the eigenstates yields

$$\eta_{-1} = \zeta_1\chi_1 + \zeta_{-1}\chi_{-1}$$

with

$$\zeta_1 = \chi_1^{+}\cdot\eta_{-1} = e^{-i\Phi/2}\sin\frac{\Theta}{2} \quad,$$

$$\zeta_{-1} = \chi_{-1}^{+}\cdot\eta_{-1} = e^{-i\Phi/2}\cos\frac{\Theta}{2} \quad.$$

The time-dependent state is obtained as

$$\varphi(t) = \zeta_1\xi_1(t) + \zeta_{-1}\xi_{-1}(t)$$

$$= e^{-i\Phi/2}\sin\frac{\Theta}{2}e^{-i\omega_1 t}\chi_1 + e^{-i\Phi/2}\cos\frac{\Theta}{2}e^{-i\omega_{-1}t}\chi_{-1}$$

with the angular frequencies

$$\omega_r = E_r/\hbar \quad, \qquad r = 1, -1 \quad.$$

The probability to find the system (originally in the state $\eta_{-1}$) in the state $\eta_1$ is given by

$$P_{1,-1} = \sin^2\Theta\,\sin^2\frac{|h|}{\hbar}t \quad,$$

and, of course, the probability to find it in the state $\eta_{-1}$ is

$$P_{-1,-1} = 1 - P_{1,-1} \quad.$$

# C. Analyzing Amplitude

## C.1 Classical Considerations: Phase-Space Analysis

We consider a detector which is capable of measuring position and momentum of a particle simultaneously with certain accuracies. If the result of a measurement is the pair $x_D$, $p_D$ of values we may assume that the true values $x$, $p$ of the quantities to be measured are described by the uncorrelated bivariate Gaussian (cf. Section 3.5) probability density

$$\rho_D(x,p,x_D,p_D) = \frac{1}{2\pi\,\sigma_{xD}\sigma_{pD}} \exp\left\{-\frac{1}{2}\left[\frac{(x-x_D)^2}{\sigma_{xD}^2} + \frac{(p-p_D)^2}{\sigma_{pD}^2}\right]\right\} \quad .$$

That is to say, the probability for the true values $x$, $p$ of the particle to be the intervals between $x$ and $x+dx$ and between $p$ and $p+dp$ is

$$dP = \rho_D(x,p)\,dx\,dp \quad .$$

The particle to be measured by the detector possesses position and momentum values $x$ and $p$. The particle may have been produced by a source which does not define exactly the values of $x$ and $p$ but according to a probability density

$$\rho_S(x,p,x_S,p_S) = \frac{1}{2\pi\,\sigma_{xS}\sigma_{pS}} \exp\left\{-\frac{1}{2}\left[\frac{(x-x_S)^2}{\sigma_{xS}^2} + \frac{(p-p_S)^2}{\sigma_{pS}^2}\right]\right\} \quad .$$

This is an uncorrelated bivariate Gaussian probability density with the expectation values $x_S$ and $p_S$ and the variances $\sigma_{xS}^2$ and $\sigma_{pS}^2$.

We now describe how much information can at best be obtained about the probability density $\rho_S(x,p)$ using the above detector. The probability for a particle prepared by the source to be detected within the intervals $(x_D, x_D + dx_D)$ and $(p_D, p_D + dp_D)$ is given by

$$dP = w^{cl}(x_D,p_D,x_S,p_S)\,dx_D\,dp_D$$

with the probability density in phase space

S. Brandt and H.D. Dahmen, *The Picture Book of Quantum Mechanics*,
DOI 10.1007/978-1-4614-3951-6, © Springer Science+Business Media New York 2012

$$w^{\text{cl}}(x_{\text{D}}, p_{\text{D}}, x_{\text{S}}, p_{\text{S}})$$

$$= \int_{-\infty}^{+\infty} \int_{-\infty}^{+\infty} \rho_{\text{D}}^{\text{cl}}(x, p, x_{\text{D}}, p_{\text{D}}) \rho_{\text{S}}^{\text{cl}}(x, p, x_{\text{S}}, p_{\text{S}}) \, dx \, dp$$

$$= \frac{1}{2\pi \sigma_x \sigma_p} \exp\left\{ -\frac{1}{2} \left[ \frac{(x_{\text{D}} - x_{\text{S}})^2}{\sigma_x^2} + \frac{(p_{\text{D}} - p_{\text{S}})^2}{\sigma_p^2} \right] \right\} .$$

Here the variances $\sigma_x^2$ and $\sigma_p^2$ are obtained by summing up the variances of the detector and source distribution,

$$\sigma_x^2 = \sigma_{x\text{D}}^2 + \sigma_{x\text{S}}^2 \quad , \qquad \sigma_p^2 = \sigma_{p\text{D}}^2 + \sigma_{p\text{S}}^2 \quad .$$

The quantity $w^{\text{cl}}(x_{\text{D}}, p_{\text{D}}, x_{\text{S}}, p_{\text{S}})$ is the result of analyzing the phase-space probability density of the source $\rho_{\text{S}}(x, p, x_{\text{S}}, p_{\text{S}})$ with the help of the phase-space probability density $\rho_{\text{D}}(x, p, x_{\text{D}}, p_{\text{D}})$. The function $w^{\text{cl}}(x_{\text{D}}, p_{\text{D}}, x_{\text{S}}, p_{\text{S}})$ is itself a phase-space probability density and obtained through a process we shall call *phase-space analysis*.

This distribution can be measured in principle if the source consecutively produces a large number of particles which are observed in the detector. For a detector of arbitrarily high precision,

$$\sigma_{x\text{D}} \to 0 \quad , \qquad \sigma_{p\text{D}} \to 0 \quad ,$$

the distribution $w^{\text{cl}}$ approaches the source distribution $\rho_{\text{S}}(x = x_{\text{D}}, p = p_{\text{D}}, x_{\text{S}}, p_{\text{S}})$.

We have seen that with a detector of high precision and with a sufficiently high number of measurements the source distribution can be measured with arbitrary high accuracy. We now assume that the minimum-uncertainty relations,

$$\sigma_{x\text{D}} \sigma_{p\text{D}} = \frac{\hbar}{2} \quad , \qquad \sigma_{x\text{S}} \sigma_{p\text{S}} = \frac{\hbar}{2} \quad ,$$

hold for the widths characterizing the detector and the source. Besides this restriction we stay within the framework of classical physics. Now it is no longer possible to measure the source distribution exactly.[1] However, we may still measure the distribution in position alone or the distribution in momentum alone with arbitrary accuracy. To show this we construct the marginal distributions of $w^{\text{cl}}$ in the variables $x_{\text{D}} - x_{\text{S}}$, and $p_{\text{D}} - p_{\text{S}}$, respectively,

$$w_x^{\text{cl}}(x_{\text{D}}, x_{\text{S}}) = \frac{1}{\sqrt{2\pi} \sigma_x} \exp\left\{ -\frac{1}{2} \frac{(x_{\text{D}} - x_{\text{S}})^2}{\sigma_x^2} \right\}$$

and

---

[1] We could, however, compute the source distribution $\rho_{\text{S}}$ by unfolding it from $w^{\text{cl}}$.

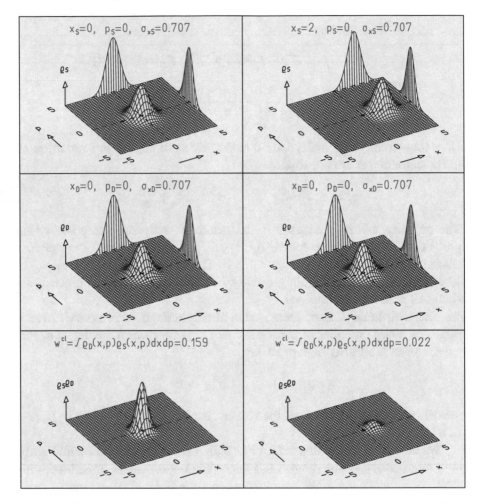

**Fig. C.1.** Phase-space distributions $\rho_S$ **(top)**, $\rho_D$ **(middle), and their product $\rho_S\rho_D$ (bottom) together with the marginal distributions of $\rho_S$ and $\rho_D$. The two columns differ only in the spatial mean $x_S$ of $\rho_S$. Units are used in which $\hbar = 1$.**

$$w_p^{cl}(p_D, p_S) = \frac{1}{\sqrt{2\pi}\,\sigma_p} \exp\left\{-\frac{1}{2}\frac{(p_D - p_S)^2}{\sigma_p^2}\right\} \quad .$$

The first distribution approaches the corresponding marginal position distribution of the source,

$$\rho_{Sx}(x, x_S) = \int_{-\infty}^{+\infty} \rho_S(x, p, x_S, p_S)\,dp = \frac{1}{\sqrt{2\pi}\,\sigma_{xS}} \exp\left\{-\frac{1}{2}\frac{(x - x_S)^2}{\sigma_{xS}^2}\right\}$$

in the case of $\sigma_{xD} \to 0$. However, because of the minimum uncertainty relations, $\sigma_{pD}$ as well as $\sigma_p$ approach infinity. Therefore, the second distribution

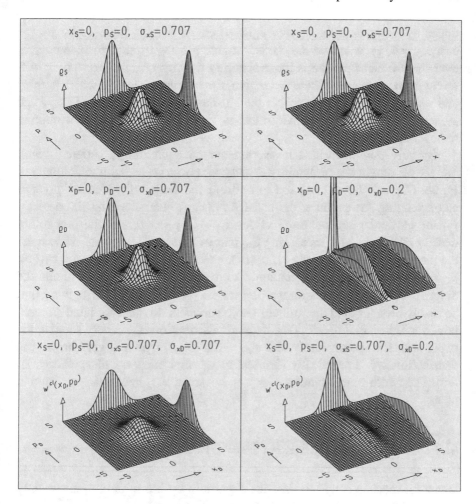

**Fig. C.2.** The phase-space distribution $\rho_S$ (top), the distribution $\rho_D$ for a particular point $(x_D, p_D)$ of mean values (middle), and convolution of $\rho_S$ with $\rho_D$ for all possible mean values (bottom). Also shown are the marginal distributions. The two columns differ in the widths of $\rho_S$. Units $\hbar = 1$ are used.

$w_p^{cl}$ becomes so wide – and actually approaches zero – that no information about the momentum distribution can be obtained from it.

Conversely, for $\sigma_{pD} \to 0$ the momentum distribution of the source can be measured accurately. However, then the information about the position distribution is lost.

We illustrate the concepts of this section in Figures C.1 and C.2. We begin with the discussion of the result of a single measurement, yielding the pair $x_D$, $p_D$ of measured values. In the two columns of Figure C.1 we show (from top to bottom) the probability density $\rho_S(x, p)$ characterizing the particle as produced by the source, the density $\rho_D(x, p)$ characterizing the detector for the case $x_D = 0$, $p_D = 0$, and the product function $\rho_S(x, p)\rho_D(x, p)$. The integral over the product function is the probability density $w^{cl}$. It is essentially different from zero only if there is a region, the overlap region, in which both $\rho_S$ and $\rho_D$ are different from zero. In the left-hand column of Figure C.1 $\rho_S$ and $\rho_D$ were chosen to be identical, so that $w^{cl}$ is large. In the right-hand column the overlap is smaller.

By very many repeated measurements, each yielding a different result $x_D$, $p_D$ we obtain the probability density $w^{cl}(x_D, p_D)$. In the two columns of Figure C.2 we show (from top to bottom) the probability density $\rho_S(x, p)$ characterizing the particle, the density $\rho_D(x, p)$ characterizing the detector for the particular set measured values $x_D = 0$, $p_D = 0$, and the probability density $w^{cl}(x_D, p_D)$ for measuring the pair of values $x_D$, $p_D$. Also shown are the marginal distribution $\rho_{Sx}(x)$, $\rho_{Dx}(x)$, and $w_x^{cl}(x_D)$ in position, and $\rho_{Sp}(p)$, $\rho_{Dp}(p)$, and $w_p^{cl}(p_D)$ in momentum. Comparing in the left-hand column the diagram of $\rho_S$ with the diagram of $w^{cl}$ we see that the latter distribution is appreciably broader than the former in both variables. In the right-hand column, however, the spatial width of the detector distribution $\sigma_{xD}$ is very small at the expense of the momentum width $\sigma_{pD} = \hbar/(2\sigma_{xD})$, which is very large. The distribution $w^{cl}$ is practically identical to $\rho_S$ what concerns its spatial variation. The width in momentum of $w^{cl}$ is, however, very much larger than that of $\rho_S$.

## C.2 Analyzing Amplitude: Free Particle

Quantum-mechanically we describe a particle by the minimum-uncertainty wave packet

$$\varphi_S(x) = \varphi_S(x, x_S, p_S) = \frac{1}{(2\pi)^{1/4}\sigma_{xS}^{1/2}} \exp\left\{-\frac{(x - x_S)^2}{4\sigma_{xS}^2} + \frac{i}{\hbar}p_S(x - x_S)\right\}.$$

We consider this wave packet as having been prepared by some physical apparatus, the source. The question now arises how the phase-space analysis of the particle as discussed in the last section can be described in quantum mechanics.

If, in a particle detector with position-measurement uncertainty $\sigma_{xD}$ and momentum-measurement uncertainty $\sigma_{pD} = \hbar/(2\sigma_{xD})$, the values $x_D$, $p_D$ are measured, we want to interpret the result as in Section C.1. The same probability density

$$\rho_D(x, p, x_D, p_D) = \frac{1}{2\pi\sigma_{xD}\sigma_{pD}} \exp\left\{-\frac{1}{2}\left[\frac{(x - x_D)^2}{\sigma_{xD}^2} + \frac{(p - p_D)^2}{\sigma_{pD}^2}\right]\right\}$$

describes the probability density of position $x$ and momentum $p$ of the particle. Quantum-mechanically this probability density is the phase-space distribution introduced by Eugene P. Wigner in 1932 of the wave packet

$$\varphi_D(x, x_D, p_D) = \frac{1}{(2\pi)^{1/4}\sigma_{xD}^{1/2}} \exp\left\{-\frac{(x-x_D)^2}{4\sigma_{xD}^2} + \frac{i}{\hbar}p_D(x-x_D)\right\} \quad.$$

Therefore, $\rho_D(x, p, x_D, p_D)$ is also called *Wigner distribution* of $\varphi_D$ (cf. Appendix D).

Let us construct the *analyzing amplitude*

$$a(x_D, p_D, x_S, p_S) = \frac{1}{\sqrt{h}} \int_{-\infty}^{+\infty} \varphi_D^*(x, x_D, p_D)\varphi_S(x, x_S, p_S)\,dx$$

representing the *overlap* between the particle's wave function $\varphi_S$ and the detecting wave function $\varphi_D$. It turns out to be

$$\begin{aligned} a(x_D, p_D, x_S, p_S) &= \frac{1}{\sqrt{2\pi\sigma_x\sigma_p}} \exp\left\{-\frac{(x_D-x_S)^2}{4\sigma_x^2} - \frac{(p_D-p_S)^2}{4\sigma_p^2}\right. \\ &\quad \left.- \frac{i}{\hbar}\frac{\sigma_{xD}^2 p_D + \sigma_{xS}^2 p_S}{\sigma_x^2}(x_D - x_S)\right\} \quad, \end{aligned}$$

where, like in Section C.1,

$$\sigma_x^2 = \sigma_{xD}^2 + \sigma_{xS}^2 \quad, \qquad \sigma_p^2 = \sigma_{pD}^2 + \sigma_{pS}^2 \quad.$$

The absolute square of the analyzing amplitude

$$\begin{aligned} |a|^2 &= \frac{1}{2\pi\sigma_x\sigma_p} \exp\left\{-\frac{1}{2}\left[\frac{(x_D-x_S)^2}{\sigma_x^2} + \frac{(p_D-p_S)^2}{\sigma_p^2}\right]\right\} \\ &= w^{cl}(x_D, p_D, x_S, p_S) \end{aligned}$$

is identical to the probability density $w^{cl}(x_D, p_D, x_S, p_S)$ of Section C.1.

We conclude that the probability amplitude analyzing the values $x_D$ and $p_D$ of position and momentum of a particle as a result of the interaction of a particle with a detector is given by

$$a(x_D, p_D, x_S, p_S) = \frac{1}{\sqrt{h}} \int_{-\infty}^{+\infty} \varphi_D^*(x, x_D, p_D)\varphi_S(x, x_S, p_S)\,dx \quad.$$

Here, $\varphi_S$ is the wave function of the particle and $\varphi_D$ the *analyzing wave function*. The probability to observe a position in the interval between $p_D$ and $p_D + dp_D$ is

$$dP = |a(x_D, p_D, x_S, p_S)|^2\, dx_D\, dp_D \quad.$$

In analogy to the classical case we may now ask whether we can still recover the original quantum-mechanical spatial probability density

$$\rho_S(x) = |\varphi_S(x)|^2 = \frac{1}{\sqrt{2\pi}\,\sigma_{xS}}\exp\left\{-\frac{1}{2}\frac{(x-x_S)^2}{\sigma_{xS}^2}\right\}$$

from $|a|^2$. Information about the position of the particle only is obtained by integrating $|a|^2$ over all values of $p_D$, i.e., by forming the marginal distribution with respect to $x_D$,

$$|a|_x^2 = \int_{-\infty}^{+\infty}|a|^2\,dp_D = w_x^{cl}(x_D, x_S) \quad .$$

The result is the same as in the classical case. Again in the limit $\sigma_{xD} \to 0$ we find that the function $|a|_x^2$ approaches the quantum-mechanical probability density $\rho_S(x)$ which is equal to the classical distribution $\rho_{Sx}(x)$.

In Figures C.3 and C.4 we demonstrate the construction of the analyzing amplitude using particular numerical examples. Each column of three plots in the two figures is one example. At the top of the column the particle wave function is shown as two curves depicting $\mathrm{Re}\,\varphi_S(x)$ and $\mathrm{Im}\,\varphi_S(x)$ together with the numerical values of the parameters $x_S$, $p_S$, $\sigma_{xS}$ which define $\varphi(x)$. Likewise, the middle plot shows the detector wave function, given by $\mathrm{Re}\,\varphi_D(x)$ and $\mathrm{Im}\,\varphi_D(x)$. The bottom plot contains the real and imaginary parts of the product function

$$\varphi_D^*(x)\varphi_S(x) \quad ,$$

which after integration and absolute squaring yields the probability density

$$|a|^2 = \frac{1}{h}\left|\int_{-\infty}^{+\infty}\varphi_D^*(x)\varphi_S(x)\,dx\right|^2$$

of detection. Also given in the bottom plot is the numerical value of $|a|^2$. Four different situations are shown in the two figures. In each case the same detector function $\varphi_D$ is used. Only the particle wave function $\varphi_S$ changes from case to case.

(i) In the left-hand column of Figure C.3 $\varphi_S$ and $\varphi_D$ are identical. For that case we know that the overlap integral is explicitly real and that $\int_{-\infty}^{+\infty}\varphi_S^*\varphi_S\,dx = 1$, so that $|a|^2 = 1/h = 1/2\pi$ in the units $\hbar = 1$ used.

(ii) In the right-hand column of Figure C.3 the particle wave packet is moved to a position expectation value $x_S \neq x_D$, but we still have $p_S = p_D$, $\sigma_{xS} = \sigma_{xD}$. By construction, the overlap function is different from zero in that $x$ region where both $\varphi_S$ and $\varphi_D$ are sizably different from zero. As expected, the value of $|a|^2$ is considerably smaller than in case (i).

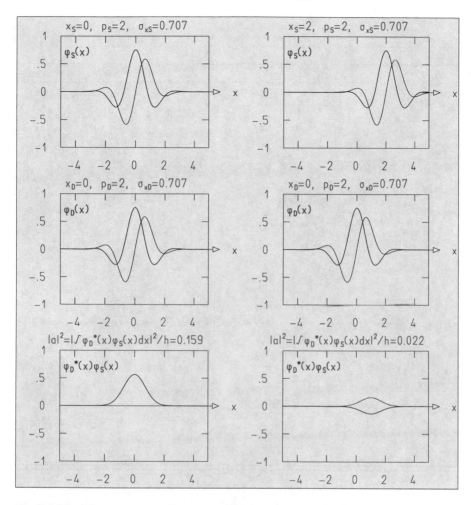

**Fig. C.3. Wave function $\varphi_S$ (top) and $\varphi_D$ (middle) and the product function $\varphi_D^*\varphi_S$ (bottom). Real parts are drawn as thick lines, imaginary parts as thin lines. The two columns differ in the mean value $x_S$ of $\varphi_S$. Units $\hbar = 1$ are used.**

(iii) In the left-hand column of Figure C.4 the position expectation values and the widths of particle and detector wave function are identical, $x_S = x_D$, $\sigma_{xS} = \sigma_{xD}$, but the momentum expectation values differ, $p_S \neq p_D$. As in case (i) the product function $\varphi_D^*\varphi_S$ is different from zero in the region $x \approx x_0$ but due to the different momentum expectation values it oscillates. Therefore, the value of $|a|^2$ is much smaller than in case (i) since positive and negative regions of the product function nearly cancel when the integration is performed.

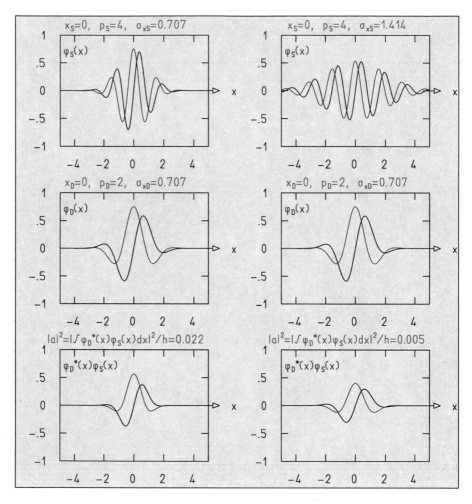

**Fig. C.4. As Figure C.3 but for different functions $\varphi_S$. The two columns differ only in the value of $\sigma_{xS}$.**

(iv) In the right-hand column of Figure C.4 the particle wave packet has a larger width $\sigma_{xS} > \sigma_{xD}$. All other parameters are as in case (iii). The product function is similar to that for case (iii) and is concentrated in the region $x_D - \sigma_{xD} \leq x \leq x_D + \sigma_{xD}$ where both wave functions are appreciably different from zero. However, the amplitude of the product function is smaller than in case (iii) since the amplitude of $\varphi_S(x)$ is smaller in the overlap region. Therefore, the value of $|a|^2$ is also smaller.

## C.3 Analyzing Amplitude: General Case

The lesson learned in the last section can be generalized to the analysis of an arbitrary normalized wave function $\varphi(x)$ describing a single particle in terms of an arbitrary complete or overcomplete set of normalized wave functions $\varphi(x)$ or $\varphi(x, q_1, \ldots, q_N)$. The functions $\varphi_n(x)$ can in particular be eigenfunctions of a Hermitean operator, e.g., the energy. Examples for a set of overcomplete functions $\varphi(x, q_1, \ldots, q_n)$ are

- free wave packets $\varphi_D(x, x_D, p_D)$ as in the last section,

- coherent states of the harmonic oscillator $\varphi(x, x_0, p_0)$ as we shall study in detail in the next section, and

- minimum-uncertainty states of a set of noncommuting operators like the operators $L_x$, $L_y$, $L_z$ of angular momentum or $S_x$, $S_y$, $S_z$ of spin as investigated in Sections 10.5 and 17.2.

The analyzing amplitude for the different cases is given by

$$a = \frac{1}{N_1} \int_{-\infty}^{+\infty} \varphi_n^*(x)\varphi(x)\,dx$$

or

$$a = \frac{1}{N_2} \int_{-\infty}^{+\infty} \varphi^*(x, q_1, \ldots, q_N)\varphi(x)\,dx \quad .$$

Of course, a mutual analysis of two sets of analyzing functions is also of interest, e.g.,

$$a = \frac{1}{N_3} \int_{-\infty}^{+\infty} \varphi_n^*(x)\varphi(x, q_1, \ldots, q_N)\,dx \quad .$$

The normalization constants have to be individually determined for every type of analyzing amplitude.

## C.4 Analyzing Amplitude: Harmonic Oscillator

For the harmonic oscillator of frequency $\omega$ we have discussed in Sections 6.3 and 6.4 two sets of states in particular:

(i) The eigenstates $\varphi_n$ corresponding to the energy eigenvalues $E_n = (n + \frac{1}{2})\hbar\omega$,

$$\varphi_n(x) = (\sqrt{2\pi}\,2^n n!\sigma_0)^{-1/2} H_n\left(\frac{x}{\sigma_0}\right) \exp\left\{-\frac{x^2}{2\sigma_0}\right\} \quad ,$$

with the ground-state width

$$\sigma_0 = \sqrt{2}\sigma_x \quad , \qquad \sigma_x = \sqrt{\frac{\hbar}{2m\omega}} \quad .$$

Plots of the $\varphi_n$ are shown in Figure 6.5.

(ii) The coherent states,

$$\psi(x,t,x_0,p_0) = \sum_{m=0}^{\infty} a_m(x_0,p_0)\varphi_m(x)\exp\left\{-\frac{\mathrm{i}}{\hbar}E_m t\right\} \quad ,$$

where the complex coefficients $a_n$ are given by

$$a_n(x_0,p_0) = \frac{z^n}{\sqrt{n!}}\exp\left\{-\frac{1}{2}z^*z\right\} \quad , \qquad n = 0, 1, 2,\ldots \quad .$$

The variable $z$ is complex and a dimensionless linear combination of the initial expectation values $x_0$ of position and $p_0$ of momentum,

$$z = \frac{x_0}{2\sigma_x} + \mathrm{i}\frac{p_0}{2\sigma_p} \quad , \qquad \sigma_p = \frac{\hbar}{2\sigma_x} \quad .$$

Plots of the coherent states $\psi(x,t)$ are shown in Figure 6.6c.

The set of energy eigenfunctions $\varphi_n(x)$ is complete, the set of coherent states is overcomplete. We can form four kinds of analyzing amplitudes.

## Eigenstate – Eigenstate Analyzing Amplitude

We analyze the energy eigenfunctions using energy eigenfunctions as analyzing wave functions. Thus, we obtain as analyzing amplitude

$$a_{mn} = \int_{-\infty}^{+\infty} \varphi_m(x)\varphi_n(x)\,\mathrm{d}x = \delta_{mn} \quad ,$$

which yields as probability

$$a_{mn}^2 = \delta_{mn} \quad .$$

This result, based on the orthonormality of the eigenfunctions $\varphi_n(x)$, is illustrated in Figure C.5 which shows the functions $\varphi_m\varphi_n$. Whereas $\varphi_m^2$ is non-negative everywhere so that the integral over $\varphi_m^2$ cannot vanish it is qualitatively clear from the figure that the integral over $\varphi_m\varphi_n$ vanishes for $m \neq n$.

The analysis of an eigenstate $\varphi_n(x)$ with all eigenstates $\varphi_m(x)$ thus yields with probability $a_{mn}^2 = 1$ the answer that the original wave function was indeed $\varphi_n$, and with probability $a_{mn} = 0$ the result that the original wave function was $\varphi_m$ with $m \neq n$. Such an analysis can also be considered as an energy determination which with certainty yields the energy eigenvalue $E_n$.

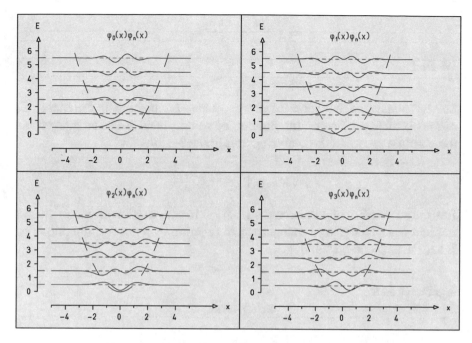

**Fig. C.5. Product $\varphi_m(x)\varphi_n(x)$ of the wave functions of the harmonic oscillator for $\hbar\omega = 1$. The long-dash curve indicates the potential energy $V(x)$, the short-dash lines show the energy eigenvalue $E_n$ of the functions $\varphi_n$. These lines also serve as zero lines for the product functions.**

### Eigenstate – Coherent-State Analyzing Amplitude

The function to be analyzed is the time-dependent wave function $\psi(x, t, x_0, p_0)$ of the coherent state. The analyzing function is the energy eigenfunction $\varphi_n(x)$. As analyzing amplitude we obtain

$$
\begin{aligned}
a(n, x_0, p_0) &= \int_{-\infty}^{+\infty} \varphi_n(x)\psi(x, t, x_0, p_0)\, dx \\
&= \frac{z^n}{\sqrt{n!}} \exp\left\{-\frac{1}{2}z^*z\right\} \exp\left\{-\frac{i}{\hbar}E_n t\right\} \quad .
\end{aligned}
$$

The corresponding probability is given by

$$
|a(n, x_0, p_0)|^2 = \frac{(|z|^2)^n}{n!}e^{-|z|^2} \quad , \qquad |z|^2 = \frac{x_0^2}{4\sigma_x^2} + \frac{p_0^2}{4\sigma_p^2} \quad .
$$

The probabilities $|a(n, x_0, p_0)|^2$ for fixed $x_0$, $p_0$ are distributed in the integer $n$ according to a Poisson distribution, cf. Appendix G. Its physical interpretation can be understood if we express $|z|^2$ in terms of the expectation value of the total energy of an oscillator with initial values $x_0$ and $p_0$,

$$E_0 = \frac{p_0^2}{2m} + \frac{m}{2}\omega^2 x_0^2 \quad .$$

We find

$$|z|^2 = E_0/(\hbar\omega) = n_0 \quad ,$$

i.e., $|z|^2$ equals the number $n_0$ of energy quanta $\hbar\omega$ making up the energy $E_0$ of the classical oscillator. This number, of course, need not be an integer. For the absolute square of the analyzing amplitude we thus find

$$|a_n(x_0, p_0)|^2 = \frac{n_0^n}{n!} e^{-n_0} \quad .$$

It is the probability of a Poisson distribution for the number of energy quanta $n$ found when analyzing a coherent wave function with the eigenfunctions $\varphi_n$. It has the expectation value

$$\langle n \rangle = n_0$$

and the variance

$$\mathrm{var}(n) = n_0 \quad .$$

## Coherent-State – Eigenstate Analyzing Amplitude

Analyzing the eigenstate wave functions $\varphi_n(x)$ with the coherent state wave functions for $t = 0$,

$$\varphi_D(x, x_D, p_D) = \sum_{n=0}^{\infty} a_n(x_D, p_D)\varphi_n(x) \quad ,$$

with the coefficients

$$a_n(x_D, p_D) = \frac{z_D^n}{\sqrt{n!}} \exp\left\{-\frac{1}{2}z_D^* z_D\right\} \quad , \qquad z_D = \frac{x_D}{2\sigma_x} + \mathrm{i}\frac{p_D}{2\sigma_p} \quad ,$$

we find as the analyzing amplitude

$$a(x_D, p_D, n) = \frac{1}{\sqrt{h}} \int_{-\infty}^{+\infty} \varphi_D^*(x, x_D, p_D)\varphi_n(x)\,\mathrm{d}x$$

$$= \frac{1}{\sqrt{h}} \frac{z_D^n}{\sqrt{n!}} \exp\left\{-\frac{1}{2}z_D^* z_D\right\} \quad ,$$

and for its absolute square

$$|a(x_D, p_D, n)|^2$$
$$= \frac{1}{2\pi(\sqrt{2}\sigma_x)(\sqrt{2}\sigma_p)} \frac{1}{n!} \left(\frac{x_D^2}{4\sigma_x^2} + \frac{p_D^2}{4\sigma_p^2}\right)^n \exp\left\{-\left(\frac{x_D^2}{4\sigma_x^2} + \frac{p_D^2}{4\sigma_p^2}\right)\right\} \quad .$$

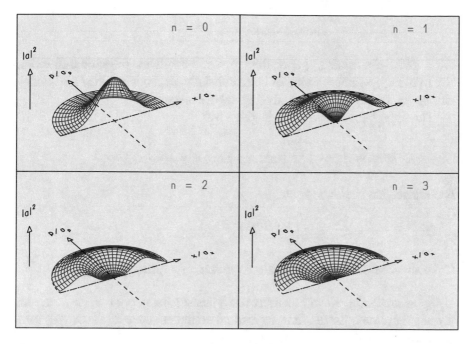

**Fig. C.6. Absolute square** $|a(x_D, p_D, n)|^2$ **of the amplitude analyzing the harmonic-oscillator eigenstate** $\varphi_n(x)$ **with a coherent state of position and momentum expectation value** $x_D$ **and** $p_D$.

For a given quantum number $n$ of the eigenstate, $|a|^2$ is a probability density in the $x_D, p_D$ phase space of the analyzing coherent state which is shown in Figure C.6 for a few values of $n$. It has the form of a ring wall with the maximum probability at

$$|z_D|^2 = \frac{x_D^2}{4\sigma_x^2} + \frac{p_D^2}{4\sigma_p^2} = n \quad .$$

In terms of the energy

$$E_D = \frac{p_D^2}{2m} + \frac{m}{2}\omega^2 x_D^2$$

of a classical particle of mass $m$ with position $x_D$ and momentum $p_D$ in a harmonic oscillator of angular frequency $\omega$, we have

$$|z_D|^2 = \frac{x_D^2}{4\sigma_x^2} + \frac{p_D^2}{4\sigma_p^2} = n_D \quad ,$$

where $n_D$ is the average number of energy quanta $\hbar\omega$ in the analyzing wave function $\varphi_D(x, x_D, p_D)$. We find

$$|a(x_D, p_D, n)|^2 = \frac{1}{h} e^{-n_D} \frac{n_D^n}{n!} \quad .$$

For a given eigenstate $\varphi_n(x)$ of the harmonic oscillator the probability density in the $x_D, p_D$ phase space of coherent states depends only on the average number $n_D$ of quanta in the analyzing coherent state.

The expectation value of $n_D$ is given by

$$\langle n_D \rangle = \int_{-\infty}^{+\infty} \int_{-\infty}^{+\infty} n_D |a(x_D, p_D, n)|^2 \, dx_D \, dp_D = n + 1 \quad .$$

Its variance has the same value:

$$\text{var}(n_D) = n + 1 \quad .$$

## Coherent-State – Coherent-State Analyzing Amplitude

Using as analyzing wave functions the coherent states $\varphi_D(x, x_D, p_D)$, the analyzing amplitude for the time-dependent coherent states $\psi(x, t, x_0, p_0)$ turns out to be

$$
\begin{aligned}
&a(x_D, p_D, x_0, p_0, t) \\
&= \frac{1}{\sqrt{h}} \int_{-\infty}^{+\infty} \varphi_D^*(x, x_D, p_D) \psi(x, t, x_0, p_0) \, dx \\
&= \frac{1}{\sqrt{h}} \exp\left\{ -\frac{1}{2} \left( z_D^* z_D + 2 z_D^* z(t) + z^*(t) z(t) \right) \right\} \exp\left\{ -\frac{i}{2} \omega t \right\}
\end{aligned}
$$

with

$$z(t) = z e^{-i\omega t} \quad , \qquad z = \frac{x_0}{2\sigma_x} + i \frac{p_0}{2\sigma_p} \quad .$$

The absolute square yields

$$
\begin{aligned}
&|a(x_D, p_D, x_0, p_0, t)|^2 \\
&= \frac{1}{2\pi(\sqrt{2}\sigma_x)(\sqrt{2}\sigma_p)} \exp\left\{ -\frac{1}{2} \left[ \frac{(x_D - x_0(t))^2}{2\sigma_x^2} + \frac{(p_D - p_0(t))^2}{2\sigma_p^2} \right] \right\}
\end{aligned}
$$

with

$$
\begin{aligned}
x_0(t) &= x_0 \cos \omega t + \frac{p_0}{m\omega} \sin \omega t \quad , \\
p_0(t) &= -m\omega x_0 \sin \omega t + p_0 \cos \omega t
\end{aligned}
$$

representing the expectation values of position and momentum of the coherent state $\psi(x, t, x_0, p_0)$ at time $t$. It is a bivariate Gaussian in the space of $x_D$ and $p_D$ centered about the classical positions $x_0(t)$, $p_0(t)$ of the oscillator. The probability density $|a(x_D, p_D, x_0, p_0, t)|^2$ shows the same behavior as that of a classical particle. The expectation values of the position $x_D$ and of momentum $p_D$ are simply given by the classical values

$$\langle x_D \rangle = x_0(t) \quad , \qquad \langle p_D \rangle = p_0(t) \quad .$$

The variances of $x_D$ and $p_D$ are

$$\mathrm{var}(x_D) = 2\sigma_x^2 \quad ,$$
$$\mathrm{var}(p_D) = 2\sigma_p^2 \quad .$$

This is twice the values of the ones of the coherent state itself, a consequence of the broadening caused by the analyzing wave packet $\varphi_D(x)$ having itself the variances $\sigma_x^2$ and $\sigma_p^2$.

The classical Gaussian phase-space probability density corresponding to $\psi(x, t, x_0, p_0)$ is of the same form as $|a(x_D, p_D, x_0, p_0, t)|^2$. It possesses, however, the widths $\sigma_x$ and $\sigma_p$, and has the explicit form

$$\rho^{\mathrm{cl}}(x, p, x_0, p_0, t)$$
$$= \frac{1}{2\pi \sigma_x \sigma_p} \exp\left\{ -\frac{1}{2}\left[ \frac{(x - x_0(t))^2}{\sigma_x^2} + \frac{(p - p_0(t))^2}{\sigma_p^2} \right] \right\} \quad .$$

By the same token the classical phase-space density corresponding to the detecting wave packet is

$$\rho_D^{\mathrm{cl}}(x, p, x_D, p_D)$$
$$= \frac{1}{2\pi \sigma_x \sigma_p} \exp\left\{ -\frac{1}{2}\left[ \frac{(x - x_D)^2}{\sigma_x^2} + \frac{(p - p_D)^2}{\sigma_p^2} \right] \right\} \quad .$$

The functions $\rho^{\mathrm{cl}}$ and $\rho_D^{\mathrm{cl}}$ are equal to the Wigner distributions (cf. Appendix D) of $\varphi$ and $\varphi_D$, respectively. The analyzing probability density $|a|^2$ can again be written as

$$|a(x_D, p_D, x_0, p_0, t)|^2$$
$$= \int_{-\infty}^{+\infty} \int_{-\infty}^{+\infty} \rho_D^{\mathrm{cl}}(x, p, x_D, p_D) \rho^{\mathrm{cl}}(x, p, x_0, p_0, t) \, dx_D \, dp_D \quad ,$$

which once more shows the reason for the broadening of $|a|^2$ relative to $\rho$.

# D. Wigner Distribution

The quantum-mechanical analog to a classical phase-space probability density is a distribution introduced by Eugene P. Wigner in 1932. In the simple case of a one-dimensional system described by a wave function $\varphi(x)$ the *Wigner distribution* is defined by

$$W(x,p) = \frac{1}{h} \int_{-\infty}^{+\infty} \exp\left\{\frac{i}{\hbar}py\right\} \varphi(x - \frac{y}{2})\varphi^*(x + \frac{y}{2})\,dy \quad .$$

For an uncorrelated Gaussian wave packet with the wave function

$$\varphi(x,x_0,p_0) = \frac{1}{\sqrt[4]{2\pi}\sqrt{\sigma_x}} \exp\left\{-\frac{(x-x_0)^2}{4\sigma_x^2} + i\frac{p_0}{\hbar}(x-x_0)\right\}$$

it has the form of a bivariate normalized Gaussian:

$$W(x,p,x_0,p_0) = \frac{1}{2\pi\sigma_x\sigma_p} \exp\left\{-\frac{1}{2}\left[\frac{(x-x_0)^2}{\sigma_x^2} + \frac{(p-p_0)^2}{\sigma_p^2}\right]\right\} \quad ,$$

where $\sigma_x$ and $\sigma_p$ fulfill the minimum-uncertainty relation

$$\sigma_x\sigma_p = \hbar/2 \quad .$$

The expression obtained for $W(x,p,x_0,p_0)$ coincides with the classical phase-space probability density for a single particle introduced in Section 3.6. The marginal distributions in $x$ or $p$ of the Wigner distribution are

$$W_p(x) = \int_{-\infty}^{+\infty} W(x,p)\,dp = \varphi^*(x)\varphi(x) = \rho(x)$$

and

$$W_x(p) = \int_{-\infty}^{+\infty} W(x,p)\,dx = \widetilde{\varphi}^*(p)\widetilde{\varphi}(p) = \rho_p(p) \quad ,$$

where $\widetilde{\varphi}(p)$ is the Fourier transform

S. Brandt and H.D. Dahmen, *The Picture Book of Quantum Mechanics*,
DOI 10.1007/978-1-4614-3951-6, © Springer Science+Business Media New York 2012

$$\widetilde{\varphi}(p) = \frac{1}{\sqrt{2\pi\hbar}} \int_{-\infty}^{+\infty} \exp\left\{-\frac{i}{\hbar}px\right\}\varphi(x)\,dx$$

of the wave function $\varphi(x)$, i.e., $\widetilde{\varphi}(p)$ is the wave function in momentum space.

For the case of the Gaussian wave packet we find for the marginal distributions

$$W_p(x) = \frac{1}{\sqrt{2\pi}\sigma_x}\exp\left\{-\frac{1}{2}\frac{(x-x_0)^2}{\sigma_x^2}\right\} \quad,$$

$$W_x(p) = \frac{1}{\sqrt{2\pi}\sigma_p}\exp\left\{-\frac{1}{2}\frac{(p-p_0)^2}{\sigma_p^2}\right\} \quad.$$

An alternative representation for the Wigner distribution can be obtained by introducing the wave function in momentum space into the expression defining $W(x,p)$ with the help of

$$\varphi(x) = \frac{1}{\sqrt{2\pi\hbar}} \int_{-\infty}^{+\infty} \exp\left\{i\frac{p}{\hbar}x\right\}\widetilde{\varphi}(p)\,dp \quad.$$

We find

$$W(x,p) = \frac{1}{h}\int_{-\infty}^{+\infty} \exp\left\{-\frac{i}{\hbar}xq\right\}\widetilde{\varphi}\left(p-\frac{q}{2}\right)\widetilde{\varphi}^*\left(p+\frac{q}{2}\right)dq \quad.$$

A note of caution should be added: For a general wave function $\varphi(x)$ the Wigner distribution is not a positive function everywhere. Thus, in general it cannot be interpreted as a phase-space probability density. It is, however, a real function

$$W^*(x,p) = W(x,p) \quad.$$

As an example we indicate the Wigner distribution $W(x,p,n)$ for the eigenfunction $\varphi_n(x)$ of the harmonic oscillator,

$$W(x,p,n) = \frac{(-1)^n}{\pi\hbar}L_n^0\left(\frac{x^2}{\sigma_x^2}+\frac{p^2}{\sigma_p^2}\right)\exp\left\{-\frac{1}{2}\left(\frac{x^2}{\sigma_x^2}+\frac{p^2}{\sigma_p^2}\right)\right\} \quad.$$

Here, the widths $\sigma_x$, $\sigma_p$ are given by

$$\sigma_x = \sqrt{\hbar/(2m\omega)} \quad, \qquad \sigma_p = \hbar/(2\sigma_p) \quad,$$

and $L_n^0(x)$ is the Laguerre polynomial with upper index $k=0$ as discussed in Section 13.4.

Figure D.1 shows the Wigner distributions for the lowest four eigenstates of the harmonic oscillator $n = 0, 1, 2, 3$ plotted over the plane of the scaled variables $x/\sigma_x$, $p/\sigma_p$. Accordingly, the plots are rotationally symmetric about the $z$ axis of the coordinate frame. The nonpositive regions of $W(x,p,n)$ can be clearly seen. The corresponding plots for the absolute square of the analyzing amplitude are shown in Figure C.6.

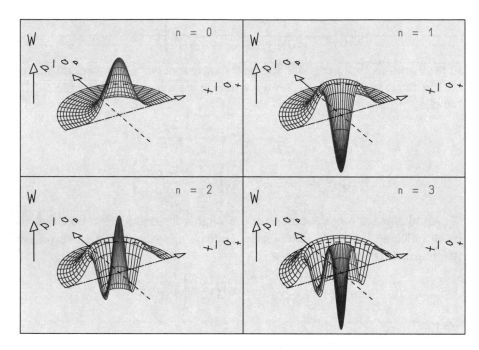

**Fig. D.1. Wigner distributions** $W(x, p, n)$ **of the harmonic-oscillator eigenstates** $\varphi_n(x)$, $n = 0, 1, 2, 3$.

The relation to the analyzing amplitude can easily be inferred from the following observation. For an arbitrary analyzing wave function $\varphi_D(x)$ we form the Wigner distribution

$$W_D(x, p) = \frac{1}{h} \int_{-\infty}^{+\infty} \exp\left\{\frac{i}{\hbar} p y'\right\} \varphi_D\left(x - \frac{y'}{2}\right) \varphi_D^*\left(x + \frac{y'}{2}\right) dy' \quad .$$

Then, the integral over $x$ and $p$ of the product of $W_D$ and $W$ yields

$$\int_{-\infty}^{+\infty} \int_{-\infty}^{+\infty} W_D(x, p) W(x, p) \, dx \, dp = \frac{1}{h} \left| \int_{-\infty}^{+\infty} \varphi_D^*(x) \varphi(x) \, dx \right|^2 = |a|^2 \quad .$$

This is to say, analyzing the Wigner distribution $W(x, p)$ of a wave function $\varphi(x)$ with the Wigner distribution $W_D(x, p)$ of an (arbitrary) analyzing wave function $\varphi_D(x)$ yields exactly the absolute square of the analyzing amplitude

$$a = \frac{1}{\sqrt{h}} \int_{-\infty}^{+\infty} \varphi_D^*(x) \varphi(x) \, dx$$

introduced in Appendix C.

The temporal evolution of a Wigner distribution

$$W(x,p,t) = \frac{1}{\hbar} \int_{-\infty}^{+\infty} e^{\frac{i}{\hbar}py} \psi^*(x+\frac{y}{2},t)\psi(x-\frac{y}{2},t)\,dy$$

corresponding to a time-dependent solution $\psi(x,t)$ of the Schrödinger equation with the Hamiltonian $H = p^2/2m + V(x)$, $p = (\hbar/i)d/dx$, is governed by the Wigner–Moyal equation. It is the quantum-mechanical analog of the Liouville equation for a classical phase-space distribution. For potentials $V(x)$ which are constant, linear, or quadratic in the coordinate $x$ or linear combinations of these powers the two equations of Wigner and Moyal, and of Liouville are identical. For these types quantum-mechanical and classical phase-space distributions that coincide at one instant $t$ in time, say the initial one, coincide at all the times.

# E. Gamma Function

The *gamma function* $\Gamma(z)$ introduced by Leonhard Euler is a generalization of the *factorial function* for integers $n$,

$$n! = 1 \cdot 2 \cdot 3 \cdot \ldots \cdot n \quad , \qquad 0! = 1! = 1 \quad ,$$

to noninteger and eventually complex numbers $z$. It is defined by *Euler's integral*

$$\Gamma(z) = \int_0^\infty t^{z-1} e^{-t} \, dt \quad , \qquad \mathrm{Re}(z) > 0 \quad .$$

By partial integration of

$$\int_0^\infty t^z e^{-t} \, dt = \Gamma(1+z)$$

we find the recurrence formula

$$\Gamma(1+z) = -t^z e^{-t} \Big|_0^\infty + z \int_0^\infty t^{z-1} e^{-t} \, dt = z\Gamma(z)$$

valid for complex $z$.

From Euler's integral we obtain

$$\Gamma(1) = 1$$

and, thus, with the help of the recurrence relation for non-negative integer $n$,

$$\Gamma(1+n) = n! \quad .$$

Euler's integral can also be computed in closed form for $z = 1/2$,

$$\Gamma\left(\frac{1}{2}\right) = \sqrt{\pi} \quad ,$$

so that – again through the recurrence relation – it is easy to find the gamma function for positive half-integer arguments.

S. Brandt and H.D. Dahmen, *The Picture Book of Quantum Mechanics*,
DOI 10.1007/978-1-4614-3951-6, © Springer Science+Business Media New York 2012

For nonpositive integer arguments the gamma function has poles as can be read off the reflection formula

$$\Gamma(1-z) = \frac{\pi}{\Gamma(z)\sin(\pi z)} = \frac{\pi z}{\Gamma(1+z)\sin(\pi z)} \quad .$$

In Figure E.1 we show graphs of the real and the imaginary part of $\Gamma(z)$ as surfaces over the complex $z$ plane. The most striking features are the poles for nonpositive real integer values of $z$. For real arguments $z = x$ the gamma function is real, i.e., $\mathrm{Im}(\Gamma(x)) = 0$. In Figure E.2 we show $\Gamma(x)$ and $1/\Gamma(x)$. The latter function is simpler since it has no poles. The gamma function for purely imaginary arguments $z = iy, y$ real, is shown in Figure E.3.

For complex argument $z = x + iy$ an explicit decomposition into real and imaginary part can be given,

$$\Gamma(x+iy) = (\cos\theta + i\sin\theta)|\Gamma(x)| \prod_{j=0}^{\infty} \frac{|j+x|}{\sqrt{y^2+(j+x)^2}} \quad ,$$

where the angle $\theta$ is determined by

$$\theta = y\psi(x) + \sum_{j=0}^{\infty}\left[\frac{y}{j+x} - \arctan\frac{y}{j+x}\right] \quad .$$

Here $\psi(x)$ is the *digamma function*

$$\psi(x) = \frac{\mathrm{d}}{\mathrm{d}x}(\ln\Gamma(x)) = \frac{\Gamma'(x)}{\Gamma(x)} \quad .$$

For integer $n$ the following formula follows from the recurrence formula:

$$\begin{aligned}\Gamma(n+z) &= \Gamma(1+(n-1+z)) \\ &= (n-1+z)(n-2+z)\cdot\ldots\cdot(1+z)z\Gamma(z) \quad .\end{aligned}$$

For purely imaginary $z = iy$ we find

$$\Gamma(n+iy) = (n-1+iy)(n-2+iy)\cdot\ldots\cdot(1+iy)iy\Gamma(iy) \quad .$$

The gamma function of a purely imaginary argument can be obtained by specialization of the argument $x + iy$ to $x = 0$ in $\Gamma(x+iy)$ to yield

$$\Gamma(iy) = (\sin\theta - i\cos\theta)\sqrt{\frac{\pi}{y\sinh y}}$$

with

$$\theta = -\gamma y + \sum_{j=1}^{\infty}\left[\frac{y}{j} - \arctan\frac{y}{j}\right] \quad ,$$

where *Euler's constant* $\gamma$ is given by

$$\gamma = -\psi(1) = \lim_{n\to\infty}\left(\sum_{k=1}^{n-1}\frac{1}{k} - \ln n\right) = 0.5772156649\ldots \quad .$$

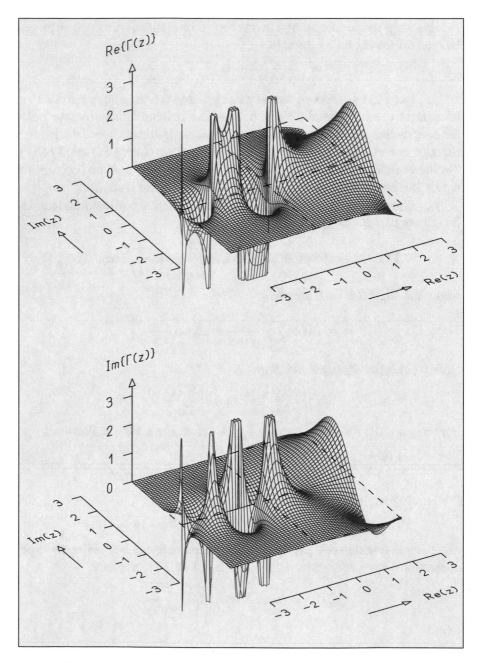

**Fig. E.1. Real part (top) and imaginary part (bottom) of $\Gamma(z)$ over the complex $z$ plane.**

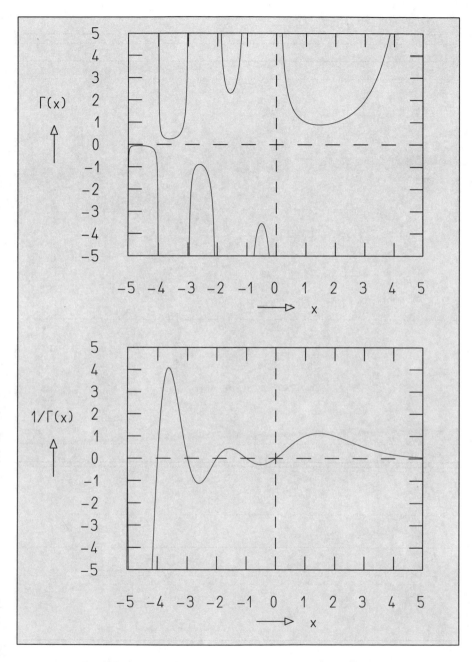

**Fig. E.2. The functions** $\Gamma(x)$ **and** $1/\Gamma(x)$ **for real arguments** $x$.

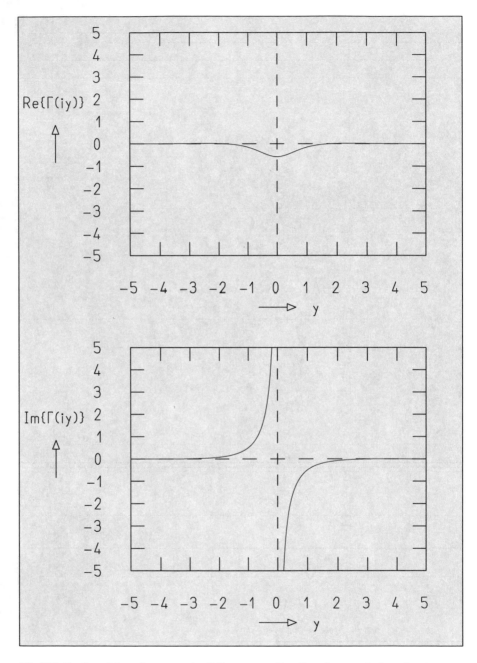

Fig. E.3. Real and imaginary parts of the gamma function for purely imaginary arguments.

# F. Bessel Functions and Airy Functions

Bessel's differential equation

$$x^2 \frac{d^2 Z_\nu(x)}{dx^2} + x \frac{dZ_\nu(x)}{dx} + (x^2 - \nu^2) Z_\nu(x) = 0$$

is solved by the *Bessel functions* of the first kind $J_\nu(x)$, of the second kind (also called *Neumann functions*) $N_\nu(x)$, and of the third kind (also called *Hankel functions*) $H_\nu^{(1)}(x)$ and $H_\nu^{(2)}(x)$ which are complex linear combinations of the former two. The Bessel functions of the first kind are

$$J_\nu(x) = \left(\frac{x}{2}\right)^\nu \sum_{k=0}^\infty \frac{(-1)^k}{k! \Gamma(\nu+k+1)} \left(\frac{x^2}{4}\right)^k \quad,$$

where $\Gamma(z)$ is Euler's gamma function.

The Bessel functions of the second kind are

$$N_\nu(x) = \frac{1}{\sin \nu\pi} [J_\nu(x) \cos \nu\pi - J_{-\nu}(x)] \quad.$$

For integer $\nu = n$ one has

$$J_{-n}(x) = (-1)^n J_n(x) \quad.$$

The *modified Bessel functions* are defined as

$$I_\nu(x) = \left(\frac{x}{2}\right)^\nu \sum_{k=0}^\infty \frac{1}{k! \Gamma(\nu+k+1)} \left(\frac{x^2}{4}\right)^k \quad.$$

The Hankel functions are defined by

$$H_\nu^{(1)}(x) = J_\nu(x) + iN_\nu(x) \quad,$$

$$H_\nu^{(2)}(x) = J_\nu(x) - iN_\nu(x) \quad.$$

The following relations hold for the connections of the functions just discussed and the spherical Bessel, Neumann, and Hankel functions, cf. Section 10.8. The *spherical Bessel functions* of the first kind are

S. Brandt and H.D. Dahmen, *The Picture Book of Quantum Mechanics*,
DOI 10.1007/978-1-4614-3951-6, © Springer Science+Business Media New York 2012

$$j_\ell(x) = \sqrt{\frac{\pi}{2x}} J_{\ell+1/2}(x) \quad .$$

The spherical Bessel functions of the second kind (also called *spherical Neumann functions*) are given by

$$n_\ell(x) = -\sqrt{\frac{\pi}{2x}} N_{\ell+1/2}(x) = (-1)^\ell j_{-\ell-1}(x) \quad ,$$

and the spherical Bessel functions of the third kind (also called *spherical Hankel functions* of the first and second kind) are

$$h_\ell^{(+)}(x) = n_\ell(x) + i j_\ell(x) = i[j_\ell(x) - i n_\ell(x)] = i\sqrt{\frac{\pi}{2x}} H_{\ell+1/2}^{(1)}(x) \quad ,$$

$$h_\ell^{(-)}(x) = n_\ell(x) - i j_\ell(x) = -i[j_\ell(x) + i n_\ell(x)] = -i\sqrt{\frac{\pi}{2x}} H_{\ell+1/2}^{(2)}(x) \quad .$$

In Figures F.1 and F.2 we show the functions $J_\nu(x)$ and $I_\nu(x)$ for $\nu = -1, -2/3, -1/3, \ldots, 11/3$. The features of these functions are simple to describe for $\nu \geq 0$. The functions $J_\nu(x)$ oscillate around zero with an amplitude that decreases with increasing $x$, whereas the functions $I_\nu(x)$ increase monotonically with $x$. At $x = 0$ we find $J_\nu(0) = I_\nu(0) = 0$ for $\nu > 0$. Only for $\nu = 0$ we have $J_0(0) = I_0(0) = 1$. For $\nu > 1$ there is a region near $x = 0$ in which the functions essentially vanish. The size of this region increases with increasing index $\nu$. For negative values of the index $\nu$ the functions may become very large near $x = 0$.

Closely related to the Bessel functions are the *Airy functions* Ai(x) and Bi(x). They are solutions of the differential equation

$$\left( \frac{d^2}{dx^2} - x \right) f(x) = 0$$

and are given by

$$\mathrm{Ai}(x) = \begin{cases} \dfrac{1}{3}\sqrt{x}\left\{ I_{-1/3}\left(\dfrac{2}{3}x^{3/2}\right) - I_{1/3}\left(\dfrac{2}{3}x^{3/2}\right) \right\} & , \quad x > 0 \\[2ex] \dfrac{1}{3}\sqrt{x}\left\{ J_{-1/3}\left(\dfrac{2}{3}|x|^{3/2}\right) + J_{1/3}\left(\dfrac{2}{3}|x|^{3/2}\right) \right\} & , \quad x < 0 \end{cases} \quad ,$$

and

$$\mathrm{Bi}(x) = \begin{cases} \sqrt{\dfrac{x}{3}}\left\{ I_{-1/3}\left(\dfrac{2}{3}x^{3/2}\right) + I_{1/3}\left(\dfrac{2}{3}x^{3/2}\right) \right\} & , \quad x > 0 \\[2ex] \sqrt{\dfrac{x}{3}}\left\{ J_{-1/3}\left(\dfrac{2}{3}|x|^{3/2}\right) - J_{1/3}\left(\dfrac{2}{3}|x|^{3/2}\right) \right\} & , \quad x < 0 \end{cases} \quad .$$

Graphs of these functions are shown in Figure F.3. Both functions oscillate for $x < 0$. The wavelength of the oscillation decreases with decreasing $x$. For $x > 0$ the function Ai(x) drops fast to zero whereas Bi(x) diverges.

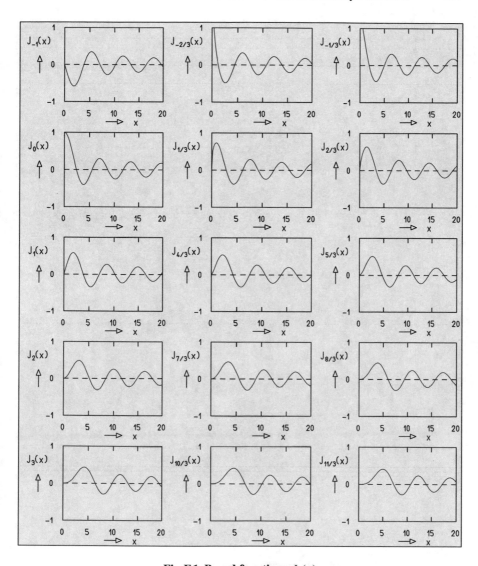

**Fig. F.1. Bessel functions** $J_\nu(x)$.

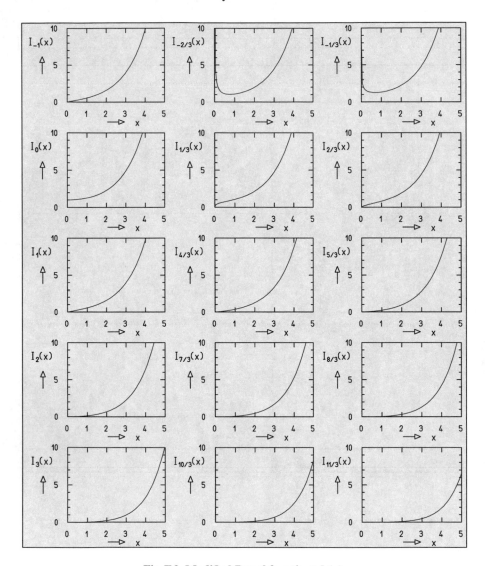

**Fig. F.2. Modified Bessel functions** $I_\nu(x)$.

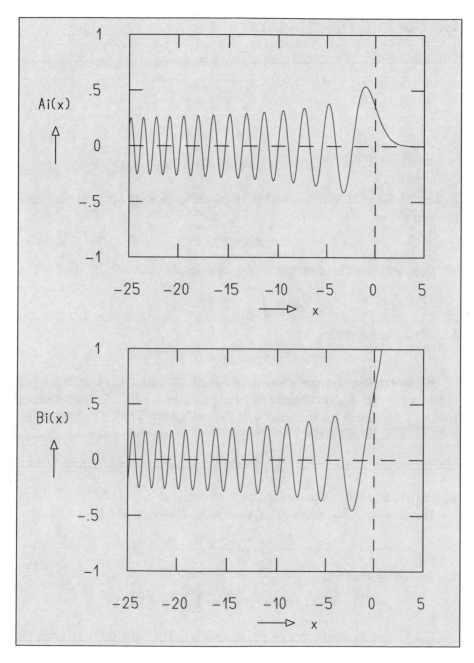

**Fig. F.3. Airy functions** $\mathrm{Ai}(x)$ **and** $\mathrm{Bi}(x)$.

# G. Poisson Distribution

In Section 3.3 we first introduced the *probability density* $\rho(x)$, which is normalized to one,

$$\int_{-\infty}^{+\infty} \rho(x)\,\mathrm{d}x = 1 \quad .$$

We also introduced the concepts of the *expectation value* of $x$,

$$\langle x \rangle = \int_{-\infty}^{+\infty} x\rho(x)\,\mathrm{d}x \quad ,$$

and of the *variance* of $x$,

$$\mathrm{var}(x) = \sigma_x^2 = \left\langle (x - \langle x \rangle)^2 \right\rangle \quad .$$

We now replace the *continuous variable* $x$ by the *discrete variable* $k$ which can assume only certain discrete values, e.g., $k = 0, 1, 2, \ldots$ . In a statistical process the variable $k$ is assumed with the *probability* $P(k)$. The total probability is normalized to one,

$$\sum_k P(k) = 1 \quad ,$$

where the summation is performed over all possible values of $k$.

The *average value*, *mean value*, or *expectation value* of $k$ is

$$\langle k \rangle = \sum_k k P(k) \quad ,$$

and the *variance* of $k$ is

$$\mathrm{var}(k) = \sigma^2(k) = \left\langle (k - \langle k \rangle)^2 \right\rangle = \sum_k (k - \langle k \rangle)^2 P(k) \quad .$$

The simplest case is that of an *alternative*. The variable only takes the values

$$\kappa = 0, 1 \quad .$$

The process yields with probability $p = P(1)$ the result $\kappa = 1$ and with probability $P(0) = 1 - p$ the result $\kappa = 0$. Therefore, the expectation value of $\kappa$ is

$$\langle \kappa \rangle = 0 \cdot (1 - p) + 1 \cdot p = p \quad .$$

S. Brandt and H.D. Dahmen, *The Picture Book of Quantum Mechanics*,
DOI 10.1007/978-1-4614-3951-6, © Springer Science+Business Media New York 2012

We now consider a process which is a sequence of $n$ independent alternatives each yielding the result $\kappa_i = 0, 1$, $i = 1, 2, \ldots, n$. We characterize the result of the process by the variable

$$k = \sum_{i=1}^{n} \kappa_i \quad ,$$

which has the range

$$k = 0, 1, \ldots, n \quad .$$

A given process yields the result $k$ if $\kappa_i = 1$ for $k$ of the $n$ alternatives and $\kappa_i = 0$ for $(n - k)$ alternatives. The probability for the sequence

$$\kappa_1 = \kappa_2 = \cdots = \kappa_k = 1 \quad , \qquad \kappa_{k+1} = \cdots = \kappa_n = 0$$

is $p^k (1 - p)^{n-k}$. But this is only one particular sequence leading to the result $k$. In total there are

$$\binom{n}{k} = \frac{n!}{k!(n-k)!}$$

such sequences where

$$n! = 1 \cdot 2 \cdot 3 \cdots \cdots n \quad , \qquad 0! = 1! = 1 \quad .$$

Therefore, the probability that our process yields the result $k$ is

$$P(k) = \binom{n}{k} p^k (1 - p)^{n-k} \quad .$$

This is the *binomial probability distribution*. The expectation value can be computed by introducing $P(k)$ into the definition of $\langle k \rangle$ or, even simpler, from

$$\langle k \rangle = \sum_{i=1}^{n} \langle \kappa_i \rangle = np \quad .$$

In Figure G.1 we show the probabilities $P(k)$ for various values of $n$ but for a fixed value of the product $\lambda = np$. The distribution changes drastically for small values of $n$ but seems to approach a limiting distribution for very large $n$. Indeed, we can write

$$P(k) \;=\; \frac{n!}{k!(n-k)!} \left(\frac{\lambda}{n}\right)^k \frac{\left(1 - \frac{\lambda}{n}\right)^n}{\left(1 - \frac{\lambda}{n}\right)^k}$$

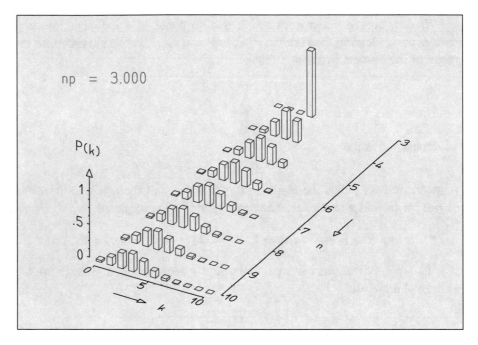

**Fig. G.1. Binomial distributions for various values of $n$ but fixed product $np = 3$.**

$$= \frac{\lambda^k}{k!} \left(1 - \frac{\lambda}{n}\right)^n \frac{n(n-1)\cdot \ldots \cdot (n-k+1)}{n^k \left(1 - \frac{\lambda}{n}\right)^k}$$

$$= \frac{\lambda^k}{k!} \left(1 - \frac{\lambda}{n}\right)^n \frac{\left(1 - \frac{1}{n}\right)\left(1 - \frac{2}{n}\right)\cdot \ldots \cdot \left(1 - \frac{k-1}{n}\right)}{\left(1 - \frac{\lambda}{n}\right)^k} \quad .$$

In the limit $n \to \infty$ every term in brackets in the last factor approaches one, and since

$$\lim_{n \to \infty} \left(1 - \frac{\lambda}{n}\right)^n = e^{-\lambda}$$

we have

$$P(k) = \frac{\lambda^k}{k!} e^{-\lambda} \quad .$$

This is the *Poisson probability distribution*. It is shown for various values of the parameter $\lambda$ in Figure G.2. The expectation value of $k$ is

$$\langle k \rangle = \sum_{k=0}^{\infty} k \frac{\lambda^k}{k!} e^{-\lambda} = \sum_{k=1}^{\infty} k \frac{\lambda^k}{k!} e^{-\lambda}$$

$$= \sum_{k=1}^{\infty} \frac{\lambda \lambda^{k-1}}{(k-1)!} e^{-\lambda} = \lambda \sum_{j=0}^{\infty} \frac{\lambda^j}{j!} e^{-\lambda} = \lambda \quad .$$

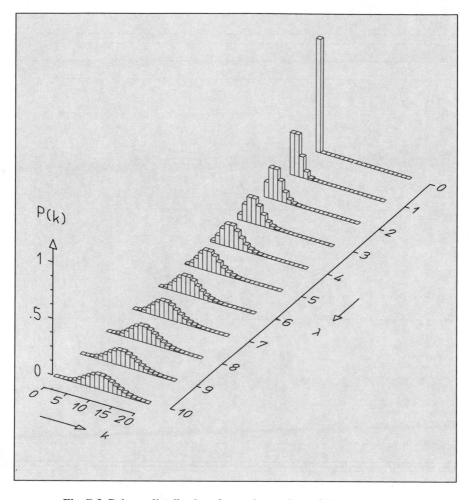

**Fig. G.2. Poisson distributions for various values of the parameter $\lambda$.**

In a similar way one finds

$$\langle k^2 \rangle = \lambda(\lambda + 1)$$

and therefore, also the variance of $k$ is equal to $\lambda$,

$$
\begin{aligned}
\text{var}(k) &= \langle (k - \langle k \rangle)^2 \rangle = \langle k^2 - 2k\langle k \rangle + \langle k \rangle^2 \rangle \\
&= \langle k^2 \rangle - 2\langle k \rangle^2 + \langle k \rangle^2 = \langle k^2 \rangle - \langle k \rangle^2 \\
&= \lambda(\lambda + 1) - \lambda^2 = \lambda \quad .
\end{aligned}
$$

The Poisson distribution is markedly asymmetric for small values of $\lambda$. For large $\lambda$, however, it becomes symmetric about its mean value $\lambda$ and in that case its bell shape resembles that of the Gaussian distribution.

# Index

Acceleration, sudden, 78
Airy functions, 72, 456
$\alpha$ particle, 86, 386, 395
Analyzing amplitude, 430, 435
– coherent-state – coherent-state, 444
– coherent-state – eigenstate, 442
– eigenstate – coherent-state, 441
– eigenstate – eigenstate, 440
– free particle, 434
– general case, 439
– harmonic oscillator, 439
Angular
– distribution, 205
– frequency, 11
– momentum, 191
– – variances, 200
– – expectation values, 199
– – in polar coordinates, 192
– – interpretation of eigenfunctions, 201
– – intrinsic , 353
– – of free wave packet, 219
– – of plane waves, 214
– – vector, semiclassical, 207, 358
– – $z$ component, 195
Antiquarks, 389
Antisymmetry, 172
Argand diagram, 94, 247, 314
Atomic
– beam, 399
– number, 381
Average value, 38, 460

Azimuth, 192

Balmer series, 378
Band spectra, 133, 380
Base
– states, 425
– vectors, 416
Bessel functions, 455
– modified, 455
– spherical, 210, 211, 455
Binomial distribution, 461
Bivariate Gaussian, 47
Bloch equations, 404
Bloch, Felix, 400
Boersch, Hans, 6
Bohm, David, 153
Bohr
– magneton, 397
– radius, 268
Bohr, Niels, 268
Born, Max, 37
Bose–Einstein particles, 172
Boson, 172, 298
Bound states
– and resonances, 322
– in atoms, 377
– in crystals, 377
– in deep square well, 111
– in harmonic oscillator, 117
– in nuclei, 377
– in periodic potential, 133
– in spherical square well, 251
– in square well, 129

S. Brandt and H.D. Dahmen, *The Picture Book of Quantum Mechanics*,
DOI 10.1007/978-1-4614-3951-6, © Springer Science+Business Media New York 2012

# Frequently Used Symbols

| | | | |
|---|---|---|---|
| $a$ | Bohr radius | $H$ | Hamiltonian |
| $a(x_0, p_0, x_S, p_S)$ | analyzing amplitude | $j$ | probability current density |
| $A_I, A_{II}, \ldots$ | amplitude factors in regions | $j_\ell(\rho)$ | spherical Bessel function |
| $B_I, B_{II}, \ldots$ | I, II, ... of space | $k$ | wave number |
| $\mathbf{B}$ | magnetic-induction field | $\ell$ | angular-momentum |
| $c$ | speed of light | | quantum number |
| $c$ | correlation coefficient | $L_n^\alpha$ | Laguerre polynomials |
| $d_{mm'}^\ell$ | Wigner function | $\mathbf{L}$ | angular momentum |
| $D_{mm'}^\ell$ | Wigner function | $\hat{\mathbf{L}}$ | angular-momentum operator |
| $d\sigma/d\Omega$ | differential scattering cross section | $m$ | quantum number of $z$ component of angular momentum |
| $e$ | elementary charge | | |
| $E_c$ | complex electric field strength | $m, M$ | mass |
| $E$ | energy | $M(x,t)$ | amplitude function |
| $E_n$ | energy eigenvalue | $\mathbf{M}$ | magnetization |
| $E_{kin}$ | kinetic energy | $n$ | refractive index |
| $\mathbf{E}$ | electric field strength | $n$ | principal quantum number |
| $f(k)$ | spectral function with respect to wave number | $n_\ell(\rho)$ | spherical Neumann function |
| | | $\mathbf{n}$ | unit vector |
| $f(p)$ | spectral function with respect to momentum | $p$ | momentum |
| | | $\hat{p}$ | momentum operator |
| $f(\vartheta)$ | scattering amplitude | $\langle p \rangle$ | momentum expectation value |
| $f_\ell$ | partial scattering amplitude | $\mathbf{p}$ | momentum vector |
| $f_{\ell m}(\Theta, \Phi)$ | directional distribution | $\hat{\mathbf{p}}$ | vector operator of momentum |
| $g$ | gyromagnetic factor | $P_\ell$ | Legendre polynomial |
| $h$ | Planck's constant | $P_\ell^m$ | associated Legendre function |
| $\hbar = h/(2\pi)$ | | | |
| $h_\ell^{(+)}(\rho)$ | spherical Hankel function of the first kind | $r$ | relative coordinate |
| | | $r$ | radial distance |
| $h_\ell^{(-)}(\rho)$ | spherical Hankel function of the second kind | $\mathbf{r}$ | position vector |
| | | $R$ | center-of-mass coordinate |

| | | | | |
|---|---|---|---|
| $R(r)$, $R_\ell(k,r)$ | radial wave function | $\eta_\ell(\mathbf{r})$ | scattered partial wave |
| $R_{n\ell}$ | radial eigenfunction | $\vartheta$, $\Theta$ | polar angle |
| $s$ | spin quantum number | $\vartheta$ | scattering angle |
| $S = (S_1, S_2, S_3)$ | spin-vector operator | $\lambda$ | wavelength |
| $S_\ell$ | scattering-matrix element | $\mu_0$ | vacuum permeability |
| $t$ | time | $\mu$ | reduced mass |
| $T$ | oscillation period | $\boldsymbol{\mu}$ | magnetic moment |
| $T$ | transmission coefficient | $\rho$ | probability density |
| $T$ | kinetic energy | $\rho^{\mathrm{cl}}$ | classical phase-space probability density |
| $T_\mathrm{T}$, $T_\mathrm{R}$ | transition-matrix elements | |
| $U$ | voltage | $\sigma_0$ | width of ground state of harmonic oscillator |
| $v_0$ | group velocity | |
| $v_\mathrm{p}$ | phase velocity | $\sigma_1$, $\sigma_2$, $\sigma_3$ | Pauli matrices |
| $V$ | potential (energy) | $\sigma_k$ | width in wave number |
| $V_\ell^{\mathrm{eff}}$ | effective potential | $\sigma_\ell$ | partial cross section |
| $w$ | average energy density | $\sigma_\mathrm{p}$ | width in momentum |
| $W$ | Wigner distribution | $\sigma_x$ | width in position |
| $W$ | energy | $\sigma_{\mathrm{tot}}$ | total cross section |
| $W_\ell$, $W_{\ell m}$ | coefficients in the angular decomposition of a wave packet | $\varphi(x)$ | stationary wave function |
| | | $\varphi_\mathbf{p}(\mathbf{r})$ | stationary harmonic wave function |
| $x$ | position | |
| $\langle x \rangle$ | position expectation value | $\varphi$ | state vector of stationary state |
| $Y_{\ell m}$ | spherical harmonic | $\phi$, $\Phi$ | azimuthal angle |
| $Z$ | atomic number | $\chi$ | magnetic susceptibility |
| $\alpha$ | fine-structure constant | $\boldsymbol{\chi}$ | general spin state |
| $\delta_\ell$ | scattering phase shift | $\psi(x,t)$ | time-dependent wave function |
| $\Delta k$ | wave-number uncertainty | $\psi_\mathbf{p}(\mathbf{r},t)$ | harmonic wave function |
| $\Delta p$ | momentum uncertainty | $\boldsymbol{\psi}$ | state vector |
| $\Delta x$ | position uncertainty | $\omega$ | angular frequency |
| $\varepsilon_0$ | vacuum permittivity | $\Omega$ | solid angle |
| $\eta_1$, $\eta_{-1}$ | spin $\frac{1}{2}$ base states | $\nabla$ | nabla (or del) operator |
| $\eta_\mathbf{k}(\mathbf{r})$ | scattered wave | $\nabla^2$ | Laplace operator |

# Basic Equations

**de Broglie wave**

$$\psi_{\mathbf{p}}(\mathbf{r},t) = \frac{1}{(2\pi\hbar)^{3/2}} \exp\left(-\frac{i}{\hbar}Et\right) \exp\left(\frac{i}{\hbar}\mathbf{p}\cdot\mathbf{r}\right)$$

**time-dependent Schrödinger equation**

$$i\hbar\frac{\partial}{\partial t}\psi(\mathbf{r},t) = \left[-\frac{\hbar^2}{2M}\nabla^2 + V(\mathbf{r})\right]\psi(\mathbf{r},t)$$

**stationary Schrödinger equation**

$$\left[-\frac{\hbar^2}{2M}\nabla^2 + V(\mathbf{r})\right]\varphi_E(\mathbf{r}) = E\varphi_E(\mathbf{r})$$

**momentum operators**

$$\hat{\mathbf{p}} = (\hat{p}_x, \hat{p}_y, \hat{p}_z) = \frac{\hbar}{i}\left(\frac{\partial}{\partial x}, \frac{\partial}{\partial y}, \frac{\partial}{\partial z}\right) = \frac{\hbar}{i}\nabla$$

**angular-momentum operators**

$$\hat{\mathbf{L}} = \mathbf{r}\times\hat{\mathbf{p}} = \frac{\hbar}{i}\mathbf{r}\times\nabla$$

**radial Schrödinger equation for spherically symmetric potential**

$$-\frac{\hbar^2}{2M}\left[\frac{1}{r}\frac{d^2}{dr^2}r - \frac{\ell(\ell+1)}{r^2} - \frac{2M}{\hbar^2}V(r)\right]R_\ell(k,r) = E\,R_\ell(k,r)$$

**stationary scattering wave**

$$\varphi_{\mathbf{k}}^{(+)}(\mathbf{r})\xrightarrow[kr\gg 1]{} e^{i\mathbf{k}\cdot\mathbf{r}} + f(\vartheta)\frac{e^{ikr}}{r}$$

**scattering amplitude**

$$f(\vartheta) = \frac{1}{k}\sum_{\ell=0}^{\infty}(2\ell+1)f_\ell(k)P_\ell(\cos\vartheta)$$

**differential, partial, and total cross sections**

$$\frac{d\sigma}{d\Omega} = |f(\vartheta)|^2\,,\quad \sigma_\ell = \frac{4\pi}{k^2}(2\ell+1)|f_\ell(k)|^2\,,\quad \sigma_{\text{tot}} = \sum_{\ell=0}^{\infty}\sigma_\ell$$

# Physical Constants

**Planck's constant**

$$h = 4.136 \cdot 10^{-15}\,\text{eV}\,\text{s} = 6.626 \cdot 10^{-34}\,\text{J}\,\text{s}$$
$$\hbar = h/(2\pi) = 6.582 \cdot 10^{-16}\,\text{eV}\,\text{s}$$
$$= 1.055 \cdot 10^{-34}\,\text{J}\,\text{s}$$

**speed of light**

$$c = 2.998 \cdot 10^{8}\,\text{m}\,\text{s}^{-1}$$

**elementary charge**

$$e = 1.602 \cdot 10^{-19}\,\text{C}$$

**fine-structure constant**

$$\alpha = \frac{e^2}{4\pi\varepsilon_0\hbar c} = \frac{1}{137.036}$$

**electron mass**

$$m_e = 0.5110\,\text{MeV}/c^2 = 9.110 \cdot 10^{-31}\,\text{kg}$$

**proton mass**

$$m_p = 938.3\,\text{MeV}/c^2 = 1.673 \cdot 10^{-27}\,\text{kg}$$

**neutron mass**

$$m_n = 939.6\,\text{MeV}/c^2 = 1.675 \cdot 10^{-27}\,\text{kg}$$

# Conversion Factors

**mass**   $1\,\text{kg} = 5.609 \cdot 10^{35}\,\text{eV}/c^2\,,$   $1\,\text{eV}/c^2 = 1.783 \cdot 10^{-36}\,\text{kg}$

**energy**   $1\,\text{J} = 6.241 \cdot 10^{18}\,\text{eV}\,,$   $1\,\text{eV} = 1.602 \cdot 10^{-19}\,\text{J}$

**momentum**   $1\,\text{kg}\,\text{m}\,\text{s}^{-1} = 1.871 \cdot 10^{27}\,\text{eV}/c\,,$   $1\,\text{eV}/c = 5.345 \cdot 10^{-28}\,\text{kg}\,\text{m}\,\text{s}^{-1}$

Printed in the United States
By Bookmasters